Constructing Mathematical Knowledge

Studies in Mathematics Education Series

Series Editor
Paul Ernest
School of Education
University of Exeter
Exeter

1 The Philosophy of Mathematics Education
Paul Ernest
2 Understanding in Mathematics
Anna Sierpinska
3 Mathematics Education and Philosophy: An International Perspective
Edited by Paul Ernest
4 Constructing Mathematical Knowledge: Epistemology and Mathematics Education
Edited by Paul Ernest
5 Investigating Mathematics Teaching: A Constructivist Enquiry
Barbara Jaworski
6 Radical Constructivism: A Way of Knowing and Learning
Ernst von Glasersfeld

Studies in Mathematics Education Series: 4

Constructing Mathematical Knowledge: Epistemology and Mathematics Education

Edited by

Paul Ernest

Routledge
Taylor & Francis Group
LONDON AND NEW YORK

First published 1994 by RoutledgeFalmer

2 Park Square, Milton Park, Abingdon, Oxon OX14 4RN
711 Third Avenue, New York, NY 10017, USA

Routledge is an imprint of the Taylor & Francis Group, an informa business

First issued in hardback 2016

Transferred to Digital Printing 2004

A catalogue record for this book is available from the British Library

Library of Congress Cataloging-in-Publication Data are available on request

ISBN 0 7507 0354 7 cased
ISBN 0 7507 0570 1 limp

Jacket design by Caroline Archer

Typeset in 9.5/11pt Bembo by
Graphicraft Typesetters Ltd, Hong Kong

ISBN 978-0-7507-0570-7 (pbk)
ISBN 978-1-138-16297-6 (hbk)

Contents

List of Tables and Figures

Preface by Series Editor

Mathematics education is established worldwide as a major area of study, with numerous dedicated journals and conferences serving national and international communities of scholars. Research in mathematics education is becoming more theoretically orientated. Vigorous new perspectives are pervading it from disciplines and fields as diverse as psychology, philosophy, logic, sociology, anthropology, history, feminism, cognitive science, semiotics, hermeneutics, post-structuralism and post-modernism. The series *Studies in Mathematics Education* consists of research contributions to the field based on disciplined perspectives that link theory with practice. It is founded on the philosophy that theory is the practitioner's most powerful tool in understanding and changing practice. Whether the practice is mathematics teaching, teacher education, or educational research, the series intends to offer new perspectives to assist in clarifying and posing problems and to stimulate debate. The series *Studies in Mathematics Education* will encourage the development and dissemination of theoretical perspectives in mathematics education as well as their critical scrutiny. It aims to have a major impact on the development of mathematics education as a field of study into the twenty-first century.

Unusually for the series this book (Volume 4) and Volume 3 are edited collections. Instead of the sharply focused concerns of a research monograph the books offer a panorama of complementary and forward-looking perspectives. In the spirit of the philosophy of the series Volumes 3 and 4 illustrate between them the breadth of theoretical and philosophical perspectives that can fruitfully be brought to bear on mathematics and education. The companion to the present volume is *Mathematics, Education and Philosophy: An International Perspective.* It offers a reconceptualization of mathematics from a range of philosophical, educational and social perspectives, as well as philosophical reflections on mathematics education itself. The present volume provides a complementary focus. Its emphasis is on epistemological issues, encompassing multiple perspectives on the learning of mathematics, as well as broader philosophical reflections on the genesis of knowledge. The two books aim to set a research agenda for the philosophy of mathematics education, a rapidly developing area of enquiry. Together they survey research, providing a report on advances made so far, as well as indicating orientations for potentially fruitful work in the future.

Paul Ernest
School of Education
University of Exeter
March 1994

Introduction

Paul Ernest

> To introduce epistemological considerations into a discussion of education has always been dynamite. Socrates did it, and he was promptly given hemlock. Giambattista Vico did it in the 18th century, and the philosophical establishment could not bury him fast enough. (von Glasersfeld, 1983, p. 41)

This book is a contribution to the philosophy of mathematics education, that loosely defined cluster of interests at the intersection of mathematics, education and philosophy. These interests indicate an attitude of mind, the desire to enquire into philosophical and reflective aspects of mathematics and mathematics education, as well as a relish for discussion and argument over deep-seated epistemological issues. They also embody the move towards interdisciplinarity that is sweeping the social sciences and humanities, which is driven by the desire to break down barriers between disparate fields of knowledge, and to apply some of the exciting new methods and perspectives from one field to another. What all of these interests share is the potential to ignite controversy, as the above quote from von Glasersfeld suggests.

Central to the philosophy of mathematics education are two problems. First, there is of course the problem of the nature of mathematics itself. Second, there is the problematic relationship between philosophy and epistemology and the learning and teaching of mathematics. Traditionally the first problem belongs to the philosophy of mathematics. But this field is changing, reflecting a controversy and a new interdisciplinarity. The philosophy of mathematics is in the midst of a 'Kuhnian revolution'. The Euclidean paradigm of mathematics as an objective, absolute, incorrigible and rigidly hierarchical body of knowledge is increasingly under question. One reason is that the foundations of mathematics are not as secure as was claimed. Technical results such as Gödel's theorems have shown that formal axiomatic systems can never be regarded as ultimate. Another reason is a growing dissatisfaction amongst mathematicians, philosophers, educators and multidisciplinary scholars with the traditional narrow focus of the professional philosophy of mathematics, usually limited to questions of the foundations of pure mathematical knowledge and of the existence of mathematical objects (Kitcher and Aspray, 1988).

A revolutionary new tradition in the philosophy of mathematics has been emerging termed quasi-empiricist (Lakatos, 1976; Kitcher, 1984; Tymoczko; 1986), maverick (Kitcher and Aspray, 1988) and post-modernist (Tiles, 1991). This is primarily naturalistic, concerned to describe the nature of mathematics and the practices of mathematicians, both current and historical. It is quasi-empiricist and fallibilist in its epistemology, thus displacing mathematics from its place as the secure cornerstone of absolutism. A number of philosophers and mathematicians are contributors to this new tradition, including Wittgenstein, Lakatos, Putnam, Wang, Davis and Hersh, Kitcher, Tymoczko. These authors have proposed that the task of the philosophy of mathematics is to account for mathematics more fully, including the practices of mathematicians, its history and applications, the place of mathematics in human culture, perhaps even including issues of values and education, in short, describing the human face of mathematics.

This concern with external, social dimensions of mathematics including its applications and uses, has given rise to the desire to see a multidisciplinary account of mathematics drawing inspiration from many currents of thought. These include ethnomathematics (Ascher, D'Ambrosio, Zaslavsky), social constructivism and the rhetoric of science (Billig, Knorr-Cetina, Latour), post-structuralism (Foucault, Walkerdine), post-modernism (Derrida, Lyotard), semiotics (Peirce, Eco), social constructionist psychology (Gergen, Harré, Shotter), feminism (Harding, Rose), Critical Theory (Habermas, Marcuse), externalist philosophy of science (Feyerabend, Hacking, Kuhn, Laudan), social epistemology (Fuller, Toulmin), and philosophy in general (Rorty, Bernstein).

Consequently, there is a number of researchers who are drawing on other disciplines to account for the nature of mathematics, including Bloor and Restivo, from social constructivism in sociology; Wilder and Livingston, from cultural studies and ethno-methodology; Rotman from semiotics, Aspray and Kitcher, Joseph, Kline and Gillies from the history of mathematics.

Thus a growing number of scholars shares a common concern with external, social dimensions of mathematics including its history, applications and uses. Many share a desire to see a multidisciplinary account of mathematics that accommodates ethnomathematics, mathematics education studies, and feminist and multicultural critiques. What drives this for many is a sense of the social responsibility of mathematics. For once mathematics is reconceptualized as a social construction, then the social function of mathematics in society must be examined. Its relation with broader issues of power, social structure and values, needs to be considered to see whose interests it serves. The question must be asked: who in the world economy gains by mathematics, and who loses? An ethics of mathematics is called for, once it is seen as an instrument and product of values and power. For one of the conclusions to be drawn from a radical social view of mathematics is that it plays a key role in the distribution of life chances. For example, there is widespread concern with how mathematics acts as a 'critical filter' in depriving minority and women students of equal opportunities in employment. Philosophical and ethical considerations like these thus have important implications for mathematics, and especially for educational theory and practice.

Within mathematics education there is an increasing awareness of the significance of epistemological and philosophical issues for important traditional areas of inquiry. Theories of learning have themselves been epistemologically orientated for some time, with widespread discussion of the philosophical assumptions of constructivism and various forms of social constructivism and socio-cultural cognition. The present volume makes a significant contribution to this area. A growing number of scholars is trying to accommodate fallibilist views of mathematics and knowledge in their accounts of learning and teaching. Issues and concerns that now draw on the philosophy of mathematics and philosophical perspectives in general include mathematical thinking, problem-solving, problem-posing and investigational pedagogy, curriculum theories and ideologies, teacher education and development, teacher beliefs, learner conceptions and beliefs, and applications of such theories as the 'Perry Scheme' (Perry, 1970) and *Women's Ways of Knowing* (Belenky *et al.*, 1986). Scholars are also reflecting on broader social issues including the culture of mathematics and its politics, its philosophy, its social context, its language, and issues of gender, race and class.

Philosophical considerations are also central to empirical research in mathematics education and its methods of inquiry, with researchers becoming increasingly aware of the epistemological foundations of their methodologies and inquiries, and referring to them explicitly. Constructivist researchers in particular have helped raise awareness of this, and the present volume also makes a contribution on these issues. Multiple research paradigms are now widely used, with proponents of the scientific, interpretative and critical theoretic research paradigms discussing and comparing their philosophical and methodological bases (when not engaged in internecine paradigm war, *à la* Gage, 1989).

Problems of the Philosophy of Mathematics Education

This introduction outlines some of the areas of activity in the philosophy of mathematics education. A more systematic insight into its central questions or *problématique* arises from considering Schwab's (1978) four 'commonplaces' of teaching. These are the subject (mathematics), the learner, the teacher, and the milieu of teaching. Each gives rise to a characteristic set of problems and questions, providing a research agenda for the philosophy of mathematics education.

Aims and the Social Context

What are the aims of mathematics education? Are these aims valid? Whose aims are they? For whom are they intended? What values are they based on? Who gains and who loses? How do the social, cultural and historical contexts relate to mathematics, the aims of teaching, and the teaching and learning of mathematics?

The Nature of Learning

What philosophical assumptions, possibly implicit, underpin theories of the learning of mathematics? Are these assumptions valid? Which epistemologies and learning theories are assumed? How can student learning in mathematics be studied empirically? How can the social context of learning be accommodated? This is the central set of problems addressed in this volume.

The Nature of Teaching

What philosophical assumptions, possibly implicit, does mathematics teaching rest on? Are these assumptions valid? What means are adopted to achieve the aims of mathematics education? Are the ends and means consistent? What is the role of language and mediating tools in teaching, such as computers, and what is their impact?

The Nature of Mathematics

What is mathematics, and how can its nature be accounted for? What philosophies of mathematics have been developed? Whose? What features of mathematics do they pick out as significant? What is their impact on the teaching and learning of mathematics?

The Nature of Mathematics and its Relation to Teaching

> All mathematical pedagogy, even if scarcely coherent, rests on a philosophy of mathematics. (Thom, 1973, p. 204)

> The issue, then, is not: What is the best way to teach? but, What is mathematics really all about? . . . Controversies about . . . teaching cannot be resolved without confronting problems about the nature of mathematics. (Hersh, 1979, p. 34)

A central problem of the philosophy of mathematics education is the issue of the relationship between philosophies of mathematics and mathematics education.[1] The question is: What is their impact on the teaching and learning of mathematics? Thom and Hersh claim that different philosophical positions have significantly different educational implications. Steiner elaborates this claim as follows.

> **Thesis 1**. Generally speaking, all more or less elaborated conceptions, epistemologies, methodologies, philosophies of mathematics (in the large or in part) contain — often in an implicit way — ideas, orientations or germs for theories on the teaching and learning of mathematics.

> **Thesis 2**. Concepts for the teaching and learning of mathematics — more specifically: goals and objectives (taxonomies), syllabi, textbooks, curricula, teaching methodologies, didactical principles, learning theories, mathematics education research designs (models, paradigms, theories, etc.), but likewise teachers' conceptions of mathematics and mathematics teaching as well as students' perceptions of mathematics — carry with them or even rest upon (often in an implicit way) particular philosophical and epistemological views of mathematics. (Steiner, 1987, p. 8)

Thus the claim is that the association is bi-directional. First, that all learning and teaching practices in mathematics rest upon possibly implicit epistemologies or philosophies of mathematics. Some of these issues are treated in this volume. Second, that any philosophy of mathematics (including personal philosophies) has powerful implications for social and educational issues, and many educational and pedagogical consequences. However, it must be stressed that such consequences are not strict logical implications of a philosophy, and additional values, aims and other assumptions are required to reach such conclusions (Ernest 1991, 1994). Because the link is not one of logical implication, it is theoretically possible to consistently associate a philosophy of mathematics with almost any educational practice or approach. A neo-behaviourist or cognitivist (such as Ausubel) and a radical constructivist may both be concerned to ascertain what a child knows before commencing teaching, despite having diametrically opposite epistemologies. Likewise a traditional purist mathematician and a social constructivist may both favour a multicultural approach to mathematics, but for different reasons (the former perhaps to humanize mathematics, the latter to show it as the social construction of all of humanity for social-justice reasons).

Although there is no logical necessity for, e.g., a transmission-style pedagogy to be associated with an absolutist, objectivist epistemology and philosophy of mathematics, such associations often are the case (Ernest, 1988, 1991). They may be due to the resonances and sympathies between different aspects of a person's philosophy, ideology, values and belief-systems. These form links and associations and often become restructured in moves towards maximum coherence, consistency, and the integration of the personality.[2]

> In particular, the observed consistency between the teachers' professed conceptions of mathematics and the way they typically presented the content strongly suggests that the teachers' views, beliefs and preferences about mathematics do influence their instructional practice. (Thompson, 1984, p. 125)

> Research over the past several years on teacher's beliefs gives strong testimony that teachers' conceptions make a difference in how mathematics is taught. (Cooney, 1988, p. 356)

Much work in the philosophy of mathematics education pertains to exploring the link between the philosophies of mathematics implicit in teachers' beliefs, in texts and the mathematics curriculum, in systems and practices of mathematical assessment and in mathematics classroom practices and the results with learners. Whilst much progress has been made, much work remains to be done in the area.

The Philosophy of Mathematics Education Network

The burgeoning of this area of shared interest led to the founding of a philosophy of mathematics education network in 1990. This is a network of interested persons revolving around an organizing group, a newsletter and periodic conference symposia.

The Philosophy of Mathematics Education Organising Group is an informal collective of interested and cooperating scholars, who have made significant contributions to the area in their own personal research. The members are the following: Raffaella Borasi (USA), Stephen I. Brown (USA), Leone Burton (UK), Paul Cobb (USA), Jere Confrey (USA), Thomas S. Cooney (USA), Kathryn Crawford (Australia), Ubiratan D'Ambrosio (Brazil), Philip J. Davis (USA), Sandy Dawson (Canada), Paul Ernest (UK), Group Chair, Ernst von Glasersfeld (USA), David

Henderson (USA), Reuben Hersh (USA), Christine Keitel-Kreidt (Germany), Stephen Lerman (UK), John Mason (UK), Marilyn Nickson (UK), David Pimm (UK), Sal Restivo (USA), Leo Rogers (UK), Anna Sfard (Israel), Ole Skovsmose (Denmark), Francesco Speranza (Italy), Leslie P. Steffe (USA), Hans-Georg Steiner (Germany), John Volmink (South Africa), Yuxin Zheng (Peoples Republic of China).

The Philosophy of Mathematics Education Newsletter is edited in rotation by members of this organizing group, including, to date: Sandy Dawson, Paul Ernest, Stephen Lerman, Marilyn Nickson and Leo Rogers. The newsletter is mailed to several hundred subscribers in many countries including all continents of the globe. Its aims are to foster awareness of philosophical aspects of mathematics and mathematics education, understood broadly and inclusively; to disseminate news of events and new thinking in these topics; and to encourage international cooperation and informal communication and dialogue between teachers, scholars and others engaged in research in the area. The newsletter has carried discussions on many themes, including radical constructivism, ethnomathematics, the popular image of mathematics, post-modernism, revolutions in mathematics and the philosophy of mathematics education itself.

There are also periodic Philosophy of Mathematics Education symposia at national and international conferences. The most significant of these was the Philosophy of Mathematics Education Topic Group at the 7th International Congress of Mathematical Education, Québec, August 1992, which played an important part in the genesis of this book. There have also been philosophy of mathematics education symposia and discussion groups at the First British Congress of Mathematics Education (Loughborough, 1991), the British Society for Research into Learning Mathematics (Bath, 1990), and the conferences of the International Group for the Psychology of Mathematics Education numbers 14 (Mexico, 1990), 16 (USA, 1992) and 17 (Japan, 1993).

This Volume

This book represents some of the most interesting aspects of current work in the philosophy of mathematics education, as well as an indication of some of the more exciting new departures. The aim is both to survey the field and to try to predict future areas of fruitful research. A few of the chapters began as contributions to the Philosophy of Mathematics Education symposium at the ICME-7 conference in Québec, 1992, but most have now changed beyond recognition. Almost all of the chapters were specially written for the book, and were solicited from leading figures as well as from promising young researchers in the field. The response to the invitation to contribute was so overwhelming, that even when only the very best chapters were retained, there was enough to make two good-sized volumes. Therefore this book is published at the same time as a companion volume entitled *Mathematics, Education and Philosophy: An International Perspective*. Together the two volumes survey current research in the philosophy of mathematics and mathematics education. As editor, one of the most exciting features of putting this collection together is that many of the most important contributors to the field are themselves represented in it.

The contents of this volume have been divided into four sections.

Constructivism and the Learning of Mathematics. This is the largest section in the book. It explores constructivist and social theories of learning mathematics, and their relationships, viewed from multiple perspectives. It also includes discussions of the role and impact of the computer in the learning of mathematics. Overall, this continues to be one of the most central areas of philosophical research in mathematics education, and has a great deal to yield in the way of insight, as well as controversy. How we view learners' interactions with mathematical situations has to be one of the most important issues that mathematics educators face.

Psychology, Epistemology and Hermeneutics. This section ties in both with the preceding section, and with that on post-modernist and post-structuralist approaches in Volume 1. It brings new analyses from psycho-analysis, Hermeneutics and other perspectives to bear on issues of mathematics and learning.

Enquiry in Mathematics Education. This is concerned with both the nature and the outcomes of reflective research that illuminates mathematics in some significant way. One of the themes in this and the preceding section is that of the role of language in mathematics education. This is beginning to emerge more widely as a central issue of concern in research in mathematics education, not least because it is implicated in a number or leading social theories of learning.

History, Mathematics and Education. This relates the historical development of mathematics to the teaching and learning of mathematics; a vital and growing area of research. This section exemplifies the increasing attention that is being paid to the parallel between cultural and personal epistemological developments in the mathematics education community.

As the title of the book indicates, each of these areas concerns epistemological issues in mathematics education research. The primary focus is the construction of mathematical knowledge and meaning during the process of learning mathematics, whether accounted for in individual terms or socially. Also included are an enquiry in mathematics and mathematics education, and the construction of knowledge in these fields, and over the course of history. The volume can thus legitimately claim to embody the epistemology of mathematics education in its concerns with the nature, genesis and development of knowledge.[3]

The four themes in the other volume are as follows.

Reconceptualizing the Philosophy of Mathematics. This includes contributions on fresh approaches to the philosophy of mathematics from fallibilist and social perspectives.

Post-modernist and Post-structuralist Approaches. This is the largest section, made up of a number of innovative chapters applying post-structuralist and post-modernist perspectives to both mathematics and mathematics education. It offers a challenging and controversial but illuminating set of theoretical perspectives applied to mathematics education. Overall, it represents one of the areas of growth in philosophical research on mathematics and education, and it is expected to bear significant fruits in the future. It also ties in with the section on Psychology, Epistemology and Hermeneutics in this volume.

The Human Face of Mathematics. In varying ways, this section shows how the human face of mathematics can be revealed by contrasting individual mathematicians, disciplines, or even cultures.

The Social Context of Mathematics and Education. This adds to the issues raised in the previous section, but considers particularly social aspects of mathematics, including the crucial issues of race and gender. Because of the political implications, and the challenges offered to traditional ways of thinking, this remains one of the most controversial and important areas of research.

The titles and groupings indicate some of the central themes in both volumes. However there are other themes shared by the chapters too, some remarked above and some not. These include the relationship of mathematics with art, computers, history, gender, race, social critique, language and curriculum. Shared perspectives in the chapters include those of social theory, sociology of knowledge, and various insider, outsider and educational viewpoints. All these recurring themes serve to illustrate the rich complexity of the chapters and their interconnections, and the value of interdisciplinary perspectives on mathematics and mathematics education.

Notes

1. This issue remains controversial, as a reading of, e.g., Kroon (1994) shows.
2. Of course 'splitting' and the simultaneous adoption of inconsistent social roles are also possible.
3. The one aspect of epistemology that is understressed is the issue of the warranting of knowledge. See the other volume, Ernest (1993) and Noddings (1991) for a discussion of this issue.

References

Cooney, T.J. (1988) 'The issue of reform', *Mathematics Teacher*, 80, pp. 352–63.

Belenky, M.F., Clinchy, B.M., Goldberger, N.R. and Tarule, J.M. (1986) *Women's Ways of Knowing*, New York, Basic Books.

Ernest, P. (1988) 'The impact of beliefs on the teaching of mathematics', in Keitel, C., Damerow, P., Bishop, A. and Gerdes, P. (Eds) *Mathematics, Education and Society*, Paris, UNESCO, 1989, pp. 99–101.

Ernest, P. (1991) *The Philosophy of Mathematics Education*, London, The Falmer Press.

Ernest, P. (1993) 'Epistemology and the relationship between Subjective and Objective knowledge of Mathematics', Paper presented at the Conference on the Cultural Context of the Mathematics Classroom, Osnabrück, Germany, 11–15 October.

Ernest, P. (1994) 'The philosophy of mathematics and mathematics education', in Biehler, R. Scholz, R.W., Straesser, R. and Winkelmann, B. (Eds) *The Didactics of Mathematics as a Scientific Discipline*, Dordrecht, Kluwer, pp. 335–49.

Gage, N.L. (1989) 'The paradigm wars and their aftermath: A "Historical" sketch of research on teaching since 1989', *Teachers College Record*, 91, 2, pp. 135–50.

Hersh, R. (1979) 'Some proposals for reviving the philosophy of mathematics', *Advances in Mathematics*, 31, pp. 31–50.

KITCHER, P. (1984) *The Nature of Mathematical Knowledge*, New York, Oxford University Press.

KITCHER, P. and ASPRAY, W. (1988) 'An opinionated introduction', in ASPRAY, W. and KITCHER, P. (Eds) *History and Philosophy of Modern Mathematics*, Minneapolis, University of Minnesota Press, pp. 3–57.

KROON, F. (1994) 'Review of Ernest (1991)' *Science and Education*, 3, 1, pp. 77–85.

LAKATOS, I. (1976) *Proofs and Refutations*, Cambridge, Cambridge University Press.

NODDINGS, N. (1990) 'Constructivism in mathematics education', in DAVIS, R.B., MAHER, C.A. and NODDINGS, N. (Eds) *Constructivist Views on the Teaching and Learning of Mathematics*, Reston, Virginia, National Council of Teachers of Mathematics, pp. 7–18.

PERRY, W.G. (1970) *Forms of Intellectual and Ethical Development in the College Years: A Scheme*, New York, Holt, Rinehart and Winston.

SCHWAB, J. (1978) *Science, Curriculum, and Liberal Education*, WESTBURY, I. and WILKOF, N. (Eds) Chicago, University of Chicago Press.

STEINER, H.G. (1987) 'Philosophical and epistemological aspects of mathematics and their interaction with theory and practice in mathematics education', *For the Learning of Mathematics*, 7, 1, pp. 7–13.

THOM, R. (1973) 'Modern mathematics: Does it exist?', in HOWSON, A.G. (Ed) *Developments in Mathematical Education*, Cambridge, Cambridge University Press, pp. 194–209.

THOMPSON, A.G. (1984) 'The relationship between teachers conceptions of mathematics and mathematics teaching to instructional practice', *Educational Studies in Mathematics*, 15, pp. 105–27.

TILES, M. (1991) *Mathematics and the Image of Reason*, London, Routledge.

TYMOCZKO, T. (Ed) (1986) *New Directions in the Philosophy of Mathematics*, Boston, Birkhauser.

VON GLASERSFELD, E. (1983) 'Learning as a constructive activity', in BERGERON, J. and HERSCOVICS, N. (Eds) *Proceedings of the 5th Conference of the North American Chapter of the International Group for the Psychology of Mathematics Education*, Montréal, Québec, PME-NA, 1, pp. 41–69.

Part 1

Constructivism and the Learning of Mathematics

We can know nothing that we have not made. (Vico, 1710, p. 76)

Until recently, mathematics has supported the deceptive appearance of being fully formed and perfectly finished knowledge. Hence only a few decades ago, the dominant view in mathematics education assumed that the teaching and learning of mathematics only required the effective transmission of mathematical knowledge. After all, did not Shannon and Weaver's Communication Theory indicate that all that is necessary is that the message of mathematics be adequately coded, transmitted, received by the addressee, decoded and represented internally? Although the 1960s brought a softening of this view, with its emphasis on 'discovery learning', underpinning it was still an empiricist view of mind. Instead of receiving the message transmitted by the teacher, the learner now looked for mathematical knowledge in the world, albeit a carefully arranged and orchestrated corner of the world. Discovery learning also assumes that mathematical knowledge is pre-existing: it just needs to be actively noticed to be discovered. Furthermore, because it is 'truth', when discovered, it is recognized unproblematically.

The outlooks I have just described are not the products of naive error. To assert this would be false and unjust. Instead, they are a function of an empirical scientific or positivistic epistemology and methodological paradigm. This was the legacy of mathematics education, emerging from the disciplines of mathematics and experimental psychology. Even the influence of Piaget on the 1960s' discovery learning approach was assimilated into this scientific research paradigm. However in the decade that followed a different paradigm emerged better reflecting Piaget's psychology and his clinical interview method. Other strands combined with this, drawing on Kelly's Personal Construct Theory, Problem-solving research from the *Gestalt* psychologists and early cognitive scientists such as Newell and Simon, as well as advances in cybernetics and other areas. The outcome was a new research paradigm for mathematics education, that of constructivism.

Piaget's constructivism has its roots in an evolutionary biological metaphor, according to which the evolving organism must adapt to its environment in order to survive. Likewise, the developing human intelligence also undergoes a process of adaptation in order to fit with its circumstances and remain viable. Personal theories are constructed as constellations of concepts, and are adapted by the twin processes of assimilation and accommodation in order to fit with the human organism's world of experience. Indeed Piaget claims that the human intelligence is ordering the very world it experiences in organizing its own cognitive structures. *L'intelligence organise le monde en s'organisant elle-même* (Piaget, 1937, cited in von Glasersfeld, 1989a, p. 162). Epistemologically, what is of tremendous significance in Piaget's constructivism is (1) the notion that knowing is embodied, and essentially implicates interaction with the world, and (2) his 'Genetic Epistemology' that sees knowledge of the individual and group as historical and evolutionary, growing and changing to meet challenges and contradictions. This first aspect directly challenges the separation of mind and body in Cartesian Dualism that has so long dominated western thought. In their survey Varela *et al.* (1991) distinguish the three stages of 'Cognitivism, Emergence and Enactive' in the development of cognitive science, and locate Piaget's contribution already in the last stage.

As part of the broader interdisciplinary movement of Structuralism, Piaget was seduced by the Bourbakian account of mathematics as logically constituted by three mother structures. Like many thinkers before him, Piaget afforded a privileged place to mathematical knowledge in his scheme. However, a scholar who has been influential in freeing constructivism from these constraints is von Glasersfeld. He has taken the scepticism of Piaget's constructivism further, and argued in a radical version of constructivism that all knowledge, including mathematics, is constructed and fallible. That conclusion embroiled radical constructivism in a great deal of controversy in mathematics education, where until recently absolutist views of knowledge prevailed. However, the debate over the nature of mathematical knowledge, and in particular, that between absolutism and fallibilism, is not reflected in this present section on constructivist theories of the learning of mathematics.[1] For the contributors to this section, this debate is largely settled. Instead a new set of controversies has set in concerning such things as individual versus social construction. Epistemological issues, such as those raised by constructivism do seem to be dynamite in education!

Ultimately, the import of any theory of learning mathematics consists in facilitating interventions in the processes of its teaching and learning. Constructivism accounts for the individual idiosyncratic constructions of meaning, for systematic errors, misconceptions, and alternative conceptions in the learning of mathematics. Thus it facilitates diagnostic teaching, and the diagnosis and remediation of errors. A growing number of instructional projects is based on a constructivist sensitivity to children's sense making and on the spontaneous strategies they have been observed to develop (Grouws, 1992). Until recently, the available literature on constructivist learning theory in mathematics has been limited to one or two monographs like Steffe *et al.* (1983). However, a growing number of publications is appearing offering a discussion of constructivist views on the learning and teaching of mathematics (e.g., Davis *et al.*, 1990; Steffe, 1991; von Glasersfeld, 1991, 1994). In addition, there is a growing literature about the relationship between radical and individualistic versions of constructivism and social constructivism (e.g., Ernest, 1991, 1993; Steffe, in press). The present section adds to this literature, and begins to explore a number of significant issues for the future, such as the role of microcomputers in helping learners to construct mathematical knowledge and meaning. Microcomputers have great potential here, because they encourage children to think 'outside their heads', providing direct evidence of children's learning and thought processes.

Note

1. But see the section on 'Reconceptualizing the Philosophy of Mathematics' in the companion volume *Mathematics, Education and Philosophy*.

References

Davis, R.B., Maher, C.A. and Noddings, N. (Eds) (1990) *Constructivist Views on the Teaching and Learning of Mathematics* (J.R.M.E. Monograph 4), Reston, Virginia, National Council of Teachers of Mathematics.

Ernest, P. (1991) *The Philosophy of Mathematics Education*, London, The Falmer Press.

Ernest, P. (1993) 'Constructivism, the psychology of learning, and the nature of mathematics: Some critical issues', *Science and Education*, 2, 2, pp. 87–93.

Grouws, D.A. (Ed) (1992) *Handbook of Research on Mathematics Teaching and Learning*, New York, Macmillan.

Steffe, L.P. (Ed) (1991) *Epistemological Foundations of Mathematical Experience*, New York, Springer-Verlag.

Steffe, L.P. and Gale, J. (Eds) (in press) *Constructivism in Education*, London, Erlbaum.

Steffe, L.P., von Glasersfeld, E., Richards, J. and Cobb, P. (1983) *Children's Counting Types: Philosophy, Theory, and Application*, New York, Praeger.

Varela, F.J., Thompson, E. and Rosch, E. (1991) *The Embodied Mind*, Cambridge, Massachusetts, MIT Press.

Vico, G. (1710) *De antiquissima Italorum sapientia*, Nicolini, F. (Ed) Bari, Latzera (1914).

von Glasersfeld, E. (Ed) (1989a) 'Constructivism in Education', in Husen, T. (Ed) *International Encyclopedia of Education* (supplementary volume), Oxford, Pergamon.

von Glasersfeld, E. (Ed) (1991) *Radical Constructivism in Mathematics Education*, Dordrecht, Kluwer.

von Glasersfeld, E. (1994) *Radical Constructivism: A Way of Knowing and Learning*, London, The Falmer Press.

Chapter 1

A Radical Constructivist View of Basic Mathematical Concepts

Ernst von Glasersfeld

I am not a mathematician. My interest is in conceptual analysis, and mathematical concepts are extremely interesting — especially the seemingly simple ones that are linked to the basic elements of arithmetic. The most basic of these elements are the symbols that we call 'numerals'.

There is an old statement that mathematics has to do with symbols and the manipulation of symbols. Frequent repetition of this statement has encouraged the belief that it is only the symbols that matter and that their conceptual referents need not be examined — presumably because to adults who have become used to doing arithmetic, there seems to be no difference between numerals and the concepts they refer to. But symbols do not generate the concepts that constitute their referents; they have to be linked to them by a thinking agent, even when this linkage has become automatic. It is, indeed, a ground rule of semiotics that a sound or a mark on paper becomes a symbol only when it is deliberately associated with a conceptual meaning.

I trust that you will agree that mathematics could not have happened if the concepts of 'unit' and 'Plurality of units' had not somehow been generated. How this was done may not be quite so obvious.

Thinkers as diverse as Edmund Husserl, Albert Einstein, and Jean Piaget have stated very clearly that the concept of 'unit' is derived from the construction of 'objects' in our experiential world.

Einstein's description of this construction is one of the clearest:

> I believe that the first step in the setting of a 'real external world' is the formation of the concept of bodily objects and of bodily objects of various kinds. Out of the multitude of our sense experiences we take, mentally and arbitrarily, certain repeatedly occurring complexes of sense impressions (partly in conjunction with sense impressions which are interpreted as signs for sense experiences of others), and we correlate to them a concept — the concept of the bodily object. Considered logically this concept is not identical with the totality of sense impressions referred to; but it is a free creation of the human (or animal) mind. (Einstein, 1954, p. 291)

I have elsewhere shown that the concept of 'plurality', (unlike that of unitary physical object) cannot be derived from 'sense impressions', but only from the awareness that the recognition of a particular physical object is being repeated.

In Piaget's terms, it is not an empirical abstraction from sensory-motor experience, but a reflective abstraction from the experiencer's own mental operations. This is an important distinction. Let me try to make it as clear as possible. Initially, the operations of constructing or recognizing physical objects do require sense impressions; but the realization that one is carrying out the same recognition procedure more than once, arises from one's own operating and not from the particular sensory material that furnishes the occasions for this way of operating.

'Units' and 'pluralities', however, do not yet constitute a basis for the development of mathematics or even arithmetic. At least one other concept is needed: the concept of 'number'.

In our book on counting types, Steffe, Richards, Cobb, and I have presented a model of how an abstract concept of 'Number' may be derived from the activity of counting (Steffe *et al.*, 1983). So far, I have not seen a more plausible model. But irrespective of the adequacy of that particular model, I have no doubt that both the notions of ordinal and cardinal number derive from the operations an active subject carries out and not from any specific sensory material.

If you agree that the concept of 'Number' requires both the concepts of 'unit' and 'plurality' — and if you further agree that without a concept of 'number' there can be no development of a mathematics of numbers, then it is clear that this mathematics is an affair of mental operations that have to be carried out by an active subject. As Hersh says:

> Symbols are used as aids to thinking just as musical scores are used as aids to music. The music comes first, the score comes later. (Hersh, 1986, p. 19)

If this is so, it would seem to follow incontrovertibly that a string of mathematical symbols remains meaningless until someone has associated specific mental operations with the symbols. These operations, because they are mental operations, cannot be witnessed by anyone else. What can be witnessed, are the symbols that an acting subject produces in a spoken or written form as the result of his or her mental operations — and one can then examine whether or not they are compatible with the symbols one would have produced oneself.

This involves a two-fold process of interpretation:

- First, the other subject's interpretation of the string of symbols he or she has produced;
- Second, one's own interpretation of these as one perceives them.

I submit that this awareness of processes of interpretation would have far-reaching consequences for any theory of proof. But this is not my present concern.

Here, I want to emphasize that the conceptual analyses I have suggested would have far-reaching consequences also for the teaching of arithmetic. If mathematical symbols have to be interpreted in terms of mental operations, the teacher's task is to stimulate and prod the student's mind to operate mathematically. Sensory-motor material, graphic representations, and talk can provide occasions for the abstraction of mathematical operations, but they cannot convey them ready-made to the student. The frequently used expression 'mathematical objects' is misleading, because the meaning of mathematical symbols is never a static object but rather a particular way of operating. This goes for plurality and number, for point and line, and consequently for all the more complex abstractions of mathematics. The teacher's task, then, is to orient the students' mental operating; and to do this, the teacher needs at least a hypothetical model of how the student's mind operates at the outset of the lesson (von Glasersfeld and Steffe, 1991).

To conclude, I would summarize these consequences by saying that, if we really want to teach arithmetic, we have to pay a great deal of attention to the mental operations of our students. Teaching has to be concerned with *understanding* rather than performance and the rote-learning of, say, the multiplication table, or training the mechanical performance of algorithms — because *training* is suitable only for animals whom one does not credit with a thinking mind.

References

EINSTEIN, A. (1954) 'Physics and reality', *Ideas and Opinions*, New York, Bonanza Books.

HERSH, R. (1986) 'Some proposals for revising the philosophy of mathematics', in TYMOCZKO, T. (Ed) *New Directions in the Philosophy of Mathematics* (9–28), Boston/Basel, Birkhauser.

STEFFE, L.P., VON GLASERSFELD, E. RICHARDS, J. and COBB, P. (1983) *Children's Counting Types: Philosophy, Theory, and Application*, New York, Praeger, pp. 112–23.

VON GLASERSFELD, E. and STEFFE, L.P. (1991) 'Conceptual models in educational research and practice', *Journal of Educational Thought*, 25, 2, pp. 91–103.

Chapter 2

Interaction and Children's Mathematics

Leslie P. Steffe and Ron Tzur

> Knowledge implies interaction, and we cannot step out of our domain of interactions, which is closed. We live, therefore, in a domain of subject-dependent knowledge and subject-dependent reality . . . We literally create the world in which we live by living it. (Maturana, 1978, pp. 60–1)

In recent years, the social interaction involved in the construction of the mathematics of children has been brought into the foreground in order to specify its constructive aspects (Bauersfeld, 1988; Yackel, *et al.*, 1990). One of the basic goals in our current teaching experiment is to analyse such social interaction in the context of children working on fractions in computer microworlds. In our analyses, however, we have found that social interaction does not provide a full account of children's mathematical interaction. Children's mathematical interaction also includes enactment or potential enactment of their operative mathematical schemes. Therefore, we conduct our analyses of children's social interaction in the context of their mathematical interaction in our computer microworlds.[1]

We interpret and contrast the children's mathematical interaction from the points of view of radical constructivism and of Soviet activity theory. We challenge what we believe is a common interpretation of learning in radical constructivism by those who approach learning from a social-cultural point of view. Renshaw (1992), for example, states that 'In promulgating an active, constructive and creative view of learning, . . . the constructivists painted the learner in close-up as a solo-player, a lone scientist, a solitary observer, a meaning maker in a vacuum.' In Renshaw's interpretation, learning is viewed as being synonymous with construction in the absence of social interaction with other human beings. To those in mathematics education who use the teaching experiment methodology, this view of learning has always seemed strange because we emphasize social interaction as a primary means of engendering learning and of building models of children's mathematical knowledge (Cobb and Steffe, 1983; Steffe, 1993).

We see at least three possible reasons why constructivist learning could be interpreted as a univocal process. The first is that mathematics may be viewed as being innate, a view similar to Chomsky's (1980) view of language as being innate. Among the neo-Vygotskians, however, this assumption would be in conflict with Vygotsky's general genetic law of cultural development. A second more plausible reason is the radical constructivists' view of human beings as self-organizing, self-reproducing, living systems. This second reason, when coupled with the principle of subject-dependent knowledge and reality, may very well lead to interpreting learning as a univocal process. A third reason for the belief that radical constructivists omit social interaction in their accounts of learning is seemingly based on a belief that Piaget did not regard social interaction as being essential in logical-mathematical development. However, Piaget's work can be legitimately interpreted as a socio-cultural approach even though those who concentrate specifically on social interaction interpret Piaget as taking an almost exclusive biological approach to genetic epistemology (Shotter, 1994).

We view learning as the capability of an individual to change his or her conceptual structures in response to perturbation (Konold and Johnson, 1991). To the extent that these conceptual structures are operative mathematical schemes, mathematical learning occurs as a result of using the schemes in some type of interaction. By working intensively with children in teaching experiments, we have come to believe that children's learning of mathematics is much more complex than we ever imagined. In view of our concept of the mathematics of children as a dynamic and living subject, it is too much to expect that the grand theories of Piaget and Vygotsky would include an account of mathematical learning. Our purpose in this paper is to begin formulating a model of mathematical learning that supersedes the meanings of learning in Piaget's and Vygotsky's theories. We first delineate some basic Piagetian and Vygotskian notions for the purposes of comparison and contrast.

Piaget's and Vygotsky's Views on Learning

Piaget (1964) regarded learning as being subordinate to development. He viewed the development of knowledge as a spontaneous process tied to the whole process of embryogenesis.

> Embryogenesis concerns the development of the body, but it concerns as well the development of the nervous system, and the development of mental functions. In the case of the development of knowledge in children, embryogenesis ends only in adulthood. It is a total developmental process that we must re-situate in its general biological and psychological context.
>
> Learning presents the opposite case. In general, learning is provoked by situations — provoked by a psychological experimenter; or by a teacher with respect to some didactic point; or by an external situation. It is provoked in general, as opposed to spontaneous. In addition, it is a limited process — limited to a single problem, or to a single structure. (Piaget, 1964, p. 8)

Piaget's view of learning as being provoked by situations is compatible with our understanding of human beings as interactive organisms. Piaget did not regard the situations in the framework of the classical stimulus–response paradigm. He believed that a stimulus was a stimulus only when it was assimilated into a structure, and it was the structure which set off the response. Because the conceptual structures setting off the response were the products of development, learning could be understood only in terms of development. In the case of mathematics, a particular structure might be provoked or learned if it is based on an existing, simpler developmental mathematical structure.

Piaget regarded the basic mental operations involved in mathematics learning as products of spontaneous reconstruction.

> This brings us back to the child, since within the space of a few years he spontaneously reconstructs operations and basic structures of a logico-mathematical nature, without which he would understand nothing of what he will be taught in school . . . He reinvents for himself, around his seventh year, the concepts of reversibility, transitivity, recursion, reciprocity of relations, class inclusion, conservation of numerical sets, measurements, organization of spatial references. (Piaget, 1980, p. 26)

Piaget's notion of development as a spontaneous reconstruction is in contrast to learning as being provoked. In our superseding model of learning, two aspects of spontaneous reconstruction are of importance. The first is the unintentionality of spontaneous reconstruction. Children certainly do not intend to reconstruct operations and basic structures of a logical mathematical nature. The intentionality of other human beings with regard to children's reconstructions is not as certain. But we believe that only rarely do other human beings with whom children interact intend for the children to reconstruct operations and structures of a logical mathematical nature. The second is the independence of spontaneous reconstruction. Children contribute the operations and basic structures of a logico-mathematical nature to their interactions in their milieu. But we do not interpret Piaget as having meant that the child is a solo player, a lonely voyager, or a meaning maker in a vacuum even though he viewed mathematics as a contribution of the child. Piaget (1964) made it clear that the mathematical operations and basic structures he identified are reconstructed by children as a product of their interactions within their physical, social, and cultural environments. In this sense, Piaget explained mathematical development as a result of children interacting in their environments as self-regulating and autonomous organisms.

Vygotsky (1978) criticized the view of learning being subordinate to development. He also rejected the view of learning being development; in which learning is perceived as the process of elaborating innate structures. Instead, Vygotsky (1978) emphasized the essential influence of learning on development. In his work, he concentrated on means by which children acquire language as a result of social and natural processes. We appeal to Vygotsky's general genetic law of cultural development for insight into how he understood the influence of learning on development (Vygotsky, 1978; quoted in Wertsch and Toma, 1994):

> Any function in the child's cultural development appears twice, or on two planes. First, it appears on the social plane, and then on the psychological plane. First it appears between people as an interpsychological category. This is equally true with regard to voluntary attention, logical memory, the formation of concepts, and the development of volition . . . It goes without saying that internalization transforms the process itself and changes its structure and functions. Social relations or relations among

> people genetically underlie all higher functions and their relationships. (Wertsch and Toma, 1994, p. 162)

Consequently, social interaction would seem to play a major role in Vygotsky's approach to learning. Interpreters of Vygotsky have made this clear.

> One of the basic tenets of the Vygotskian approach to education is the assumption that individual learning is dependent on social interaction. However, it should be clear from the outset that this is not merely a statement of correlation between individual learning and social context. This thesis should be interpreted in its strongest possible form, proposing that the qualities of thinking are actually generated by the organizational features of the social interaction. (van Oers, 1992, p. 2)

Wertsch and Toma (1994) also interpreted Vygotsky's general genetic law of cultural development in its strong form. 'In an essential sense, the same mental functions appear on the social and individual planes' (p. 162).

Both van Oers and Wertsch and Toma emphasized that individual mental functioning is not a simple and direct copy of social interaction because of the genetic transformations involved in internalization. Internalization, the fundamental process involved in learning, makes it possible to hypothesize personal forms of what van Oers regarded as the cultural meanings of mathematics. The generalized knowledge and skills for dealing with the world that have been built up throughout cultural history, according to van Oers (1992), 'can be transformed into curriculum content and, as such, it can be taught. Personal meaning (sense), however, cannot be taught directly; it can only be built up by the involvement in an educative relationship' (p. 3). Through one of his major concepts — the zone of proximal development — Vygotsky stressed that cultural meanings can become intermingled with personal sense through an educational process.

By claiming that cultural meanings can be taught, we interpret van Oers as implying that through social interaction, children internalize the organizational features of the social interaction by means of their personal concepts. This may be more than van Oers would say, but based on his following comments, it is necessary to somehow conclude that the personal meanings are a contribution of the individual.

> The meaningfulness of the learning process refers to the process of attaching personal sense to the actions, rules, methods and values as provided by a school subject . . . Learning as a process of appropriation of culture, then, is meaningful as far as it encompasses both kinds of meaning. (van Oers, 1992, p. 4)

Toward a Superseding Model

In our opinion, we should retain the most fundamental and essential points of Vygotsky and Piaget while formulating a superseding model of mathematics learning. That is, we must try to formulate a model in which mathematical learning is not perceived as being provoked as opposed to spontaneous. To do this we separate the intentionality of adults and the unintentionality of children and strive to maintain the second aspect of spontaneity we identified in the preceding section. Because learning will always involve other people or external situations and will be usually intentional on the part of adults, we also strive to maintain an emphasis on something similar to what van Oers called cultural meanings of mathematics.[2] But rather than cast cultural meanings in some idealized form, independent of the knowledge of human beings, we stress our mathematical knowledge and how we use that knowledge when working with children. The crucial realization for us is that the children cannot construct our knowledge, because our knowledge is essentially inaccessible to them. The best they can do is to modify their own knowledge as a result of interacting with us and with each other.

It is always possible for a teacher to enact his or her knowledge in some form. For example, a teacher might demonstrate how to find a family of fractions while explaining and discussing his or her actions. The social interaction with the children would then be with reference to the actions of the teacher and the children's attempts to understand and reproduce these actions. In this, there would be cultural meanings of mathematics to learn (the teacher's meanings) even if a major focus of the social interaction was with respect to the children's personal concepts. Their personal concepts would be viewed from the frame of reference of the teacher's concepts, and it would be the goal of the teacher for the children to appropriate the teacher's concepts. This is how we interpret van Oers' (1992) comment that the meaningfulness of the learning process refers to the process of attaching personal sense to the actions, rules, methods, and values as provided by a school subject. Our view of mathematical learning is quite different in that we begin with the children's mathematical concepts and operations as we understand them.

When teaching children, we base our interactions with them on schemes we observe the children use independently in mathematical activity. This is our fundamental way of acknowledging the children as self-organizing, living systems who have their own mathematical knowledge. Our emphasis on children's schemes is compatible with Piaget's emphasis on spontaneous reconstruction as well as with Vygotsky's ontogenetic orientation, which asserts 'that it is possible to understand many aspects of mental functioning only if one understands their origin and the transitions they have undergone' (Wertsch and Toma, 1994, p. 161). The mental functioning to which the mathematics of children refers consists of children's use of their mathematical schemes. But it is not that we want to simply observe children's mathematical activity; rather, we want to be involved in the transitions it undergoes. In the process of doing so, we learn how we can bring forth the mathematics of children and how we can modify it. The mathematics of children is not independent of our mathematical concepts and operations because it is constructed partially through their interactions with our goals, intentions, language, and actions.

Our orientation is compatible but not identical with Vygotsky's ontogenetic orientation. It is compatible because by seeking to bring forth and to modify the mathematics of children, we are involved in its construction. It is not identical because we acknowledge that the mathematics of children is an observer's concept springing from our own conceptual operations as we attempt to explain children's language and actions. Although we regard children as having a mathematics of their own, the only access *we* have to their mathematics is through our interactions with them. We contribute our conceptual operations to our interactions just as children contribute their operations and basic structure of a logico-mathematical nature to their interactions.

The mathematics of children appropriately includes those operations and basic structures identified by Piaget. But it is not restricted to those products of spontaneous development. In contrast to Piaget's approach, in which no agent intends for children to construct reversibility or transitivity, for example, we do intend for children to construct mathematical concepts and operations of a particular kind. We imagine how the children might engage in mathematical activity before we actually work with them. We analyse the children's current cognitive schemes, plan specific situations, and discuss what we consider to be appropriate interactions in their zones of potential constructions. We also discuss how we might interact with the children so that they might form goals for action and initiate and sustain mathematical activity. In the actual teaching episodes, it is our intention to look for the children's contributions to the situations and to interpret the language and actions of the children on-the-spot. We attempt to act on those interpretations in ways we perhaps did not foresee in our planning sessions. The children's language and actions guide us as we interact with them. This is an important point and one which we believe distinguishes our approach from those who work from a Vygotskian perspective. We do our very best to give voice to the children's mathematics. Nevertheless, when observing the mathematics of children, we co-create our occasions of observation with children. This is critical for us because it is one of our goals to learn how to act when teaching children mathematics.

We plan specific situations for the children with respect to computer microworlds we designed for the teaching experiment. When designing the microworlds, it was our intention to create possible actions, which, when used by children with specific intentions and purposes, could be implementations of their mathematical operations. In this sense, the possible actions can be regarded as mathematical tools for the children to use. As tools, the possible actions are not independent of the children's mathematical operations. The use to which the possible actions are put depends on the operations of which the children are capable.

Our design of the possible actions of the microworlds is compatible with Vygotsky's belief that a defining feature of human mental functioning is its mediation by tools and signs (Wertsch and Toma, 1994). Vygotsky (1981) believed that 'by being included in the process of behavior, the psychological tool alters the entire flow and structure of mental functions' (p. 137). He went on to say that the psychological tool 'does this by determining the structure of a new instrumental act, just as a technical tool alters the process of a natural adaptation by determining the form of labor operations' (p. 137). The possible actions of a microworld can be regarded as technical tools that can be used by children to serve their goals and intentions in the context of doing mathematics. We believe Vygotsky would agree that, for a child who has constructed the mental operations of partitioning, the possible action *parts* could indeed be a useful device. But it is the child who makes the possible action useful by assimilating it and by coordinating it with his or her mental operations. To the extent that the assimilated possible actions are psychological tools, their use is dependent on the overall goals and intentions of the individual using them. In this view, psychological tools can

Figure 2.1: The Computer Screen of Sticks

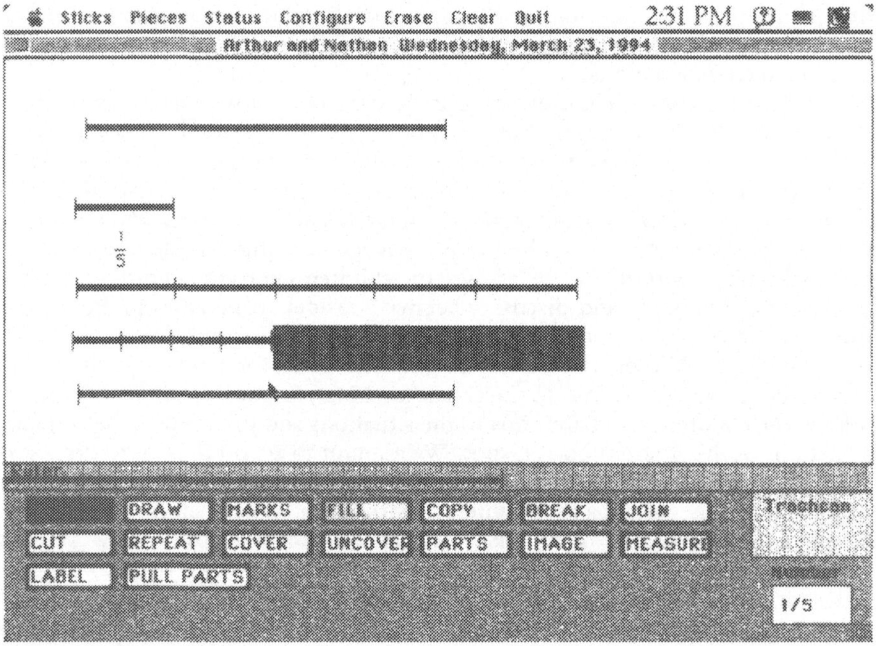

modify actions, operations, and intentions, but they do not determine the structure of those modified actions and operations because the operations were available to the subject prior to the assimilating activity.

The possible actions in the microworld Sticks (see Figure 2.1) are planned and programmed with respect to the mental operations we believe are involved in the generation of the rational numbers of arithmetic. These possible actions include: draw, copy, marks, parts, cut, join, repeat, break, image, pull parts, and measure. With 'draw' activated, a stick may be drawn by sweeping the cursor horizontally across the screen. This stick can be copied as many times as desired by using 'copy'. Another possible action is to mark a stick with a vertical hashmark by clicking on the button 'mark', positioning the cursor on the stick where the mark is to be made, and clicking on the stick. The stick can be also marked into any number of equal pieces by activating 'parts', dialing a whole number between 2 and 99, and then clicking on a stick. A stick may be cut as many times as desired using 'cut', joined to any other stick using 'join', or dragged to any position on the screen (including a trash box). A stick can be repeated any number of times and automatically joined to those preceding end-to-end using 'repeat' (a short-cut of 'copy' and 'join'). Any marked stick may be broken using 'break', and any mark can be moved or erased using menu commands. Using 'image', an image of a stick can be made and dragged to any desired position. One can also pull any number of parts out of a stick without changing its original appearance by using 'pull parts'. Any stick can be measured using any other stick by copying the stick to be used as a measuring unit into the ruler, activating 'measure', and then clicking on the stick to be measured. Two auxiliary possible actions are 'fill' and 'label'. Any stick or part of a stick can be filled in with one of ten different colours or labelled by a fraction symbol.

An Analysis of a Teaching Episode in Sticks

To illustrate our approach, we describe how Arthur and Nathan, two 10-year-old children, used their partitioning and iterating operations in the microworld Sticks. We argue that the children had already reconstructed these operations in a way Piaget described and that we brought the operations forth through social interaction. We also argue that the two children used what we refer to as an iterative partitioning scheme as they worked in the microworld Sticks in a goal-directed mathematical activity. The iterative partitioning scheme appeared in the children's mathematical activity without suggestion from their teacher.

Partitioning and Iterating Operations

In a teaching episode held on 8 October 1992, the teacher asked Arthur and Nathan to make a set of fraction sticks based on a unit stick Arthur has just made.

Protocol I (8 October 1992)

T: What I want you to work on together is to make a set of fraction sticks. What do I mean by a set of fraction sticks based on that unit stick?
A: Thirds (holds up three fingers) — it would be thirds.
N: A set of fifths would be five, sevenths would be seven, and hundredths would be a hundred.
T: Okay. What would be the largest fraction to start with?
A: One-half!
T: Start with one-half, then go to the next, the next, . . .

The teacher's primary goal was to establish a situation that could lead to independent mathematical activity. The situation as it was eventually formed could be called open-ended in that the children could use various strategies in making an indefinite number of fraction sticks.

Based on Arthur's comment, 'Thirds, it would be thirds.' and on Nathan's elaborating comment immediately following, we infer the children formed a shared goal to make fraction sticks — sticks partitioned into equal sized pieces. We also infer that the children could mentally project a specific numerosity of units of equal size into the original stick, and that they could disembed any one of these units from the stick and iterate that unit a sufficient number of times to reproduce a partitioned stick of length equal to the original.

The children seemed to have established what we regard as partitioning and iterating operations. These operations are the basic reason why the children could form a shared goal as well as a shared open-ended situation. Without these common operations, it would have been implausible for the children to form a shared goal or to engage in collaborative mathematical activity for any length of time.

An Iterative Partitioning Scheme

After the children established a shared starting point and goal, the teacher assumed the role of listener, making as few interventions as possible into their independent mathematical activity. Trying to mark the unit stick into two equal pieces, the children took turns using the mouse. The challenge of marking a stick into two equal pieces could have been circumvented by allowing them to use 'parts'. We chose, however, to configure the microworld so that 'parts' was unavailable to the children; it was our goal to find what actions the children would use in the microworld to mark the stick into its halves.

Protocol II (Continuation of Protocol I)

A: (After making a copy of the original stick) Let's see, we'll have to break it, so marks. (Clicks on 'marks' but does not mark the stick; gives the mouse to Nathan.)

N: Marks, right? But how will we make the marks?

A: Wait. (Fast) wait, wait, wait, wait (takes the mouse from Nathan and refuses to let go while smiling at Nathan). I need it.

N: (Lets go of the mouse.)

A: (Works for approximately three minutes. Draws a stick while mumbling 'one-half'. Repeats it using 'repeat', producing a stick marked in the middle that is shorter than the original stick. Starts over, drawing a stick on top of the original stick, visually estimating where to stop; repeats the new stick, producing a stick marked in the middle but shorter than the original. Draws a stick directly beneath the original stick, again visually estimating where to stop drawing; repeats this stick, producing a stick marked in the middle but this time longer than the original. Cuts the stick using 'cut' making it equal to the original. Gives the mouse to Nathan.)

Arthur's strategy of visually estimating where to stop drawing, indicates that he focused on the length of the original stick. His use of 'repeat' indicates that 'one-half' referred to one of two equal parts of the original stick, and that the parts he made were iterable parts of the whole stick. Consequently, we attribute an iterative partitioning scheme to him and the assimilating operations of partitioning and iterating.

An iterative partitioning scheme is also indicated by Nathan's actions. When Arthur gave Nathan the mouse, there were three sticks on the screen of same length; the original, the copy that Arthur had made, and the last stick that Arthur made using 'cut'.

Protocol III (Continuation of Protocol II)

N: (Works for approximately four minutes. Places two of the three sticks end-to-end and the third stick over the two, trying to gauge when the adjacent endpoints of the two sticks separate the third stick into two pieces of the same size. Marks the two sticks lying end-to-end at the endpoints of the third stick, then moves the third stick above the two others).

(After breaking the two marked sticks lying end-to-end, checks to

Figure 2.2: The Screen after Arthur's Attempt

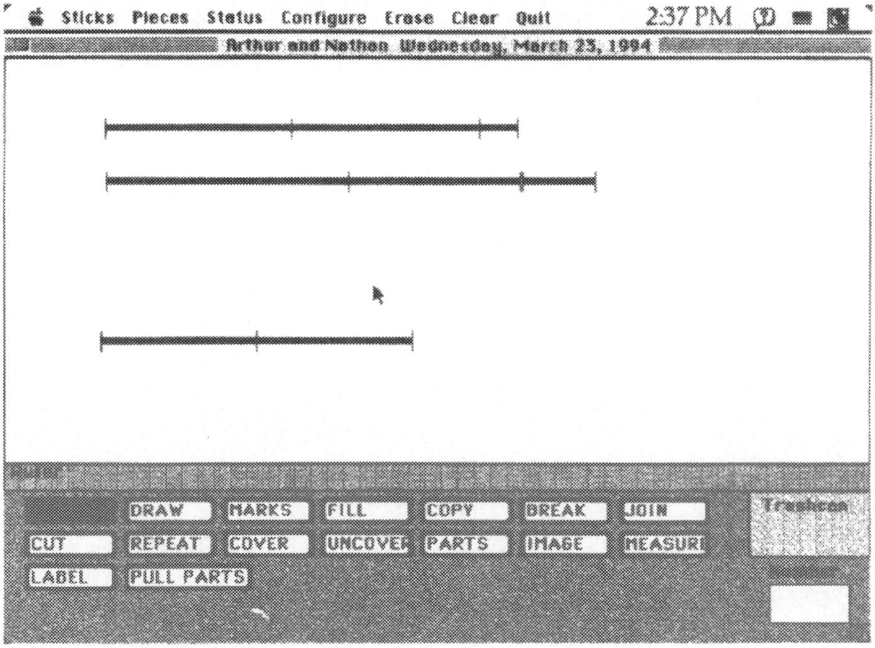

find if the two pieces of one stick are of equal size by placing them on the third stick. They are visually asymmetrical, so drags the two pieces away. Joins the two pieces of the remaining broken stick and drags the joined stick to the third stick; breaks it and cuts a small piece off the longer of the two pieces; repeats the remainder of the cut piece, producing a stick marked in the middle; he compares this stick with the third stick by placing it beneath it and finds that they are of the same length.

Nathan's actions of placing the two sticks end-to-end and the third stick on top of these two indicated the presence of partitioning operations of the same nature as Arthur's. Cutting a piece off the longer of the two pieces of a stick and then repeating that part is quite similar to Arthur's actions of repeating a drawn stick and comparing it to the original. The comparison both children made between the stick produced by using 'repeat' and the original stick illustrates the necessity they must have felt that a stick together with its replicate must produce the original stick or one of equal length. This necessity was a result of the children's conceptual operations that they enacted in the microworld.

Both Arthur and Nathan used the possible actions of the microworld to achieve their goal of marking a stick into two equal pieces. In this sense, the possible actions constituted tools. But, as tools, the possible actions were not somehow independent of the children's mental operations. Rather, they served in the enactment of the children's mental operations. For example, the children's use of 'repeat' constituted an enactment of their mental operation of unit iteration.

Figure 2.3: The Screen during Nathan's Attempt

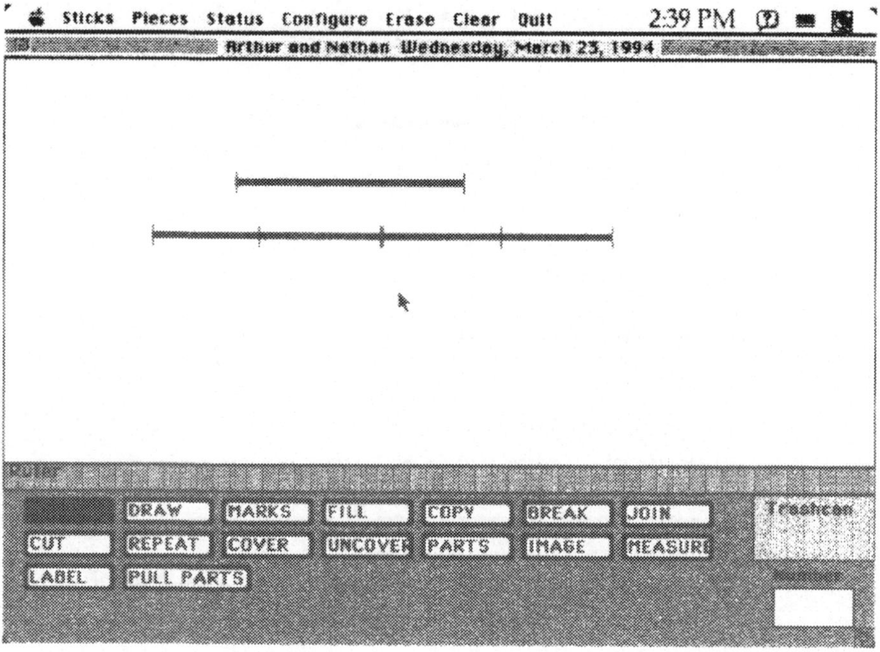

The children did not simply repeat a stick, but did so with the intention of making a stick of length equal to another stick. We would not say, however, that the possible microworld actions as assimilated by Arthur and Nathan determined the structure of their actions. Rather, their intentions as well as the conceptual operations available to them were also involved. Nevertheless, 'repeat' made it possible to act in ways that simply would not have been possible in other media. We claim that this and other possible actions became a constitutive part of their partitioning and iterating operations. Our claim provides an alternative interpretation of Vygotsky's claim that a defining feature of human mental functioning is its mediation by tools and signs.

Mathematical Interactions

Given the lack of verbal interaction in Protocol II, it might seem as if no mathematical communication had occurred between the children. To offset this misleading interpretation, it is important to interpret their mathematical activity as occurring in the context of their agreed upon open-ended situation and goal of making fraction sticks. While the children were working, there were four indications of an implicit consensual agreement between the two that they were trying to make one-half of the original stick, and further, one-third, etc. Arthur's refusal to be dominated and his subsequent collaboration with Nathan is the first indication of a consensual agreement. Arthur had just recently joined the teaching experiment, but Nathan had been a participant for one year. Therefore, we interpret Arthur's establishment of co-ownership of the mouse as an attempt to establish equality with Nathan. The next indication is the exchange of the mouse between the two children without any verbal comments of the sort that Arthur was done and that Nathan should begin. The third indication is that Nathan made no suggestions for how Arthur should proceed while he was working — Arthur had indeed established his equality. The fourth indication is that each child understood what the other was doing without interrupting the other. The two children's use of the results of the actions of the other to serve in their achievement of a goal constitutes reciprocal interactions.

Maturana and Varela (1980) suggested two cases of interactions by which an organism can affect the behaviour of another organism:

> (a) By interacting with it in a manner that directs both organisms toward each other in such a way that the ensuing behavior of each of them depends strictly on the following behavior of the other . . . A chain of interlocked behavior can thus be generated by the two organisms. (p. 27)
> (b) By orienting the behavior of the other organism to some part of its domain of interactions different from the present interaction, but comparable to the orientation of that of the orienting organism. This can take place only if the domains of interactions of the two organisms are widely coincident; in this case no interlocked chain of behaviour is elicited because the subsequent conduct of the two organisms depends on the outcome of the independent, although parallel, interactions. (Maturana and Valera, 1980, pp. 27–8)

We could not say that the ensuing behaviour of each child, as described in Protocols II and III, depended strictly on the following behaviour of the other — it was not a chain of interlocked behaviour. The relation between the two children could be more adequately understood by the second case, in which, according to Maturana and Varela (1980), the two organisms communicate. To infer that the two children were indeed communicating as explained by Maturana and Varela, we would need to establish that their interactions in the microworld were widely coincident. The indications are that (a) the children established a shared goal of making fraction sticks, and (b) each child understood and used the results of the other.

Maturana and Varela (1980) commented that 'the second case is the basis for any linguistic behavior' (p. 28). We certainly agree with them, but there was a marked absence of linguistic exchange between Arthur and Nathan during the period of their consensual activity. So, rather than use 'mathematical communication' to refer to what transpired between them, we make a distinction between two types of mathematical interactions: (a) *non-verbal* mathematical communication to emphasize their interactions in the microworld and their concomitant non-verbal interpretations of each other's interactions, and (b) *verbal* mathematical communication to emphasize their verbal interactions with each other and with their teacher as they established the goal of making fraction sticks. Both types of mathematical communication are examples of social interaction.

The verbal mathematical communication between Arthur and Nathan in Protocol I illustrates a difficulty when simply applying Vygotsky's general learning theory to explain what goes on in a consensual domain of interactions. The partitioning and iterating operations of the children were brought forth through social interaction in the context of the teaching episode. But we could not say that these operations first occurred between the children in social interaction and only then were they internalized. Rather, the operations had been already *interiorized* (not internalized) prior to the teaching episode. There was nothing that would lead us to infer that these operations first appeared in social interaction intermentally, and then, through a process of internalization, intramentally.

The non-verbal mathematical communication between Arthur and Nathan also presents a difficulty in applying Vygotsky's general learning theory to explain what goes on in a consensual domain of interactions. This communication was social, but it was dependent upon the children enacting their iterative partitioning schemes in the microworld to make fraction sticks. We could not say that the qualities of thinking were generated by the organizational features of the social interaction, as suggested by van Oers. Rather, the social interaction was dependent on the children's iterative partitioning scheme. This operative scheme served as the basis of the organization of the children's communication rather than being a result of the organization.

The question now is whether there was ever a time when the operations appeared in social interaction prior to their interiorization by Arthur and Nathan. This same question is germane to all of the operations Piaget identified as being developmental, such as reversibility or transitivity. What is the role of social interaction in children's construction of these operations? Although this interesting question lies outside of the methodology of the teaching experiment, we do have some thoughts on the more general question of the relation between social interaction and children's construction of basic mathematical operations. It is implausible, for example, that children's construction of representational space (Piaget and Inhelder, 1963) would first appear on the social plane and only then on the intramental plane. That is not to say that social interaction does not contribute to children's construction of representational space. But we fail to understand how social interaction can be the primary source of children's construction of representational space. Fundamental operations such as re-presentation (von Glasersfeld, 1991) — that is, the regeneration of a past experience in visualized imagination — are the primary sources of representational space. Like the non-verbal interaction between Arthur and Nathan, we believe these operations are indispensable in children's mathematical interactions. In particular, they form the basis of what children contribute to social interaction and in turn are modified by the social interaction. These operations also serve in more general subject–environment interactions of which social interaction is but one type. We look at these fundamental mental operations and to general subject–environment interactions for an explanation of children's construction of operations like partitioning and iterating. Limiting our explanations of higher mental functions to social interaction leaves out operations like unitizing, uniting, and re-presenting as well as subject–environment interactions that are not social. Such a limitation is unnecessary and counter-productive. But neither do we want to limit our explanations in such a way that they leave out social interaction, because social interaction is the primary means of interacting mathematically with children.

Learning a Units-Coordinating Partitioning Scheme

Based partially on our knowledge of the partitioning and iterating operations of Arthur and Nathan, we decided to guide the children toward establishing fraction families. We knew that the children had constructed a units-coordinating scheme in multiplicative contexts (Steffe, 1992). That is , the children could coordinate the production of, for example, a unit of four and a unit of five by distributing the unit of five across each element of the unit of four, producing four units of five as a result. We intended for the children to combine their units-coordinating schemes and their unit fraction schemes to produce families of equal fractions.[3] The children had not established fraction families prior to our working with them.

We decided to work with the fraction 1/3 rather than with 1/2 because children's special knowledge of doubles often makes it unnecessary for them to intentionally carry out conceptual operations to achieve their goals. It was the specific goal of the teacher to establish a situation of learning rather than to encourage independent mathematical activity. So, the situation was initially not open-ended from the point of view of the children.

Protocol IV

T: (To Arthur) could you draw a long stick for me?
A: (Draws a stick.)
T: (To Nathan) mark the stick into thirds before you make a copy.
N: (Activates 'parts', marks the stick into three parts; makes five copies; upon the teacher's request, fills 1/3 of the original stick).
T: (To Arthur) can you pull that out?
A: (Activates 'pull parts', pulls the coloured part of the original stick from the stick, leaving the original stick intact as shown in Figure 2.4).

Upon the teacher's request, the children contributed the actions involved in establishing the situation, but the situation was open-ended only from the point of view of the teacher.[4] The children had formed no goals that would sustain independent mathematical activity of the nature illustrated in Protocols II and III. But at this point, we were not interested in establishing an open-ended situation that would bring forth independent mathematical activity. Rather, we were interested in establishing a situation of learning that might lead to an accommodation of the children's fraction schemes.[5] Prior to the teaching episode, we had considered a zone of potential construction for the children by imagining a consensual domain of mathematical interactions like those in Protocol IV. We thought we knew enough about the children that we could anticipate what they might do.

Protocol V (Continuation of Protocol IV)

T: Now, what is a way to represent that one-third by another fraction?
N: Two sixths!
T: Arthur, how would you do that?
A: (After thinking for approximately 15 seconds) cut the third into one-half! (Dials '2' and clicks on the stick pulled from the original stick.)

Figure 2.4: The Screen after Arthur Had Used 'Pull Parts'

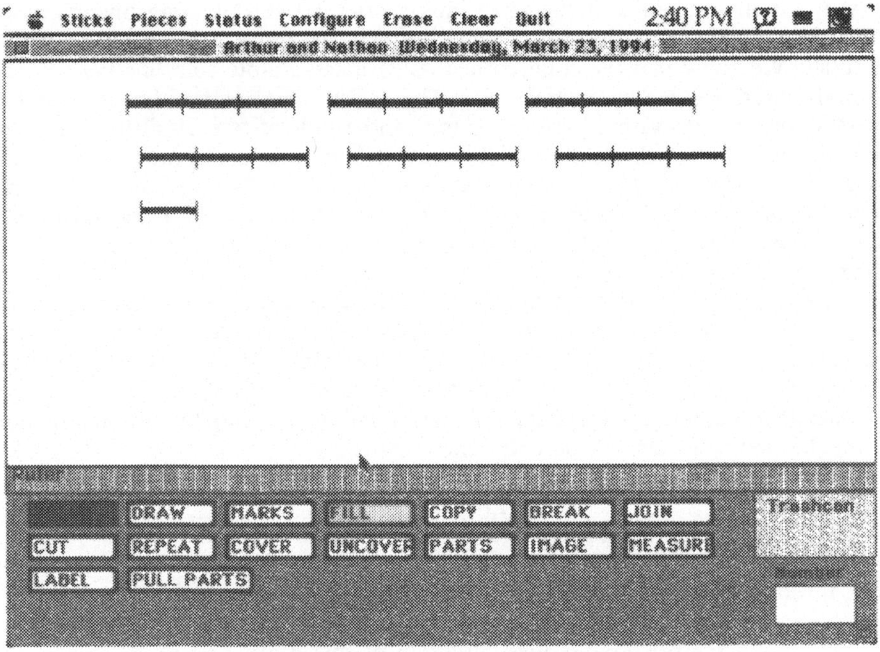

T: That would be?
A: Two-sixths (activates 'label', labels the marked stick '2/6').
T: Okay, Nathan, could you use a copy to do another one?
N: (Activates 'parts', dials '3,' and clicks on a stick marked into thirds; activates 'pull parts' and pulls the first three parts from the stick, yielding the arrangement pictured in Figure 2.5. Activates 'measure', clicks on the 3/9ths stick, yielding '1/3'. Pulls one part of the three from this stick and measures it, yielding '1/9'.)
T: Can you tell me another way to figure out ninths?
N: There are three in each third, and three times three is nine!

Nathan's utterance 'two-sixths' oriented Arthur toward finding a way to reconstitute one of three equal pieces into sixths. 'Sixths' had two meanings for Arthur. It meant six equal parts as well as any one of these six parts iterated to reconstitute a partitioned whole. However, he faced the additional constraint of finding a way to transform three equal pieces into six equal pieces. This transformation involved a coordination of breaking the pulled part into two pieces and of breaking the three parts of the original stick into six pieces. In the coordination, the pulled part had to be maintained both as a part separate from the marked stick and as a part belonging to the marked stick. In this, Arthur had to simultaneously apply partitioning operations to the three pieces to produce six pieces without destroying the three pieces. The transformation was even more demanding because the six pieces had to remain embedded in pairs within the three pieces so

Figure 2.5: Computer Screen after Nathan's and Arthur's Work

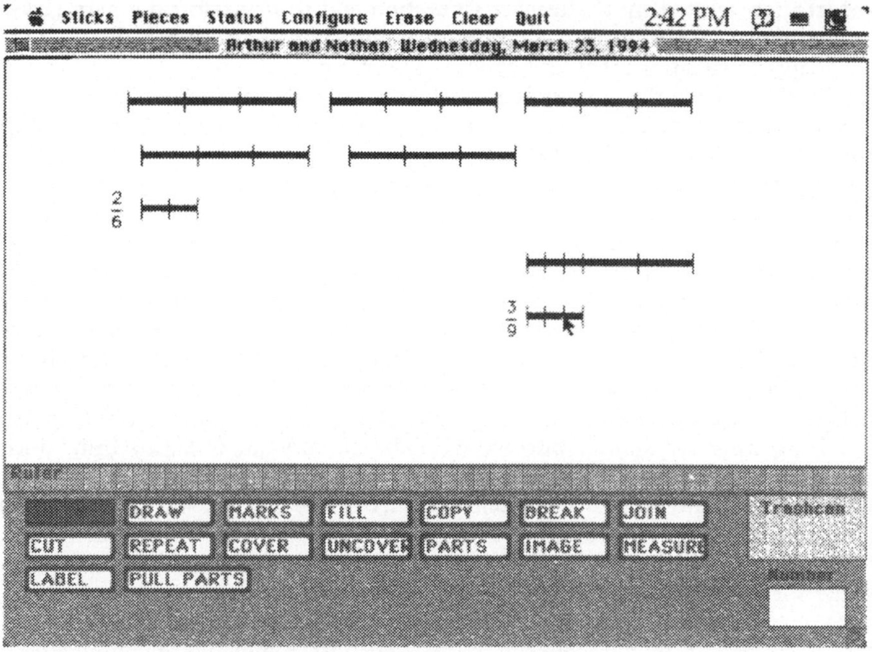

that one piece out of three yielded two smaller pieces out of six. These operations constituted the beginnings of recursive partitioning, a sophisticated form of units-coordination.

Arthur's coordination is an explanation of what he learned. But it does not fit our understanding of what it means to solve a problem because, as will be confirmed in Protocol VI, he activated the units-coordinating scheme he used in multiplying to partition the three parts of the partitioned whole. Hence, he had a coordinating scheme ready-at-hand and did not need to construct a mechanism of coordination. Nevertheless, we consider the activation of the units-coordinating scheme an event of learning because it was activated in the context of Arthur's apparently novel situation. Moreover, there were two other novelties in its application. First, he knew the results of applying it (six), as well as the numerosity of one of the composite units (three), and he had to find the numerosity of the second composite unit, which was unknown. Second, he had to use it in simultaneous partitioning rather than in distributing, which can be sequential rather than simultaneous.

The Function of Social Interaction in Learning

A primary goal of social interaction in mathematics learning is to induce perturbations of the nature experienced by Arthur. We had no foresight that Nathan would immediately say 'two-sixths', and such unexpected events will always be the norm in mathematics teaching. Nevertheless, we anticipated prior to the teaching episode that both Arthur and Nathan would be able to bear the perturbation created by the teacher's question and to neutralize the perturbation by conceptually coordinating schemes they already had available.

Planning situations that lead children to sense perturbation and to successfully modify their currently functioning schemes is an essential component of our involvement in children's construction of mathematics. It is crucial to present a situation in such a way that the children activate one or more schemes. Also, the schemes must be insufficient for the children to achieve the set goal. In this, we strive to think like the children in order that our language and actions are in harmony with the children's current knowledge. Being in harmony with the children serves as a background against which we can initiate provocations that might induce the children to modify their current mathematical knowledge.

From the perspective of an observer of the interaction in Protocol IV, it is legitimate to say that the teacher's knowledge of rational numbers constituted cultural meanings of mathematics and that Arthur and Nathan were trying to make sense of those cultural meanings. But this global analysis did not help us in specifying how we should interact with the children in order to bring forth, sustain, and modify their mathematics. For instance, the teacher could have chosen to demonstrate how to make one-sixth from one-third, while explaining and discussing what he was doing. If this had been the case, the quality of the children's thinking might have been actually generated by the organizational features of the social interactions. However, this method of teaching — demonstrating, explaining, and discussing — does not respect the mathematics of children nor does it contribute to our concept of children with mathematical knowledge of their own.

We could not say what Arthur learned (Protocol V) was actually generated by organizational features of the social interaction. Nathan did say 'two-sixths', but that served only to orient Arthur. Nathan did not perform the actions of breaking each third into sixths, nor did he indicate in any way to Arthur that this was what he would do if he were to find another fraction to represent one-third. Rather, Arthur contributed his own operations to neutralize the perturbation he seemed to experience during the 15 second-thinking time. When he said. 'Cut the third into one-half', he constructed a way of proceeding that belonged to him. He had learned how to act (interact) in the microworld with regard to the situation at hand. We claim that Arthur had learned a very general way of interacting in the microworld, and he could now communicate with Nathan concerning those potential interactions.

Still, the mathematical interactions that transpired between the teacher and the children (Protocols IV an V) did influence the quality of what Arthur (and subsequently, Nathan) learned. The children's willingness to carry out the requests of the teacher led to a situation they co-produced. This co-production of the situation kept the children active and involved, and when the teacher posed the question. 'Now, what is a way to represent that one-third by another fraction?' (Protocol V), both children could interpret the teacher's question relative to their current and immediate records of operating. The children were in a state of activation, and the teacher's question made sense to them relative to their past operating and to possible actions they could perform to make another fraction.

It wasn't as if the children were trying to make sense of a mathematical object 'out there' — another fraction. What they were trying to do was to find a way of interacting in the microworld so that they could make another fraction based on the one they had already made. 'Making sense', then, can mean (a) to construct ways and means of operating in a medium, based on current knowledge, in order to neutralize perturbations induced through social interaction, and (b) to become explicitly aware of those potential ways and means of operating through a process of symbolization. In this, the children must be able to bear the perturbations. This is how we interpret van Oers' comment that mathematics learning is a process of making sense of mathematics as brought to us by cultural history.

Symbolization in Learning

After Arthur and Nathan had taken turns making 2/6, 3/9, 4/12, and 5/15 and had labelled each marked stick with the appropriate fraction numeral, the teacher decided the children could engage in symbolic activity.[6] The goal was for Arthur and Nathan to transform the potential actions of producing equal fractions into symbolic action. Using 'label', the teacher selected '1/18' and asked the children what the numerator should be.

Protocol VI

A: Six! Because three times six is eighteen.
T: (Selects '1/66')
N: Twenty-two! Because three times twenty-two is sixty-six.

Symbolizing the potential operations involved in making and coordinating partitions of the part and the whole was possible and almost immediate. Both children abstracted the multiplicative scheme (the units-coordinating scheme) they had combined with their unit fraction schemes to produce fractions based on 1/3. It might appear as if the children had constructed an equivalence scheme for fractions on a symbolic level. However, they had yet to construct fundamental operations as indicated by the following protocol.

Protocol VII

T: (Asks Nathan to put the numeral '22/66' on the screen and to make a copy of the unmarked original stick) I want to see how you would do that one.
N: Ah, Um . . .
T: Could you do it using this one here (points to one-third of a marked stick)?
N: Yeah, there would be (can't use the teacher's suggestion, so begins dialing up to 66 using 'parts'.
T: You can't do it that way. Can you find another way to figure it out?
N: (Dials to '33' and clicks on the stick; pulls ten parts from the stick, dials '2' and marks each of the ten parts into two equal parts. Using 'repeat', repeats the stick three times, pulls six parts from it, and then joins that stick to the stick marked into sixty parts, forming a stick marked into 66 parts.)

In a subsequent task, where the children's activity was evolving into independent mathematical activity, Arthur selected '1/69' as the denominator. In his attempt to make 23/69 using an unmarked unit stick, Arthur dialed to '23', and then marked the original stick into 23 parts. He then dialed '3' and marked each of the first nine parts into thirds in an attempt to generate 23 of 69 parts. Being unsuccessful, he then pulled 23 of the 27 parts, intending to repeat this stick to make a stick with 69 parts.

Although Arthur's actions exemplify an awareness of the units-coordination of three and 23 to make 69 parts, there was still no indication of an awareness of the structure of operating with his unit-fraction scheme. Neither he nor Nathan structured their actions starting with one out of three equal parts. Had the children constructed equivalent fractions, the most likely method for making 22/66 or 23/69 would have been to partition a stick into three parts of the same size and then partition each one of the three parts into 22 or 23 parts. There is always the possibility that the children could have operated in this way, but in subsequent teaching episodes, it proved to be a conceptual obstacle for both children.

Coordinating a partition of a third pulled from the whole with the same partition of each of the three parts in the whole apparently does not imply an inversion of the partitioning operations. That is, it does not necessarily imply a coordination of uniting the subparts of the partitioned third into a composite unit and uniting the subparts of the three parts of the whole into three composite units. When the children symbolically partitioned a part pulled from the three parts of the stick, say into 22 subparts, and then each of the three parts in the whole stick, yielding 66 subparts, the results of the partitioning actions seemed to consist of composite units of subparts of the second partition. That is, the children seemed to focus on the 22 and 66 subparts rather than on partitioning the unit into three parts each partitioned into 22 subparts. They seemed to lose the first partition when focusing on the second. Symbolizing their potential operations was well within their zone of potential construction, but the symbolizing process did not reveal to the children the structure of operating that was obvious to us as we observed what the children did.

Uniting operations assemble individual unit items into a composite unit and partitioning operations break a simple unity into parts — individual unit items. These operations, when established as inversions of each other, yield an operative scheme for finding fractions equivalent to a given fraction. But even this reversible scheme is qualitatively distinct from an equivalence relation for fractions as understood by the children's teacher. Furthermore, Protocol VI provides a case of children not making sense of their teacher's helpful hints, because they could not assimilate and interpret those hints using their current schemes.

Perspectives

In the superseding model of mathematics learning we are building, we are trying to reformulate the concepts of learning and development so they are useful in social interaction with children. Learning is regarded as accommodation in the context of scheme theory (von Glasersfeld, 1991) and occurs as a product of interaction. This view of learning has two advantages.

First, learning can be regarded as being a result of intentional interaction on the part of the adult, but be non-intentional on the part of children. Although children can participate in educational contexts with an intention to learn, that intention cannot include an intention to learn a specific item. It can be only a general willingness to be involved in situations of learning in which the children know that something is going to be learned. What is being learned is regarded as the result of a particular accommodation of the children's current mathematical concepts or schemes, which usually occurs outside of the awareness of children.

Second, the teacher's interventions are essential in children's learning. But in this, we speak in terms of perturbations as well as in terms of provocations, because it is the children who must experience the perturbations. A provocation by the teacher can only induce a perturbation in the child, and there is no necessary correspondence between the two.

Neutralizing Perturbation

There are three basic ways for children to neutralize perturbation. First, it is quite possible for children to take action which neutralizes a perturbation without modifying their current schemes. This was essentially the case when Arthur and Nathan found a way to mark a stick into two pieces of equal size in Protocol II. Another way children have to neutralize perturbation is to modify their current schemes, which is what we mean by learning. This latter case is illustrated in Protocol V by Arthur modifying his unit fraction scheme to find another fraction for 1/3. His search for a way to proceed was made easier by Nathan's comment 'two-sixths', because Arthur knew how to make any fraction using possible actions of the microworld. The learning event was different from what it might have been had Nathan not made the comment, but this only illustrates differing possibilities for a teacher's interventions. What Arthur learned was not limited to the particular context of learning, because he could find fractions other than 2/6 equal to 1/3, barring practicalities. Moreover, in future teaching episodes, he demonstrated that he could find fractions equal to any proper fraction, which demonstrates what we mean by potential actions. His combination of his units-coordinating scheme and his unit-fraction scheme was indeed an accommodation. However, even this scheme was qualitatively distinct from what we regard as fractional equivalence.

Third, a child may neutralize a perturbation by a major reorganization of a current scheme. None of the accommodations illustrated in the protocols were of this nature. The most likely candidate was Arthur's construction of his units-coordinating partitioning scheme. His construction of the scheme opened a broad range of mathematical activity for him that was on a par with that in which Nathan could already engage. Nevertheless, the possibility of Arthur combining his units-coordinating scheme and his unit-fraction scheme was ready-at-hand and did not constitute a major reorganization of the latter scheme although it was a reorganization. It was not a major reorganization because, from all indications, it was constituted at the same level of interiorization as his unit-fraction scheme.

Arthur's scheme for making two-sixths was more complex than the schemes he modified. But it is somewhat misleading to say that the more complex structure might be learned if it is based on a simpler structure, because the observer can know only in retrospect what a child does learn. What the child learns may not

Figure 2.6: Social Interaction, Learning and Development

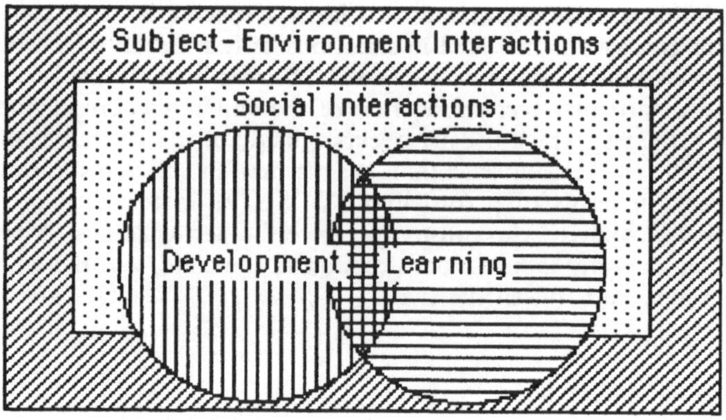

be what was intended by the teacher, and this simple realization is what blocks us from adopting the point of view that cultural meanings of mathematics may be taught in any straightforward way.

Had Arthur been able to start with 23/69 and produce 1/3 in such a way that one could infer inversions of those operations that he used when producing 23/69 starting with 1/3, then there would have been a solid indication that he was aware of how he operated in the latter case. Such an awareness indicates interiorization of operations, which is one thing we mean by mathematical development.

Social Interaction, Learning and Development

The relationships we perceive among social interaction, learning, and development is diagrammed in Figure 2.6. We view learning as engendering mathematical development of a particular kind, but there are developmental patterns that are difficult to adequately explain solely on the basis of learning. For example, mathematical learning is not sufficient to explain children's construction of number sequences, because this construction involves a metamorphosis of children's sensory-motor counting schemes (Steffe, 1991).[7] Nevertheless, we regard the process of metamorphosis to be set in motion in experiential contexts as a result of a particular kind of accommodation — an accommodation where children reprocess results of operating. We call such an accommodation engendering in that case where it leads to auto-regulation of those processes that constituted the accommodation. If auto-regulation yields a reorganization of the scheme or schemes modified in the engendering accommodation at a new level of interiorization, then we refer to the accommodation as metamorphic or developmental. This is another way to express the insufficiency of experience in the construction of mathematics. In this, we can legitimately say that learning contributes to development in a way that is compatible with Vygotsky's claim, while retaining the Piagetian notion that development contributes to as well as constrains learning.

The kind of learning in which we are most interested occurs through social

Figure 2.7: Types of Interaction

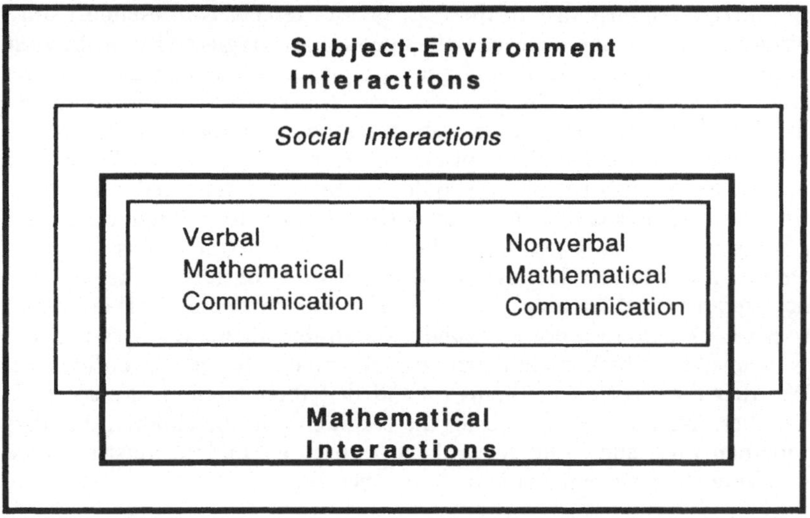

interaction, one type of the subject–environment interactions depicted in Figure 2.6. We have illustrated that mathematical interactions include mathematical communication, non-verbal and verbal, as illustrated in Figure 2.7. But they also include non-communicative mathematical interactions. One is the mathematical interactions of human beings with non-human elements in their environments and another is the mathematical interactions with others that are interlocked as explained by Maturana and Varela.

Mathematical Interactions and Independent Mathematical Activity

An important realization is that not all social interaction between two or more human beings leads to learning. We illustrated in Protocols II and III how Arthur and Nathan engaged in non-verbal mathematical communication without learning anything mathematical that was observable to us. As the two children interacted mathematically in the microworld, they implemented conceptual operations that were already available to them; partitioning unit sticks and iterating a substick to produce the original unit stick. But they did not establish a general method for finding where to mark a stick into two equal pieces. Nevertheless, each engaged in actual mathematical activity the other child could interpret. We regard such independent mathematical activity as being of central importance in the mathematical education of children because it represents, along with mathematical activity involving verbal mathematical communication, a major portion of what we regard as the mathematics of education.

We believe that it is a fundamental responsibility of mathematics teachers to learn how to bring forth independent mathematical activity of children through social interaction and how to establish mathematical communication with and among children. Although it is possible for this independent mathematical activity to engender learning, it is not the only context in which we encourage learning. Another major responsibility of the mathematics teacher is to establish situations of learning that he or she believes will engender learning activity of the children. These situations of learning should generate perturbations in what we regard as the children's range of bearable perturbations. A perturbation is likely to be bearable if a child also experiences the satisfaction of assembling those actions that neutralize the perturbation. This places the teacher in a very delicate position, because the situations of learning should not be so far removed from children's current ways and means of acting mathematically that the children cannot modify those ways and means to neutralize the experienced perturbations.

Perturbations have a dual nature that we always consider — they drive mathematical activity and learning, but they also can be debilitating in those cases where paths of action cannot be found to neutralize them. By stressing the children's independent mathematical activity, we strive to foster the confidence in the children that they can do mathematics and that their methods are acceptable to their teacher. We also strive to foster the confidence in the children that they can learn mathematics, and to do so we believe that the children must be successful in modifying their current mathematical schemes.

Notes

1. The research reported in this chapter was conducted as part of the activities of the NSF Project No. RED-8954678, Children's Construction of the Rational Numbers of Arithmetic. All opinions expressed are solely those of the authors.
2. 'External objects' means that the objects are external to the individual's present cycle of operations, but not external to the possible experiential worlds of the individual.
3. In a unit fraction scheme, starting with a fractional whole the child must be able to find n parts of the whole such that n iterations of any part yields the whole. In this, starting with the fractional whole, the child must be able to find a part such that the whole is n times the part and the part is one of n parts of the whole.
4. There are two broad types of open-ended situations. The first brings forth independent mathematical activity like that in Protocols II and III and the second brings forth problem-solving activity in which children must modify their current schemes to be successful. Situations of the second type are legitimate situations of learning. Here, we are speaking of open-ended situations of the first type.
5. We consider learning to include problem-solving activity, but we do not equate learning and problem-solving. Learning is much broader and includes modification of current schemes in their use. Such modifications can occur in situations established through interlocked mathematical interaction. The mathematical interaction in Protocol IV is mainly of this type.
6. When activating 'label', a numeric key pad appears on the screen that enables children to scroll through the unit fractions indefinitely, beginning with '1/2'. Children can select any unit fraction and any numerator using the numeric key pad, and after selecting the fraction click on the screen to form the fraction numeral.
7. Here, we consider mathematical learning to occur in experiential contexts. It is useful to distinguish these cases of learning from cases where changes occur in a scheme when, from an observer's perspective, the scheme is not in activation.

These latter cases may be regarded as developmental accommodations if they involve a metamorphosis of the scheme.

References

BAUERSFELD, H. (1988) 'Interaction, construction, and knowledge: Alternative perspectives for mathematics education', in GROUWS, D.A., COONEY, T.J., and JONES, D. (Eds) *Effective Mathematics Teaching*, Reston, Virginia, NCTM and Lawrence Erlbaum, pp. 27–46.

CHOMSKY, N. (1980) 'On cognitive structures and their development: A reply to Piaget', in PIATTELLI-PALMARINI, M. (Ed) *Language and learning: The Debate Between Jean Piaget and Noam Chomsky*, Cambridge, Massachusetts, Harvard University Press, pp. 35–54.

COBB, P. and STEFFE, L.P. (1983) 'The constructivist researcher as teacher and model builder', *Journal for Research in Mathematics Education*, 14, pp. 83–94.

KONOLD, C. and JOHNSON, D.K. (1991) 'Philosophical and psychological aspects of constructivism', in STEFFE, L.P. (Ed) *Epistemological Foundations of Mathematical Experience*, New York, Springer-Verlag, pp. 1–13.

MATURANA, H. (1978) 'Biology of language: The epistemology of reality', in MILLER, G.A. and LENNEBERG, E. (Eds) *Psychology and Biology of Language and Thought*, New York, Academic Press, pp. 24–67.

MATURANA, H. and VARELA, F. (1980) *Autopoiesis and Cognition: The Realization of the Living*, Boston, D. Reidel.

PIAGET, J. (1964) 'Development and learning', in RIPPLE, R.E. and ROCKCASTLE, V.N. (Eds) *Piaget Rediscovered: Report of the Conference on Cognitive Studies and Curriculum Development*, Ithaca, Cornell University Press, pp. 7–20.

PIAGET, J. (1980) 'The psychogenesis of knowledge and its epistemological significance', in PIATTELLI-PALMARINI. M. (Ed) *Language and Learning: The Debate Between Jean Piaget and Noam Chomsky*, Cambridge, Massachusetts, Harvard University Press, pp. 23–34.

PIAGET, J. and INHELDER, B. (1963) *The Child's Concept of Space*, London, Routledge and Kegan Paul.

RENSHAW, P.D. (1992) 'The psychology of learning and small group work', in MACLEAN, R. (Ed) *Classroom Oral Language*, Deakin, Victoria, Australia, Deakin University Press, pp. 90–4.

SHOTTER, J. (1994) 'In dialogue: Social constructionism and radical constructivism', in STEFFE, L.P. and GALE, J. (Eds) *Constructivism in Education*, Hillsdale, Lawrence Erlbaum.

STEFFE, L.P. (1984) 'The teaching experiment methodology in a constructivist research program', in ZWENG, M. *et al.* (Ed) *Proceedings of the Fourth International Congress on Mathematical Education*, Boston, Birkhauser, pp. 469–71.

STEFFE, L.P. (1991) 'The learning paradox: A plausible counterexample', in STEFFE, L.P. (Ed) *Epistemological Foundations of Mathematical Experience*, New York, Springer-Verlag, pp. 26–44.

STEFFE, L.P. (1992) 'Schemes of action and operation involving composite units', *Learning and Individual Differences*, 4, 3, pp. 259–309.

STEFFE, L.P. (1993) 'Learning an iterative fraction scheme', Paper presented at the International Working Group on the Rational Numbers of Arithmetic, University of Georgia, Athens, Georgia.

VAN OERS, B. (1992) 'Learning mathematics as a meaningful activity', Paper presented in Working Group No. 4, Theories of Mathematics Learning, Seventh International Congress on Mathematical Education, Québec City, Québec, Canada.

von Glasersfeld, E. (1991) 'Abstraction, re-presentation, and reflection: An interpretation of experience and Piaget's approach', in Steffe, L.P. (Ed) *Epistemological Foundations of Mathematical Experience*, New York, Springer-Verlag, pp. 45–67.
Vygotsky, L.S. (1978) *Mind In Society*, Boston, Harvard University.
Vygotsky, L.S. (1981) 'The instrumental method of psychology', in Wertsch, J. (Ed) *The Concept of Activity in Soviet Psychology*, Armonk, M. E. Sharpe, pp. 134–143.
Wertsch, J. and Toma, C. (1994) 'Discourse and learning in the classroom: A sociocultural approach', in Steffe, L.P. and Gale, J. (Eds) *Constructivism in Education*, Hillsdale, Lawrence Erlbaum, pp. 159–74.
Yackel, E., Cobb, P., Wood, T., Wheatly, G. and Merkel, G. (1990) 'The importance of social interaction in children's construction of mathematical knowledge', in Cooney, T.J. (Ed) *The Teaching and Learning of Mathematics in the 1990s*, Reston, Virginia, National Council of Teachers of Mathematics.

Chapter 3

Radical Constructive Criticisms of von Glasersfeld's Radical Constructivism

Robert S.D. Thomas

Introduction and First Criticism

Now that radical constructivism has been paid the high compliment of being debunked, but, not being bunk, will not go away, it may be useful to try to improve upon it; hence this attempt at constructive criticism, some of which is radical (see Suchting, 1992, pp. 223–54). Radical constructivism holds that one constructs one's own notion of the world in accordance with a confluence of genetic and evolutionary epistemologies (Piaget, 1950; Campbell, 1974; and see Rav, 1993). One of the most important consequences of this fact is that all one can claim for one's world notion is viability, more in keeping with evolving evolutionary than with genetic epistemology (Varela *et al.*, 1991). Being grateful to von Glasersfeld for having pointed out to me what I now accept, I have no desire to attack radical constructivism. There are, however, three ways in which I find it seriously, even radically, deficient, and I want in this chapter to set them out in the hope that my doing so will be some use where I, and apparently von Glasersfeld, care most about usefulness, in education. These deficiencies are a lack of due emphasis on the construction of the self, whether over against the world or as a part of it, the denial of the possibility of knowledge of the world, and von Glasersfeld's ignoring of the massive social assistance in one's construction of one's notion of the world. The latter two of these deficiencies have considerable importance for education, and that is my reason for airing my criticisms in this place.

Since it is of less educational importance, let me deal at once with my first point. While von Glasersfeld, echoing Piaget, does acknowledge that one constructs oneself while constructing the rest of the world, he puts less emphasis on this side of what is really a symmetrical matter. Rather frequent observations that one is unjustified even in concluding that there is a world that one knows should be balanced by observations that one is equally unjustified in concluding that there is a knower. (For instance, von Glasersfeld (1983, pp. 207–18) where he writes, 'words cannot refer to things that "exist" independently of an experiencer' a stronger statement even than that we cannot know anything about such things, which weaker statement does refer to such things, as (naughtily) does the stronger.) Instead, von Glasersfeld thinks that his consciousness has 'the capability to find

[him] self' (von Glasersfeld, 1991, pp. 57–68). What one clearly does have is knowings, but the world and the knower are symmetrically problematic. One might say that from Descartes' *cogito* follows logically only cogitation. The symmetry of the situation has been brought home to me by the argumentation of Varela, Thompson, and Rosch's impressive book *The embodied mind* which, against Kant's *noumena* without, sets the equally inaccessible *noumenon* within. The consequences of this are far-reaching but not specifically educational, and so I do not wish to pursue them here. Let me mention only one argument that some persons will find attractive. Descartes argued that the world is not totally illusory because God would not deceive us. A similar argument is quoted in Rupert Riedl's treatment of evolutionary epistemology: the world is not totally illusory because our adaptation to it cannot be totally mistaken or we would not be viable (Riedl, 1984; Pepper, 1958, p. 106).

Second Criticism — Abandonment of Knowledge

My second criticism of von Glasersfeld's recent writings, such as I have read, is that he seems to deduce from constructivism, which I have no wish to deny, that one cannot have knowledge of the world. Whether this is a change of ground from earlier work, as I think it may be, I think that it is both important and wrong. He seems to have begun by saying that constructivism requires that we change what we mean when we claim to know something about the world, not that we cannot do it at all. In 1985 he was attempting 'to pick apart and reconstruct the concept of knowledge', as usual of 'experiential reality', explained as follows.

> By 'experiential reality' I mean ordinary reality, which is, of course, the reality of relatively durable structures that we are able to establish, to maintain, and to *use* in the ordering and management of our actual experience. (von Glasersfeld, 1985)

Perhaps daunted by the prospect of convincing everyone of what they mean by knowing the world as well as how we learn about the world, he seems to have shifted to denying that such knowledge is possible. In one sense, this is easy to see to be true. If one's idea of knowing is altogether unrealistic, then one does not 'know' the world in that unrealistic sense. For instance, if one means by knowing the kind of certainty demanded by Descartes or even the more popular aim of belief that is both warranted and true. But one still does know the world in the way that other persons do, and with an appropriate change in one's notion of knowledge, one can be said, simply, to know the world. Since knowledge of the world is of very great educational moment and a denial of its possibility would alarm more than just Neanderthal back-to-basics parents, this is a point of some political importance and worth straightening out if possible.

Once one adopts the constructivist point of view on knowledge, one sees that epistemology since at least Kant has been heading in that direction. Knowing as the construction of a viable notion of the world, not as the 'correct' mirroring of external reality — and, even more difficult for the mirroring point of view, internal reality — relieves us of the necessity of striving for an unattainable perfect mirroring (see Rorty, 1979). The cartographers' ideal of the full-scale map of the universe can be set aside (if one can find somewhere to put it). The relation of knowledge to the world, misunderstood as a mirror image's relation to the object reflected must be replaced by the relation of reference. When one says 'cat', one refers to cats; one does not, as von Glasersfeld agrees, 'capture' or reflect anything of or about cats (von Glasersfeld, 1992, pp. 421–32). Knowledge and even lies and fiction refer; he seems to have forgotten this important but unique relationship, the existence and importance of which is widely acknowledged.[1] One routinely refers even to what one denies exists; this seems to me to refute any limitation on what one can know.[2] There seem to be no bounds to what we can refer to and so potentially have knowledge of. It needs to be said that having a viable notion of a piece of the world is to know that piece of the world. The fact that there are numerous ways in which to know even that piece of the world has to be recognized and taught. If there is any lesson to be learned from the history of science, it is that the knowledge of one time is not the knowledge of another time.

One may wish to standardize what one means by knowledge by saying that scientific knowledge is always within some shared understanding. One can only know the atomic number of oxygen within an understanding of chemical elements that assigns atomic numbers. Within an understanding, a piece of supposed knowledge can be marked right or wrong accordingly as it agrees or disagrees with the official knowledge within that understanding. Then it can be a goal of science teaching to help students to achieve various understandings within which to have knowledge.[3]

The one piece of epistemology that might with benefit be passed along with the history of science is the point that to seek perfect knowledge within a single understanding is to ignore the inherent limitations of that understanding. That such perfect knowledge is possible is an enlightenment myth that needs to be eliminated, if only to benefit the next generation of students through their teachers, since it is probably already too late for the current generation of students, whose surrounding culture clings to this myth fervently.

With a properly circumscribed claim, we can claim to know the world. What else? If we had only our own observations to go on, then we might reasonably be concerned whether our knowings were indeed of the world. But since our scientific knowledge is concordant with that of others within common understandings of the world, there is no practical room for doubt that it is the common world that we understand and have knowledge of. It is all very well for von Glasersfeld to be 'post-epistemological', but to insist that it is not the world that we know is more like the post-modernist cutting off the branch on which one sits (von Glasersfeld, 1992, pp. 421–32). Amusing as creative writing, but not to be taken seriously or taught to children. The very discussion of constructivism relies on common understanding; it is important for consistency and for teaching that it does not lapse into a self-contradictory absurdity comparable to that of a proselytizing solipsism.

Mathematical knowledge in particular needs the idea of constructivism as the means of gaining knowledge to prevent its customary appearance as the revelation of truth pre-existing human rationality but miraculously commensurate with it. My view of mathematics as the gradual exploration and working out of the consequences of rationality is a view of mathematics as knowledge in spite of its not being truth. If it were not knowledge, however, we would not work for it or teach it or value it.

Can one apply this notion of knowledge to itself? Yes, and then the question is whether it is viable to posit a world that one knows. This question is interesting because it does not have a simple answer. In quantum theory quite notoriously and in psychotherapy too, observation typically has a sufficient effect upon the observables that they cannot be held to be there independently of observation (see for example, Löfgreen, 1992). It seems to be the case therefore that the positing of a knower-independent reality to be known may be viable at certain scales and in certain circumstances but break down in the atomic and the interpersonal. At the common scales where our language does its best work (see Rosch on concepts[4]), the positing of a world does seem to be viable. With the lessening of the threat of global nuclear annihilation, we have more reason to hope so than in recent years.

I am going to conclude these remarks on the need for a redefinition of knowledge rather than its abandonment by contrasting knowledge with something more than knowledge rather than with something less, which seems to be more often done. Like von Glasersfeld and Varela, whom I have already cited, I have a relationship with psychotherapy. For the last ten years, I have enjoyed a meta-therapy session once a week with a therapist for the discussion of his practice, in particular the knowledge difficulties he encounters with his clients and knowledge difficulties that they persist in. His and their difficulties are of different sorts, but we have come to describe the events by which persons construct themselves, including their notions of the world, as 'structuring events'. Each structuring event consists of the differentiation of something from the field of possible foci of attention, the categorization of this thing in accordance with the structure of previous experience — which this categorization modifies as the current experience becomes past — the feeling that colours the event, and the action that is its result (Romeyn and Thomas, Undated). This crude analysis, while our own, appears to be a simplification of the Buddhist wheel of life, and I introduce it in order to contrast knowledge of a person, which could be the basis of such an event with no appreciable feeling component, with sympathetic understanding of the same person's situation, which would differ on account of a feeling component both appreciable and appreciated (Varela *et al., op. cit.*, Chapter 6). As I remarked in my first criticism, we are not in a position to know even ourselves in the mythical way that knowledge was once thought to work. But in the redefined knowledge, it is rather important that sympathetic understanding is based on knowledge, however far from mythic its proportions. Therapists and teachers often have and need it. And frequently sympathetic understanding is based on knowledge, although that knowledge may be incommunicable and at best is only viable (see Polanyi's 'tacit knowledge', 1964).

Third Criticism — Social Constructivism

The knowledge on which sympathetic understanding is based is much more likely to be self-deceptive than scientific knowledge is, and so for the sake of writing about knowledge that is socially viable and not just good enough for private purposes I turn to scientific knowledge, including mathematical knowledge. Much of what von Glasersfeld has written about constructivism has implicitly placed one, as knower, in a vacuum.[5] Especially in educational rather than research contexts, this is particularly strange, since in educational contexts one typically is able to check one's constructions' descriptions against the descriptions of others' constructions. It is of course an important matter of principle that one cannot check one's constructions as though they were done with straightedge and compasses but is limited to checking descriptions including consequences. Much of this checking is not even voluntary but is imposed upon one by being at school.

Some, including those that think that everyone thinks only with words, seem to consider that we can check our constructions (which they very likely do not think of as constructions) against those of others. If, however, one adopts the constructivist point of view on knowledge, then one is forced to admit that this is impossible. It is therefore impossible in principle to compare constructions. From this it is possible to conclude that one cannot know that one's knowledge is concordant with that of anyone else; hence apparently von Glasersfeld's knower in a vacuum. But such a conclusion is a mistake. With knowledge redefined as in my second criticism, one knows that one's knowledge is concordant with another's or with official knowledge merely with viable knowing. How else? This is potentially as good a piece of knowledge as any other. It is a hypothesis for which there can be evidence, quite good evidence, while admittedly being subject to disproof. I say that one's knowledge may be concordant and not identical because there seems to me no precise meaning that can be given to identity in such a context. Life experiences differ, and so what one builds one's knowledge with differs from what another's knowledge is built with; how could the results be exactly the same in any neuron-by-neuron sense?[6] And as to agreement with official knowledge, official knowledge does not even exist in the bottom-level neuron-by-neuron sense; it exists only in persons' constructions concordant with official symbolic expressions. If the symbolic expressions of one's knowledge agree physically with the symbolic expressions of another's knowledge or official knowledge, that is evidence that they are concordant. This point is not a new one, and there is even a name for the resulting view, social constructivism (see Ernest, 1991).[7] I explained elsewhere before I was aware of social constructivism, my view of the main mechanism by which individual subjective knowledge is welded into community objective knowledge, namely an assimilation that is much like Piaget's (Thomas, 1991, pp. 3–38). By this assimilation we consciously and artificially put together the separate persons' constructions concordant with official symbolic expressions. The usual word for what we do here is 'identification', but, since in a constructivist context one is concerned to emphasize that these separate constructions are neither identical nor identifiable but are on principle distinguished, I insist on using the word 'assimilation' for what we do. We know that they cannot be the same, but we insist for practical purposes on treating them as the same. As I explained in the reference given, this process is assimilable to the way we treat as 'the same' whatever we represent by common nouns for practical communicative purposes. The differences among, say, dogs, seem to me to be not wholly unlike the differences among various persons' ideas of dogs in this one respect (and probably only in this one respect), that they can safely be ignored for many purposes.[8] It is probably safer to assume that a group understands the same thing by '17' than that it understands the same thing by 'dog'. The reason why it is important to insist, as von Glasersfeld has done, on the radically personal construction of each person's notion of the world is that so much of what is accomplished in communication and therefore in education depends on the device of assimilation, which is precisely the ignoring of the distinctions he draws our attention to. The use of one's notion of the world as a stand-in for the world is — and should be recognized as — one's assimilation rather than an identity independent of oneself.[9]

Conclusion

If we are to be consistent, we must admit that any scepticism to which constructivism leads us with respect to the world of outer experience should be reflected in a similar scepticism with respect to the self. I do not suggest that either of these scepticisms is appropriate fare for school children.[10] Ontology aside, constructivism does not need to lead to scepticism with respect to the possibility of knowledge, only to a profitable redefinition of knowledge as viable rather than as mirroring nature. And for maximum profit as well as viability, that redefinition needs to be social constructivist rather than radical constructivist.

Notes

1. I do not mean this literally. Reference is mentioned, for instance, in the paper, 'A constructivist approach to experiential foundations of mathematical concepts', in S. Hills (Ed) *History and philosophy of science in science education*, Proceedings of the Second International Conference on History and Philosophy of Science and Science Teaching, Kingston, Ontario, May 11–15, 1992, Volume 2, pp. 553–71, and see von Glasersfeld, 1983, pp. 207–18.
2. To be more precise, to deny the possibility of knowing anything about X is to know that X is unknowable. One smells Russell's barber's shaving soap. Not a viable statement.
3. As Clive Sutton of the University of Leicester remarked during the constructivism discussion at the Second International Conference on History and Philosophy of Science and Science Teaching, Kingston, Ontario, May 11–15, 1992.
4. Rosch references at n54 to Chapter 8 of F.J. Varela, E. Thompson, E. Rosch, *op. cit.*
5. Not explicitly, however. In 'Reconstructing the concept of knowledge', *Archives de Psychologie*, 53 (1985), pp. 91–101, he insists, 'The consideration of Others, therefore, is no longer a requirement of ethics alone, but has become an indispensable requirement of the construction of reality', other minds lending objectivity in its intersubjective interpretation. The need for 'others' is stressed, though they are given an inadequate role, in 'Steps in the Construction of "Others" and "Reality": A Study in Self-regulation', in Robert Trappl (Ed) *Power, Autonomy, Utopia*, New York, Plenum, 1986. And in 'A constructivist approach to teaching' (Ms for meeting in Athens, Georgia, February 20–23, 1992), he quite rightly defends Piaget against the charge of 'not having considered social interaction in his theory of cognitive development' by pointing out that other persons, for Piaget, are 'part of the *environment*.' However, he claims that for 'Piaget, just as for the contemporary radical constructivists, the "others" with whom social interaction takes place' are 'no more but also no less' than other physical objects. That is not an adequate estimate of the influence of social factors, native language among them.

6. This question has only been given its due, so far as I know, by E. von Glasersfeld in many writings beginning with 'Radical constructivism and Piaget's concept of knowledge', in F.B. Murray (Ed) *Impact of Piagetian theory*, Baltimore, Maryland, University Park Press, 1979.
7. The problem of intersubjectivity is the last of Suchting's four criticisms of radical constructivism, *op. cit.*
8. This difference-ignoring basis of what is usually called identification is discussed by E. von Glasersfeld in 'The constructs of identity or the art of disregarding differences' (preprint) with the appropriate references to William James's *Psychology* and *Pragmatism.*
9. The false mental identification of what can viably be assimilated is like the identification of word and object held to be characteristic of mythical thinking by E. Cassirer, *The philosophy of symbolic forms*, New Haven, Connecticut. Yale University Press, 1955. I owe a reminder of Cassirer's side of this analogy to F. Seeger and H. Steinbring, 'The myth of mathematics', in this volume. These authors discuss revealingly the implicit view of mathematics as myth among children.
10. So I share with Michael Matthews and Peter Davson-Galle, 'Constructivism and science education: some cautions and comments', in S. Hills (Ed) *History and philosophy of science in science education*, Proceedings of the Second International Conference on History and Philosophy of Science and Science Teaching, Kingston, Ontario, 11–15, May 1992, Volume 2, pp. 135–43, the view that one need not talk about the sceptical ontology of some constructivists.

References

CAMPBELL, D. (1974) 'Evolutionary epistemology', SCHILPP, P.A. (Ed) *The philosophy of Karl Popper.* La Salle, Illinois, Open Court.

ERNEST, P. (1991) *The Philosophy of Mathematics Education*, London, The Falmer Press.

LÖFGREN, L. (1992) 'Complementarity in language: toward a general understanding', in CARVALLO, M. (Ed) *Nature, Cognition, and Systems*, 2, Dordrecht, Kluwer (preprint).

PIAGET, J. (1950) *Introduction à l'épistémologie génétique*, Presses universitaires de France.

PEPPER, S. (1958) *The Sources of Value*, Berkeley, University of California Press.

POLANYI, M. (1964) *Personal Knowledge*, New York, Harper and Row.

RAV, Y. (1989) 'Philosophical problems of mathematics in the light of evolutionary epistemology', *Philosophica*, 43, pp. 49–78, revised and reprinted in RESTIVO, S. VAN BENDEGEM, J.P. and FISCHER, R. (Eds) (1993) *Math Worlds: Philosophical and Social Studies of Mathematics and Mathematics Education*, SUNY Press, pp. 80–109.

RIEDL, R. (1984) *Biology of Knowledge: The Evolutionary Basis of Reason*, Wiley, translated by P. Foulkes (1981) from *Biologie der Erkentnis*, Verlag Paul Parey.

ROMEYN, J.A. and THOMAS, R.S.D. (Undated) 'A skeleton for psychotherapy', Unpublished manuscript and ROMEYN, J.A. and THOMAS, R.S.D. 'A note on the process underlying the effectiveness of therapeutic interventions: Rediscription of 'structuring events', Unpublished manuscript.

RORTY, R. (1979) *Philosophy and the mirror of nature*, Princeton University Press.

SUCHTING, W.A. (1992) 'Constructivism deconstructed', *Science and Education*, 1.

THOMAS, R.S.D. (1991) 'Meaning in ordinary language and in mathematics', *Philosophia Mathematica*, 2, 6.

VARELA, F.J., THOMPSON, E. and ROSCH, E. (1991) *The Embodied Mind: Cognitive Science and Human Experience*. MIT Press.

VON GLASERSFELD, E. (1983) 'On the concept of interpretation', *Poetics*, 12.

VON GLASERSFELD, E. (1985) 'Reconstructing the concept of knowledge', *Archives de Psychologie*, 53, pp. 91–101.

von Glasersfeld, E. (1991) 'Distinguishing the observer: An attempt at interpreting Maturana', *Methodologia*, 8, 5.

von Glasersfeld, E. (1992) 'Aspects of constructivism: Vico, Berkeley, Piaget', preprint in English of paper published in Italian in Ceruti, M. (Ed) *Evoluzione e Conoscenza*, Bergamo, Pierluigi Ludrina.

Chapter 4

Articulating Theories of Mathematics Learning

Stephen Lerman

The preceding chapters highlight the fascinating and controversial state of current discussions in the academic community of mathematics education regarding learning theories. Constructivism is certainly the dominant theory, but it is being subjected to much criticism. Not that this is new for constructivism; it gained in support during the 1980s despite strong attacks and even political manœuvrings in its early days. In this chapter I will attempt to create the written equivalent of a snapshot. What can be seen in the picture is a scene at one instant. By the time the snapshot has been developed, when the book is published, the scene will look different, people and places will have moved on. Yet the snapshot will have captured something, although I do not pretend that the snapshot has captured any 'truth', however temporary. It is my fiction as I write it, and the reader's fiction as it is read. It is a photo-journalist's creation: the angle, the light, the subjects, all chosen to convey the effect the photo-journalist wishes to be seen, to carry that particular story.

Radical constructivism is a neo-Piagetian position in that it draws on the theoretical foundations of Piaget's work and is also a reinterpretation of those foundations. Radical constructivism appeared on the scene of the mathematics education community at a time when there was a strong egalitarian and democratic desire to shift the locus of control from the teacher. In this sense it was in the tradition of child-centred learning (Walkerdine, 1984) and a continuation of the process of turning away from behaviourist traditions of teaching. In the UK, members of the Association of Teachers of Mathematics (ATM) had been engaged in introducing investigations to mathematics classrooms. The notion of the active creative child, building her or his own concepts, was offered a theoretical rationale by constructivism, which argued that learning is something that children can only do for themselves. At the same time there was a strong focus on, or at least great interest in, individualized learning in mathematics, again an egalitarian desire by teachers to avoid competition, seen as an anti-socialist feature of society, and to avoid labelling children as failures. Constructivism's claim that each child responds to its interactions with the world from its own individual conceptual state provided a seductive rationale for individualized learning too.

Other influences contributed to the rapid popularity of constructivism in the mathematics education community. The appearance of microcomputers in classrooms, a technology that called on just one person to interact with it directly, and perhaps two others at most to cooperate reinforced the individualistic view of learning. Many of the researchers in the academic community were inspired by the Piagetian ideas of Papert, and the notion that the computer offered each child the possibility to be a mathematician. In the philosophy of mathematics the ideas of Lakatos, Popper and Kuhn had challenged absolutism and offered a fallibilistic perspective in its place. Confrey (1980), Nickson (1981) and Lerman (1983) introduced these ideas into mathematics education, each suggesting that aspects of the teaching and learning of mathematics could be radically influenced by such views.[1] Fallibilism in mathematics too seemed to be in line with radical constructivism (Lerman, 1989; Ernest, 1991). As a final point, although I do not claim that these are exhaustive, the mathematics education community has always been influenced by developments in educational psychology, and in that community Piaget's work was still dominant and there was a growing interest in constructivism. Piaget's work had been subjected to much criticism during the 1970s for the rigidity of the perception of ages and stages and the ineffectiveness of teaching to bring children across those stages. Constructivism refocused attention on Piaget's ideas about the nature of knowledge and the need to look at a genetic epistemology in place of the inadequate epistemologies of platonism and naïve empiricism.

Thus the climate was suitable for fostering constructivism, and the mathematics education research community at large has adopted a trivial form, focusing on the active child constructing her or his knowledge and seeing the teacher's role as setting up a 'constructivist' classroom and making models of children's understanding. Some of the radical constructivists continued, and continue, the debate at the edges of the theory, as can be seen in the chapters by von Glasersfeld and Steffe (see also Confrey, Cobb etc.). In some senses, then, one can glean a perception of the state of development of constructivist theories by a reading of these chapters, not least of all because the writers are amongst the foremost thinkers and researchers in this domain.

In a recent paper by Steffe (1993), one is immediately confronted by a defence of radical constructivism against a perceived Vygotskian attack:

> We are *guided by the children's mathematical language and actions in our teaching episodes*. This is a crucial point and one which we believe distinguishes our approach from those who work in a Vygotskian perspective.
> Of course, a psychological tool modifies actions and intentions, but the psychological tool alone does not determine the structure of those modified actions and intentions.
> We contrast the interactive mathematical communication of Arthur and Nathan with Vygotsky's general genetic law of cultural development to illustrate the difficulties of simply applying a general learning theory to what goes on in a consensual domain of interactions. (Steffe, 1993)

There are instances where Steffe wants to incorporate some of Vygotsky's insights into his work: 'Our emphasis is compatible with Vygotsky's ontogenetic orientation . . .' but his general aim appears to be to present the inapplicability of fundamental aspects of Vygotsky's work as he interprets it (for a discussion of the range of interpretations of Vygotsky's ideas, see Newman and Holzman, 1993).

It appears that von Glasersfeld too is partly concerned with a defence of radical constructivism:

> Teaching has to be concerned with *understanding* rather than performance and the rote-learning of, say, the multiplication table, or training the mechanical performance of algorithms — because *training* is suitable only for animals whom one does not credit with a thinking mind. (von Glasersfeld, this volume)

One could read these words as an attack on behaviourism. However, given that this paper was written in 1992 and presented at ICME, not to experimental psychologists, one may conjecture that this is an attack on a more modern theory. If this is the case, von Glasersfeld would not be alone in ascribing a transmission metaphor of teaching to a Vygotskian view (Olivier, 1993). Indeed it has been called an absolutist theory too (Steffe, personal communication).

Steffe's (1993) paper in particular, and his chapter in this book, can be read at least partly as a defence of radical constructivism against attack from 'those who approach learning from a socio-cultural point of view'. Clearly, I am one of those (Lerman, 1992, 1993a, 1993b, 1994). In this chapter, though, I want firstly to examine how various constructivist writers, in particular those in this collection, deal with the major challenge, that of social life. Secondly, I will attempt an examination of the semiotic function of mathematical symbols, as I read von Glasersfeld's interpretation through his paper. These two critiques are foreground elements in my snapshot, the first being a growing trend which incorporates socio-cultural notions into radical constructivism resulting, I argue, in incoherence, the second an illustration of the narrow limited view offered by radical constructivism.

Radical Constructivism and the Troublesome 'Social'

It is widely recognized by radical constructivists that the role of language and social interaction is undervalued, or at least underelaborated in that theory. Many writers (e.g., Bauersfeld, Voigt, Confrey, Ernest, Cobb) have made attempts to incorporate social interactions into constructivism and the chapter by Thomas in this volume is just such an example. These writers work with the assumption that the cognizing individual is central as the meaning-maker, and in various ways have developed rationales for how the environment, including other people, plays a part in the construction of that personal meaning. Thomas achieves this by arguing that to know a piece of the world is to have a viable notion of that piece of the world, viability being achieved through comparison of one's knowledge with others. However he insists on retaining the constructivist's separation between people: 'We know that they [constructions] cannot be the same, but we insist for practical purposes on treating them as the same.' That peoples' constructions *cannot* be the same is a very strong claim and one wonders why there is a need for such strength. My concern, however, is with the argument that the search for viability through the comparison of one's ideas in social interactions overcomes the solipsism of radical constructivism, in this sense another version of the private language issue answered, in my view, by Wittgenstein (e.g., 1974). One wonders how the comparison process becomes unproblematic, so that cases of a misfit or a fit will be seen as such by the participants and observers, i.e., students and teachers. If people's knowledge is in the end individual, are not people's perceptions of misfits or fits? I may well adapt my view, given an interaction with a teacher or peer, but I may not. It is entirely my private affair.

Other social constructivists argue for a complementarity of social knowledge and personal knowledge:

> . . . we can observe that when we talk of students' constructive activities we are emphasising the cognitive aspect of mathematical learning. It then becomes apparent that we need to complement the discussion by noting that learning is also a process of acculturation. (Cobb, Yackel and Wood, 1992, p. 28)

> . . . although the primacy of focus of each of conventionalism [Ernest's term for intersubjectivity] and radical constructivism is sacrificed in social constructivism, their conjunction in it serves to compensate for their individual weaknesses . . . (Ernest, 1991, p. 86)

> The fundamental orientation of the work in our own classroom springs from the radical constructivist principles and an integrated and compatible elaboration of the role of the social dimension in these individual processes of constructing as well as the processes of social interaction in the classroom. (Bauersfeld, 1992, p. 2)

The same complication arises, though, in that if the individual is the source of meaning, then so too is he or she the source of recognition of something that someone else says as causing a cognitive conflict, or a perturbation, as Steffe terms it. The major confusion that arises from the desire to claim that knowledge is constructed by the individual but that sometimes knowledge is absorbed from culture (Cobb, *et al.*), or as social convention (Ernest) or through the role of the social dimension (Bauersfeld) is that as long as the individual is at the heart of the process, as the one who ascribes meaning, any social interaction is itself interpreted individually. The complementary role these writers desire for 'the social' has no 'bite'.

The other way of incorporating social aspects of human life into the individual's constructions is to argue that Piaget intended this theory to be understood as such a mixture. Steffe (1993) writes:

> In this sense it is legitimate to interpret Piaget's work as a social-cultural approach in which he explained the mathematical development of children as self-regulating, autonomous organisms interacting in their environments. He seemed to take the social-cultural milieu of the children as a given without attempting to alter their most general experiences. (Steffe, 1993)

Later, he characterizes the sense-making process as follows:

> 'Making sense', then can mean to construct ways and means of operating in a medium to neutralise perturbations induced through social interaction. (*ibid.*)

It seems to me that this too does not escape from the complication described above. As long as there is a separation between the subject and the world, including other people, one has to go all the way with solipsism, or give it up. When the source of knowledge and of meaning is the individual, social interactions are on the same plane as physical interactions; they are filtered, or refracted, through the perceptions of the receiver. In some senses the possibility of cognitive conflict is less likely through social interactions than physical ones. I can imagine that I could challenge my belief that I can walk through a wall and I would probably receive a rather strong perturbation. When I attempt to challenge my belief that constructivism is misguided by discussing ideas with radical constructivists, no perturbation is set up; I interpret their disagreement as their not listening to me, or that they are too entrenched in their view, and vice versa of course! The 'problem' of the social is no problem at all if one accepts that social interactions are indeed on the same plane as physical interactions and both are separate from the autonomous meaning-maker. In the context of the construction of mathematical meanings, von Glasersfeld writes here:

> Sensory-motor material, graphic representations, and talk can provide occasions for the abstraction of mathematical operations, but they cannot convey them ready-made to the student. (von Glasersfeld, *op. cit.*)

The individual constructs his or her knowledge through reflective, or refractive, abstraction and no one can ever know another's knowledge, nor that anything one offers, in particular as the teacher, will certainly create a perturbation. I take Steffe to be maintaining this position, from the (1993) paper and the chapter in this volume, and presumably von Glasersfeld too, although he doesn't address social interactions here. The incoherence which I am attempting to describe arises when the social constructivists attempt to place much greater emphasis on the social, without any satisfactory mechanism whereby it impinges on individuals *without* their choice. Where meaning is carried in social practices and people are positioned by those practices one can begin to analyse and describe the nature of people as social beings. However this is to argue that one starts from a notion of a priority of the social plane over the individual, a Vygotskian idea rejected by the radical constructivists. I have argued that social constructivism introduces incoherence into radical constructivism and I will now try to show the restricted view of communication and learning offered by radical constructivism.

The Semiotic Function of Mathematical Symbols

I want briefly to consider von Glasersfeld's description of the semiotic function of mathematical symbols. I believe that an analysis will reveal some of the limitations of the radical constructivist position. In doing so, I will be guided by the critique of Piaget's interpretation of semiotics offered by Walkerdine (1988):

> For Piaget the relationship of signifier to signified is one of representation; the semiotic function: 'consists in the ability to represent something (a signified something: object, event, conceptual scheme, etc.) by means

> of a signifier which is differentiated and which serves only a representative purpose.' [Piaget, quoted in Gruber and Voneche, 1977, p. 489]
>
> Although this view grants to the semiotic function a major role in raising thought to a representational level, it sees the signified as arising extra-discursively, from the general co-ordination of actions which form operational structures which themselves arise outside of any relationship to systems of signs. (Walkerdine, 1988, p. 3)

Von Glasersfeld argues that there are fundamental operations of the mind (common to all, *à la* Kant?) and it is these operations from which one conceptualizes unit and plurality, the former derived from the conceptual construction of 'objects' and the latter from an awareness of repetition of the recognition of objects. Neither are merely the result of sense impressions but the result of reflective abstraction on those sense impressions. A further notion is needed, von Glasersfeld argues, that of number, and he has argued elsewhere (Steffe *et al.*, 1983) that it may arise from the activity of counting. Thus, he claims, mathematics is a matter of internal mental operations, and meaning is an association of mental operations with mathematical symbols. This is a private process, which 'cannot be witnessed by anyone else'; all one has to go on is the visible results of those mental operations, the writing, speaking or other behaviour of students. Nevertheless, von Glasersfeld argues that the teacher's task is to 'stimulate and prod the student's mind to operate mathematically', and that teaching 'has to be concerned with understanding'.

What then is the nature of symbols and their referents and their connection to each other and to the mind of the individual, in von Glasersfeld's analysis? 'Symbols do not generate the concepts that constitute their referents, they have to be linked to them by a thinking agent . . . It is, indeed a ground rule of semiotics that a sound or a mark on paper becomes a symbol only when it is deliberately associated with a conceptual meaning.' Von Glasersfeld wants the listener/reader to agree that the foundations of mathematical understanding are the common mental operations of unit, plurality and number and as a consequence 'this mathematics is an affair of *mental operations that have to be carried out by an active subject.*' Thus the link between the signifier, the symbol, and the signified, the referents, is one carried out by the individual by associating a symbol with mental operations. The association, though, is between the mental operations and the symbols. The active subject constructs the concepts of unit and plurality and then associates the symbol '3', usually preceded by association with the sound 'three', with the unit repeated twice more. Repetition of this process leads to the concept 'number'. The only mention of 'meaning' is in this context; mathematical symbols remain meaningless until such association takes place and a mark on paper (or a sound) becomes a symbol only with that association. Von Glasersfeld's concern is with the psychological process of the association of mental structures with symbols. What remain unproblematized and unconsidered are: the counting process as a socio-cultural phenomenon; the role of others, parents, siblings and teachers in the activities and settings in which the practice of counting is experienced (learned?); the sign system as a system, rather than a collection of symbols each with its referents; and the sign system as a socio-cultural phenomenon through which subjects are regulated. The point I wish to make here is that it is not surprising that these aspects of the semiotic analysis are not considered. With the individual as the source of meaning and sense-making, nothing is carried in any way by the symbols. Their social origins and import are ignored. At this point, a radical constructivist reader may well indicate that I am claiming that the symbols do all the work; that the symbols convey their sense to the learner in and of themselves. It seems as though there are only two choices for the radical constructivists; either one maintains that knowledge is conveyed ready-made by symbols to the student, a naïve empiricism which they reject, or the student privately constructs the mental operations associated with those symbols, the view they maintain. In its desire to deny any form of external experience for fear of falling back into the trap of having to explain how the cognizing subject gains that experience, radical constructivism isolates the individual into a closed shell. Ultimately, meaning is personal.

They do not recognize that there is another view, namely that meaning is carried in practices and that cognition is situated. This view does not acknowledge the separation between the subject and the world, but sees consciousness as constituted through social and discursive practices. This is not to deny individuality but to set its constitution within a social focus rather than an autonomous individual one.

The Limitations of Constructivism

There appears to be an insurmountable gap between the private mental operations of the individual and the world. According to radical constructivism, meaning does not flow with language or communication of other sorts, it is the private association of mental operations with symbols. 'If mathematical symbols have to be interpreted in terms of mental operations, the teacher's task is to stimulate and prod the students' mind to operate mathematically' (von Glasersfeld, Chapter 1, this volume). The teacher does not communicate, merely somehow prods the student's mind. From behaviour, for example the comparison of one's own mathematical symbols with those produced by the acting subject, the teacher infers something undefined and unapproachable, namely understanding, although of course there would be a denial that the two are identical and reducible one to the other, namely behaviour and understanding. I would not argue, as does Suchting in an irritating and pedantic attack (Suchting, 1992), that radical constructivism as offered by von Glasersfeld is inconsistent or incoherent. On the contrary, it is a strong and consistent position. It is also, in my view, a very limited one which is unable to explain a great deal of human behaviour and relationships, and leaves unconnected with its theory such important issues for education as what actions of the teacher will prod the right mental operations so that correct (in terms of mathematical concepts accepted by the community of mathematicians) mathematical associations are formed. Texts play a role that cannot be interpreted within the theory. Regulation through power relations manifested in discourses, appropriation of cultural experience, and the notion of cognition situated in practices rather than abstracted by the cognizing subject from experience all appear to be irrelevant to the individual's construction of knowledge, which comes about by reflective abstraction on common mental operations. How can individuals arrive at the mathematical concepts generally accepted by the mathematics community or at least the mathematics education community? It seems that the radical constructivists can only argue that their analysis results in common concepts by a right intuition, as the mathematical constructivists claimed, a Kantian view of necessary features of the mind.

What makes radical constructivism incoherent is the attempt to incorporate a social view of knowledge into it, to form some kind of social constructivism. I argued above that this attempt to shore up a limited theory opens it to an inherent contradiction, namely that the individual's knowledge is private and can never be shared, but a perturbation will be seen to be a perturbation in common.

Conclusion — What Practice Am I Engaged In?

Discussions in this area often take the form of people putting ideas to each other, sometimes in quite emotionally charged settings, but ending up retaining their original points of view. What is the nature of that activity, and indeed what is the nature of the activity in which I am engaged at this moment, in replying to some other papers and putting forward my own view? At one level it is an attempt to compare theories in different theoretical frameworks. But the disputants sit in those differing situations and thus talk across each other. (I might want to use the term 'paradigms' in discussing, for example, research perspectives in these different theoretical frameworks. However I would not want to argue that the disputants are in separate incommensurable programmes and unable to communicate.) Positions adopted carry with them much more than 'ideas'; indeed it would contradict my support for the assertion that knowledge manifests in practices were I to claim that those ideas are independent of social practices. We are engaged in typical activities of academic communities including: defending the 'corners' one may have argued in meetings and conferences and in papers; identifying one's own view by comparison with another, which includes 'selective' presentation of that other view; claiming phrases of the moment as part of one's case, such as 'overcoming Cartesian dualism'; identifying with groups of people.

Can these discussions take other forms? Perhaps not, although in presenting this chapter in the setting of a self-selected snapshot it is intended to emphasize the practice in which I am engaged in writing this, as are these few final comments. Perhaps my attempt at pointing out the practice can shift the discussion into another (meta-)practice in which we can share and compare snapshots.

Note

1. Dawson (1969) preceded these writers but his work was not widely known until recently.

References

Bauersfeld, H. (1992) 'Classroom cultures from a social constructivist's perspective', *Educational Studies in Mathematics*, 23, pp. 467–81.

Cobb, P., Yackel E. and Wood T. (1992) 'A constructivist alternative to the representational view of mind in mathematics education', *Journal for Research in Mathematics Education*, 23, 1, pp. 2–33.

CONFREY, J. (1980) *Conceptual Change Analysis: Implications for Mathematics and Curriculum inquiry*, Institute for Research on Teaching, Science-Mathematics Teaching Center, Michigan State University.

DAWSON, A.J. (1969) The Implications of the Work of Popper, Polya and Lakatos for a Model of Mathematics Instruction, Unpublished doctoral dissertation, University of Alberta.

ERNEST, P. (1991) *The Philosophy of Mathematics Education*, London, The Falmer Press.

LERMAN, S. (1983) 'Problem-solving or knowledge-centred: The influence of philosophy on mathematics teaching', *International Journal of Mathematical Education in Science and Technology*, 14, 1, pp. 59–66.

LERMAN, S. (1986) Alternative Views of the Nature of Mathematics and their Possible Influence on the Teaching of Mathematics, Unpublished doctoral dissertation, University of London.

LERMAN, S. (1989) 'Constructivism, mathematics and mathematics education', *Educational Studies in Mathematics*, 20, pp. 211–23.

LERMAN, S. (1992) 'The Function Of Language In Radical Constructivism: A Vygotskian Perspective', *Proceedings of Sixteenth Meeting of the International Group for the Psychology of Mathematics Education*, 1, New Hampshire.

LERMAN, S. (1993a) 'The position of the individual in radical constructivism: In search of the subject', in MALONE, J. and TAYLOR, P. (Eds) *Constructivist Interpretations of Teaching and Learning Mathematics* Curtin University of Technology, Perth.

LERMAN, S. (1993b) *Can We Talk About Constructivism?*, *Proceedings of the British Society for Research in Learning Mathematics*, November.

LERMAN, S. (1994) 'Changing focus in the mathematics classroom', In LERMAN, S. (Ed) *Cultural Perspectives on the Mathematics Classroom*, Dordrecht, Kluwer.

NEWMAN, F. and HOLZMAN, L. (1993) *Lev Vygotsky: Revolutionary Scientist*, London, Routledge.

NICKSON, M. (1981) Social Foundations of the Mathematics Curriculum, Unpublished doctoral dissertation, University of London Institute of Education.

OLIVIER, A. (1993) 'Voluntary Interaction Groups in Problem-centred Learning', *Proceedings of Seventeenth Meeting of the International Group for the Psychology of Mathematics Education*. Tsukuba, Japan.

STEFFE, L.P. (1993) *Interaction and Children's Mathematics*, Paper presented at the American Educational Research Association, Atlanta, Georgia.

STEFFE, L.P., VON GLASERSFELD, E., RICHARDS, J. and COBB, P. (1983) *Children's Counting Types: Philosophy, Theory and Application*, New York, Praeger Scientific.

SUCHTING, W.A. (1992) 'Constructivism deconstructed', *Science and Education*, 1, pp. 223–54.

WALKERDINE, V. (1984) 'Developmental psychology and the child-centred pedagogy', in HENRIQUES, J. *et al.* (Eds) *Changing the Subject*, London, Methuen.

WALKERDINE, V. (1988) *The Mastery of Reason*, London, Routledge.

WITTGENSTEIN, L. (1974) *Philosophical Grammar*, Oxford, Basil Blackwell.

Chapter 5

Is Radical Constructivism Coherent?

Michael Otte

John Bigelow begins his review of Penelope Maddy, *Realism in mathematics*, Oxford 1990, with the following remark:

> Mathematics has three striking features. It discovers truths which are necessary, not contingent. It proceeds by *a priori* methods, and does not justify its claims by appeal to experience. It has a distinctive subject matter, not material objects, nor plants, nor animals, nor thoughts and feelings, nor societies — but numbers and other things of that sort. (Bigelow, 1992, p. 235)

Now the third point, the question whether mathematics possesses objects of its own, is itself an object of extremely controversial debates. Nominalism has always denied that there are mathematical objects at all. And since 1870 when it was formulated the following definition of mathematics has met with a lot of approval: 'Mathematics is the science which draws necessary conclusions. This definition of mathematics is wider than that which is ordinarily given, and by which its range is limited to quantitative research. The ordinary definition, like those of other sciences, is objective; whereas this is subjective' (Peirce, 1965, Vol. 4, par. 229).

Let us thus focus first on the two other points. These two features are primarily expressed in mathematical proof. Proof is what characterizes mathematics and, in particular, mathematical method. Proof as a means of truth absolute and independent of experience is inevitably confronted with the demand that it proves its effectiveness in a given sense. This means that the proof must be followed by the proof that the proof is correct, and this again by the proof that the correctness of the proof is correct, and thus *ad infinitum*. Traditionally, there have always been two ways to cut through this infinite regression. First the assertion that there is an immediate identity between thinking and being, as it is found in Parmenides and later in the Cartesian foundation of western rationality and in all subsequent forms of conceptualism and intuitionism.

All the epistemological theories derived from linguistic analysis belong here as well. It is for instance, a central point of logical positivism in the sense of Moritz Schlick and the Vienna circle of philosophers that all knowledge can only be expressed in sentences and that all epistemology must thus be logical linguistic analysis in the end. The orientation towards proof fosters this view. Proof and the analysis of proof in terms of critical discussion requires verbalization and the view that understanding is not primarily of ideas or objects but pertains to sentences. It is suggestive that this approach privileges the concept of number among the mathematical concepts and consequently strives to arithmetize all of mathematics.

The other way of avoiding the above infinite regression is to convert proof into a perfectly mechanized, as if automatical activity. No questions of meaning could arise, because a proof would just be a fact, as if reality spoke for itself. Instead of assuming that man is a cognizant being and is rational, this approach assumes man to be a being who acts, retransforming this activity, again, into a self-regulated event, quasi into a physical process. This is paradoxical insofar as the subject is, as has been said, distinguished from the object by considering man not as dead matter but as a being who acts and reasons on the one hand, and there is at the same time an intention to fit this subjective side to the objective side resp., to have it confirmed by some self-control on the other. Reasoning would fall in one with being, were it to be 'being' itself. In order to attain convergence with nature, activity itself must become naturalized, that is be mechanized.

In conclusion one finds that the two solutions, which I should like to refer to as the Cartesian and the Leibnizean solution, both fall back on philosophical nominalism. This means that both of them destroy the proof as something that is a general of reality itself. How could the proof keep its predictive power? Because of this feature, the property of being general or — what amounts to the same — of having predictive power, the proof is of a different logical or categorical type than the facts to which it refers, it is of the 'nature of a representation', that means, of the nature of 'a general principle that is operative in the real world', as Peirce (1986) said (Chapter 5, p. 105). Because of this difference in type it entitles the addressee of its argument to self-regulated and creative action. Now all this holds equally true with respect to the laws of nature; they resemble mathematical proofs to the details (see Peirce, 1986, Chapter 5, pp. 102–07). This however implies that concerns for the dignity of the individual human subject and nominalist epistemological conceptions do not go together and I believe that radical constructivism is therefore incoherent.

At first of course those foundationalist endeavours failed to perceive their own paradoxical construction, as they were based on an absolute guarantor, that is on God. How problematical this was became fully visible only at the turn of the twentieth century. Only then intuitionism was challenged with the inexplicability and a historical absoluteness of its foundations, and 'formalism' with its intention to make man into a machine. Only then it became important to distinguish both between consciousness and communication and between human, creative behaviour, and automatical, mechanized behaviour, in both cases for sociohistorical reasons.

At the same time, the persistent search for absolute foundations of rationality itself had to appear as an anachronism. The approach in that was in principle the same as in the rationalism of the seventeenth century even if there was no longer a direct evocation of God. The anachronism is, as Luhmann says, that nowadays it should be clear 'that there cannot be a Cartesian self-confirmation of rationality which can use itself as a starting point for distinctions (e.g., for that between true and false), as soon as it has been confirmed' (Luhmann, 1992, p. 86).

By this, however, knowledge loses its absolute certainty, and the concept of risk becomes a feature of rationality. Scientific fruitfulness or technological success begin to play a fundamental role in the construction of concepts and in the mathematical development of knowledge. The task thus is to transform the conceptualization of rationality from the search for absolute foundations into an orientation towards the future and towards future developments. Hermann Weyl already responded in this way to the debate on foundations (Weyl, 1968, Vol. 4, p. 334). He was also, besides Peirce, one of the first, however, to introduce another element, the problem of the human subject, the problem of the subject's self-control, the problem of values and relevance, the problem of the negative role of nominalism, etc. into the debate, thus again introducing an absolute element into epistemology.

Peirce in considering the matter from the point of view of the human subject's real life, also makes it an issue of the opposition between nominalism and realism. The flavour of this twist comes out quite nicely in the following short statement. 'Generality is, indeed, an indispensable ingredient of reality; for mere individual existence or actuality without any regularity whatever is nullity' (Chapter 5, p. 431). The meaning of a natural law, we recall for instance, is to be seen in its power to make predictions. This implies that the foundation or reality of a general, like a natural law, lies in the future and is based on the fact 'that general principles are really operative in nature' (p. 100). The same holds true with respect to the meaning of any proposition, that means, that form in which the proposition becomes applicable to human conduct, 'for future conduct is the only conduct that is subject to self-control' (p. 425).

This introduces an evolutionary perspective into epistemology and from this perspective, the third aspect named above by Bigelow, the question as to the specific objectivity of mathematics, again becomes more important, for every programme of foundations which is not absolutist relying on some self-authenticating experiences must necessarily make allowance for the fact that cognition is dependent on its content. Forms and methods of reasoning develop simultaneously with new contents of reasoning and new experience. This is the very conclusion Chaitin and others have drawn from Gödel's theorem. The question where mathematics takes its certainty from is thus most closely linked with the question as to its objects. Epistemology always seems to have been based on the distinction between cognition and object, and on the question: how does cognition get its object? Now there is the assertion (Heidegger, Luhmann) that 'the distinction between cognition and object is itself only a distinction, that is a construction used to injure, to decompose, to observe the world' (Luhmann, 1990, p. 51).

But is this really the case? For us, subjects and objects are distinct by their mode of communication. We talk with other subjects about objects, and not vice versa. To me, this seems to have implications. It is said that the problem of mathematical certainty is quite decisively linked to the problem of the infinite. Now the infinite is nothing unified. It obviously makes a difference whether we use, for instance, the principle of continuity as a heuristic principle in mathematical proofs, an approach common since the says of Euclid, or whether we rather spoke of the continuum as of something given, grasped it conceptually by definition. In his 1925 lecture 'On the Infinite', Hilbert treats these two forms of the infinite by saying that one of these forms has been clarified by Weierstrass, but that this was not yet sufficient to clarify the significance of the infinite for mathematics. 'While the infinitely small and the infinitely large have been eliminated in Weierstrass' analysis by reducing the appropriate propositions to relationships between finite quantities, the infinite still occurs in the infinite series of numbers which define the real numbers, and furthermore in the concept of the system of real numbers which is conceived of quite as a totality given in a complete and closed way.' (quote translated from *Hilbertiana*, Wissenschaftliche Buchgesellschaft, Darmstadt, 1964, pp. 79–80).

Hilbert's proposed solution to the problem consisted, finally, in the distinction and mutual relationship between mathematics and meta-mathematics. This could be the mark of realism, nominalism moving around within one overall context, presumably that of the subject's own mind. The relationship between knowledge and meta-knowledge was in a way intended to replace that between cognition and object, on which epistemology had been traditionally based. Hilbert had seen that none of the two proposed modes, neither conceptualism which identified concepts with their extensions, nor mechanism which did completely without the infinite, could do. Too much was sacrificed, in the case of mechanicism while conceptualism became entangled in the well-known paradoxes.

If mathematics and science are to have relevance for human life we have to apply them and to speak about their truths in terms of a common human world. We know, however, from the paradoxes of set theory that this common world or absolute universe of discourse cannot be conceived of in terms of a totality of individual objects. Hence results the importance of the continuum and the idea of the inexhaustible infinite represented by it. This does not mean that we have a strict separation between subject and object. We rather prefer to conceive of epistemology in terms of a complementarity of means and objects. The continuum itself can be a means, as in geometric theory, or it can be an object, as in sensation or philosophical discourse. Complementarity is a concept required to express the fact that time and evolution are the essential dimensions for certain relationships, such as the relationship between foundation and development of knowledge which is our focus here. In this sense, the concept of complementarity resembles the traditional philosophical concept of dialectics. It is different from the concept of duality and polarity in that it simultaneously strives for the unity of that which is distinguished, in our case of object and means of cognition. According to certain forms of constructivist epistemology, the unity of this distinction is 'nothing more than the blind spot which someone makes use of who produces observations and descriptions by means of this distinction. If one assumes with constructivism, however, that this is always a real process within a real environment, that is always subject to the limitations by the environment, where then is the problem? The problem could be situated in the question of how a system succeeds in reformulating such limitations into conditions for increasing its own complexity. The non-arbitrariness of cognition then would be nothing other than the selectivity of this process of transformation controlled in an evolutionary way. Other than under the auspices of idealism, constructivist cognition finds no ground and does not search for one. It reflects the change of world orientation from unity to difference and diversity. It begins with distinctions, and it ends with distinction, fully aware of the fact that this is its own affair, and nothing necessitated by that which evades it as the external world' (Luhmann, 1990).

This type of the epistemic attitude can be called, with Luhmann, systems theory, or by the term I prefer, complementarism. It is still a large step from there to radical constructivism. There is a substantial difference between the insight that one does not see what one doesn't see, or, in other terms, that one cannot see everything at the same time, and the contention that one does not really see anything. Radical constructivism may nevertheless well be classified among foundationalist conceptions of cognition. At least since Kant and his famous uncognizable *Ding an sich*, constructivism has claimed to know fundamental limits of cognition; in fact it is essentially a theory about the limits of human knowledge.

Such an assertion that there are fundamental limits of cognition is only another expression for traditional epistemological foundationalism. So radical constructivism is not radical at all and it is not a realist but a nominalist philosophy.

Newton is deemed to be the greatest scientist of modern times not least because of the fact that he was one of the first to radically advocate that it is the object to be recognized which must be spoken of in scientific cognition, and not the human subject with all his desires, representations, and prejudices. His famous *hypotheses non fingo* is the well-known expression of this attitude. Newton thereby also encouraged the separation of science from philosophy and in the long run this attitude ended up with positivism and nominalism, denying that generalizations on which theory necessarily rests are real.

This has led Gregory Bateson to state an opposition between a Newtonian world on the one hand and a world of communication on the other. The opposition 'is simply this: that the Newtonian world ascribes reality to objects and achieves its simplicity by excluding the context of the context — excluding indeed all meta-relationships — *a fortiori* excluding an infinite regress of such relations. In contrast, the theorist of communication insists upon examining the meta-relationships while achieving its simplicity by excluding all objects' (Bateson, 1973, p. 221).

This description reveals by itself that the perspective essential for understanding and knowledge development consists in combining these two worlds. Philosophically speaking the concern for the epistemic subject demands to also speak about the object of its activity and vice versa, talking about objective reality requires a consideration of the human subject. Radical constructivism sometimes claims to have combined 'several ideas that the western philosophical establishment left by the wayside in the course of history', as von Glasersfeld claimed at the ICME-7 Conference in Quebec in 1992. It would be important to first of all revive the philosophical concern in its broadest sense itself. This would imply to develop a symmetrical consideration of subject and object of human world-making. And this in turn requires to give up all kinds of epistemological foundationalism concentrating instead on the circular connections between contexts and meta-contexts.

The reference to the learning or cognizing subject has an ineliminable role within didactical or pedagogical contexts. Relevance is necessarily one of the key terms in the debate about mathematics instruction. In this vein, there is no objection against radical constructivism's pointedly insisting on a switch from indoctrination to fostering individual building. Or against the fact that so-called ethno-mathematics is anxious not to give the impression that mathematics is a well-defined area of human activity, but rather emphasizes the diversity in the mathematics both of individual groups and of the mathematics among individuals within any group.

These are all good wishes but they should not tempt us into spoiling our thinking and our epistemological insight. In the present case, this would be even more a pity as so-called radical constructivism has fostered interest in questions of epistemology and in the awareness of the relevance of epistemological considerations for mathematical education. There is no didactic and no pedagogy without a philosophy! It should become possible to make implicit epistemic assumptions explicit and to discuss their relative merits and disadvantages.

Von Glasersfeld states in his presentation for ICME-7 that human knowledge is to be evaluated according 'to its fit with the world of human experience' and not as a representation of the world as it might 'be beyond the interface of human experience'. Is not radical constructivism like logical positivism just a variant of traditional empiricism? And further on: Is there a difference between radical constructivism and traditional logical empiricism? Michael Matthews states that 'epistemologically, constructivism is the well-known old empiricist wolf in contemporary sheep's clothing'. And von Glasersfeld too considers 'constructivism an offspring of subjective empiricism'. And he continues: 'But since constructivism is explicitly instrumentalist, it holds that all conceptual construction is carried out not for the sake of representational knowledge of a given world, but to enlarge the map of viable pathways in the world constituted by the subject's experience.' This resembles logical empiricism very closely.

Are concepts considered mere schemes of action or are they also an outcome of the application of the epistemic system onto itself? Must we not even say that theoretical concepts are dispensable altogether? If knowledge is just a representation of the experiential and present theoretical terms will have the only function of establishing relationships among givens. If I know these facts the same relationships can be established without theoretical terms, so these terms are dispensable. Theoretical terms serve a meta-perspective on tendencies, developments, possible alternatives etc., etc.

Mathematics is also meta-mathematics and the meta-perspective means the thinking of one thing within the context of another thing, means to move within a particular context and at the same time within a meta-context that classifies or relates contexts to each other. This other thing is traditionally called substance, reality, nature, infinity or God.

Substance we call the sorts of thing we make subjects of predication. Substances are the real objects of knowledge. Such things do not exist in empiricism nor for radical constructivists. Empiricism denies that things have natures or essences and even that they are really substances if by substance we mean anything more than a bundle of observable qualities. Of course we can know about a thing's tendencies and powers by observing what it does, but theoretical reconstruction thereby deals with the tendencies of a thing that it perceives, rather than merely with its actual appearance. If we measure something, for instance economic value, is there something beyond the individual numerical outcomes of measurement as represented by exchange values? If we represent an idea is there something beyond the individual symbolic representation? Are numbers different from numerals? If we describe a mathematical system, are the theorems we state about it invariant in their truth content with respect to changes of the description?

A mathematical concept, such as the concept of function, does not exist independently of the totality of its possible representations, but it is not to be confused with any such representation, either. A mathematical theory does not exist independently of the entirely of its axiomatic characterizations, and it is still not to be confused with one of them (Rota *et al.*, 1988). A formal system can be represented in various ways, and still the theorems have to be invariant in their truth content with regard to changes of the representation. This does not mean that 'there is a hypostatized entity called a formal system which exists independently of any representation' (Curry, 1970, p. 30). This example is, however, somewhat queer, as in face of Gödel's theorem such a system is necessarily incomplete. In considering it with respect to its evolution one might claim that there is in fact an entity behind, which is the subject of this evolution. Gödel's incompleteness theorem would then just state that we cannot define this entity in formal terms. I think such was Gödel's own view of the matter.

The central concept of the physics of Aristotle is the concept of motion, everything moves. Mathematics, however, is the realm of static forms. Aristotle's thinking is riddled with two orientations diametrically opposed to each other. Aristotle is most often regarded as the great representative of a logic and mathematics which rests on the assumption of the possibility of clear divisions and rigorous classification. 'But this is only half the story about Aristotle; and it is questionable whether it is the more important half. For it is equally true that he first suggested the limitations and dangers of classification, and the non-conformity of nature to those sharp divisions which are so indispensable for language . . .' (Lovejoy, 1964, p. 58), and for mathematics, as one might add. How then can mathematics be applied to physics? Or to state the problem in philosophical terms: how are the principle of identity and the continuity principle related to each other?

When I had nearly finished this chapter I ran across a manuscript by Michael Matthews 'Old wine in new bottles: a problem with constructivist epistemology' which bears much resemblance to my own. In fact I completely agree with his epistemological views and with most of what he has to say about the distinction between the real and the theoretical object of theory. Matthews, in separating epistemology from questions of ontology, however takes into account one side of Aristotle only and then gets into difficulty when he has to explain how the British empiricist of the seventeenth century could be avowed opponents of Aristotle and at the same time maintain an Aristotelian paradigm of knowledge.

The seventeenth century philosopher who made most out of Aristotle's principle of continuity which was rejected by Berkeley, was certainly Leibniz. Leibniz after having read Berkeley's *Principles of Human Knowledge* notes at the end of the book that he does not accept Berkeley's refusal of the continuum and of the infinite divisibility of the latter (Breidert, 1989). The geometrical and temporal continuum is the fundamental mathematical object.

In Leibniz' philosophy, the 'principle of continuity' (Principle 2) possibly plays a part which is as central as that of the 'principle of the identity of indiscernibles' (Principle 1). The latter consists in the theory that there are no two substances which resemble each other entirely but only differ numerically because then their 'complete concepts' would coincide. However, in general it would need an infinite analysis to devise the complete concept of an individual substance that would have to contain everything which belongs or will ever belong to that substance. Therefore only God's infinite mind can establish the complete concept of a substance. While for Leibniz, as for the radical constructivists, the substance is inseparably tied to the existent and observable the complete concept of a substance i.e., the concept that characterizes it completely is only accessible to God.

To speak of both principles means, according to Gueroult, 'to assume a central perspective from which one can behold the unity of this colossal world of thought as well as the sometimes opposed components it contains'. Leibniz is also deemed to be the mathematician who developed the Cartesian unification of arithmetic and geometry further towards algebraic reasoning which conceived algebra no longer as a generalized arithmetic, but rather as a general calculus, a *characteristic a universalis*, which permits to operate with the things themselves, making algebraic calculation and conceptual analysis convergent processes. According to Leibniz' nominalist views, however, relations do not have an ontological status on a par with substances. They are merely things imagined, and no real-life entities. In this light, the 'principle of continuity' appears to assume a somewhat different role, even perhaps subordinate to Principle 1 above. The importance of the principle of continuity for the seventeenth and eighteenth centuries is seen in the following considerations dating from the mid-eighteenth century. In his work *De la Nature*, J.B. Robinet states: It is 'the first axiom of natural philosophy' that the Scale of Beings constitutes a whole infinitely graduated, with no real lines of separation; that there are only individuals, and no kingdoms or classes or genera or species . . . This great and important truth, the key to the universal system, and the basis of all true philosophy, will day by day become more evident, as we progress in the study of Nature (quoted from Lovejoy, 1964, p. 275).

The principle of continuity supports and supplements Principle 1 inasmuch as we might sometimes have the impression that two entities A and B are merely numerically two and that they are otherwise identical. In such a case, the principle of continuity will guide us in discovering intrinsically 'imperceptible differences'. And where we do perceive differences, the 'principle of continuity' will guide us in looking for intermediate elements, for 'not one example can be quoted where a property abruptly begins or disappears'. Thus, all orders of natural beings form but one single chain, just as the 'coordinates of one and the same curve'. (letter to Varignon in 1702, reprinted in *Hauptschriften zur Grundlegung der Philosophie*, 1966, edited by Cassirer, Hamburg, p. 77). There are no classes but no sharp distinctions between species either, the gradations being known distinctly only by God.

Leibniz' favourite examples to illustrate the application of Principle 2 are from geometry, for instance, those concerning the fact that the various conic sections can be continuously transformed into one another. The principle of continuity allows us to discover connections, where sense perception observes nothing but differences. *A ne considerer que la configuration externe des Paraboles, des Ellipses et des Hyperboles, on seroit tenté de croire, qu'il y a une interruption immense d'une de ces Courbes à l'autre. Cependant nous savons qu'elles sont liées intimement, de manière qu'il est impossible de ranger entre deux quelque autre espèce intermédiaire . . .* (quoted from the same letter to Varignon).

In contrast to Poncelet, Leibniz, however, would understand this connection confirmed by the 'principle of continuity' not as the reality of the various conic sections as dependent of one genus of which they are particular species nor as individual cases of a general, that were entitled to an equivalent or even superior ontological status. Leibniz treats the relation between general and particular differently, conceptually. He advocates an intensional view of concepts, similar to the so-called prototype theory of present-day cognitive psychology. This means, however, that his no-class conception is such that the individuals mentioned by Robinet could be called 'general' or prototypical individuals or monads. Such a monad is completely to be identified with its 'complete concept' or prototypical category. Leibniz does not clearly distinguish between concepts and objects, as Kant and post-Kantian epistemology do, for instance.

Research in cognitive psychology has shown that the members of categories which are considered most prototypical are those with most attributes in common with other members of the category and least attributes in common with other categories (Rosch and Mervis, 1975). Prototypes, because of the distance they keep from each other, enable us to cognitively distinguish between categories. At the same time they represent a certain continuum. In short they exemplify the interaction of Principle 1 and Principle 2.

That both principles will cooperate productively in the movement of cognition only if cognition itself is conceived of as a movement between the particular and the general, is also seen from Leibniz's conceptions concerning algebra as a general science of signs. The entire programme is best expressed in Leibniz's conception of mathematical proof. Leibniz has created the modern concept of proof (Hacking, 1984) 'Leibniz knew what a proof is. Descartes did not.' And: 'Leibniz thought that truth is constituted by proof. Descartes thought proof irrelevant to truth. This comes out nicely at the metaphorical level. Leibniz's God, in knowing a truth, knows the infinite analysis and thereby knows the truth. That is what true knowledge is' (p. 214). Leibniz's God recognizes proofs by realizing the complete concept of substances. Proofs deal with these concepts. There does in fact not exist according to Leibniz a sharp borderline between the perceptual and the conceptual nor between discursive knowledge and directly as well as apodictically perceived reality. Proof is the process which constitutes reality itself. 'But Leibniz, making proof a matter of ontology, not methodology, asserts that all true propositions have an *a priori* proof, although in general human beings cannot make those proofs' (Hacking, 1984, p. 221).

In this sense, the entire reality is anchored in God's mind, and it becomes some kind of an algebraic calculation, it being impossible to subordinate algebra to arithmetical science, as it has to do with substances, forms or qualities, that means with the different and manifold, whereas arithmetic represents calculation with the homogenous. Algebra becomes a *characteristica universalis*, by means of which truth can be rendered stable, visible and irresistible, 'so to speak, as on a mechanical basis. Algebra, which we rightly hold in such esteem, is only a part of this general device. Yet algebra accomplished this much — that we cannot err even if we wish and that truth can be grasped as if pictured on paper with the aid of a machine. I have come to understand that everything of this kind which algebra proves is due only to a higher science, which I now usually call a *combinatorial characteristic*' (Letter to Oldenbourg dated 28 December, 1675, quoted from Hacking, 1984, p. 213).

That this programme of Leibniz has failed can be expressed by stating that the situation of God, and the situation of the finite human subject of cognition, who is only a part of the world, are not really distinguished. We may illustrate this with respect to algebra itself. As Peirce mentions, algebra can be defined from the viewpoint of the mathematician or out of the meta-perspective of the logician. 'The mathematician asks what value algebra has as a calculus . . . The logician does not wish the algebra to have this character . . . He demands that the algebra shall analyze a reasoning into its last elementary steps' (Chapter 4, p. 239). The mathematician applies and develops the algebra within a certain context. A logician in the sense of Leibniz might consider it within the meta-context of 'all possible worlds'. It is obvious that Leibniz considered algebra and its use in proof from the logician's point of view and it can be shown that he did not succeed in coordinating both perspectives (see Otte, 1989, 1993).

In spite of this argumentation Leibniz's programme should not be abandoned but supplemented and corrected in the manner suggested. The questions of essence and existence of substances make up the central concern of classical philosophy. In this philosophy epistemology and metaphysics remained as unseparated as did rationality and problems related to the sense of life. I am interested in this philosophy because my concern is not only to show that so called radical constructivism espouses very old fashioned epistemological views but that it is incoherent with respect to its humane intentions as well. The reference to history adds, or better, makes aware of, a philosophical element to scientific reasoning which refers precisely to the role of the subject's self-understanding.

How in general can a substance be conceptually be represented? Leibniz's solution was by means of his famous principle of the identity of indiscernibles. Leibniz proceeded from the insight that the real substance must contain all its predicates, that is its determinateness must be the criterion of all truths about it. The substance is the subject of all of its predicates. According to the principle of indiscernibles the substance must therefore be the individual. The ultimate goal of knowledge, which is, in general only to be accomplished by God through an infinite analysis, lies in the determination of the individual substances. This seems to be in opposition to the Aristotelian theory of concepts as the modern empiricist tradition has understood it and in this respect Matthew's description is appropriate.

Leibniz's principle is a way of trying to combine strict identity with concreteness. Concrete entities, like a man or a mountain or whatever, constantly change although they are, at different occasions, referred to as being the same. Leibniz's principle amounts to the suggestion that a concrete substance, like a man, for instance, includes the sequence of occasions constituting his life from birth to death. This view essentially *spatializes* time and makes becoming a mere diversity of qualities in being. The whole conception rests on the interaction of the principle of identity and the continuity principles as was illustrated above by means of the prototype theory developed by cognitive psychology.

Bertrand Russell has concluded from the difficulties of Leibniz's principle that either the substance is only defined by its predicates and 'then it would seem to be identical with the sum of those predicates', thus losing its subject character, or that the substance cannot be defined at all. Russell concluded that in the latter case the substance would be 'wholly meaningless'. From this, Russell concludes that Leibniz's principle makes no sense. As opposed to that, I believe that this fact which sounds unacceptable at first expresses a profound philosophical truth, which has become salient in our time in particular in the subject–machine problem, but which has more or less secretly determined the development of all fundamental concepts of mathematics since the seventeenth and the eighteenth centuries. Among these, the concept of (mechanical) motion would have to be named first, and, correspondingly, the concept of function.

In the present debate on the question 'Can computers think?', the result is that the (human) subject can neither be identified with the totality of his presently accessible properties, for otherwise he could be simulated on a suitably programmed computer, nor that the subject can be conceived of as a substance beyond all his properties, because otherwise he would be inaccessible to his own self-reflection.

In this case, the human subject would lose his subject character, as human cognition differs from mere information by the fact that the subject not only knows, but knows that he knows. Knowledge and meta-knowledge are inseparably connected in human cognition.

It may thus be said that certain current problems with science and technology (the problem of meaning, the man–machine problem and others) have got all kinds of nominalism into difficulties, difficulties which have to do with the subject's self-image and that all the aporias and dualisms, the polarity of the finite and the infinite, the paradoxes of motion, of time and of the present etc. are manifested mainly in the subject, indeed in the cognizing subject's self-image as a (potentially) universal and actually always limited being.

Let us have a look at the concept of (mechanical) motion. The first historical expression of the problem inherent in this concept comes from Zenon's paradoxes. You can find this paradox in any school textbook of mathematics. Zeno's problem is a paradox of motion.

In physics, motion is understood as continuous functions of time in three-dimensional — space $g(t) = (x(t), y(t), z(t))$, with t as a time parameter: 'We talk of a movement when the coordinates of the body change over the course of time' states a randomly selected physics textbook. The continuous function as a model of motion actually very clearly reflects the double character of this concept: On the one hand, it contains discrete aspects, such as it permits me to calculate single values when it is written as a formula. On the other hand, it emphasizes continuous aspects, for example, in the illustration of the functional graph that permits me a qualitative overall idea of the function (motion). The function is simultaneously both qualitative and quantitative; conceptual and constructive. It is knowledge (overall idea) and tool (calculation formula) in one.

Now from a physical point of view the essential difference of opinions is whether motion is no more than just the occupation of successive places at successive times or whether a moving body possesses more than just a position at a given time, but also what amounts to an instantaneous velocity. Mathematically the latter is represented by a vector or a derivative. So the function that describes the motion is taken either in extension or in intension.

The function as a model of motion (or rather the relative movement of the tortoise to the 'standing' position of Achilles) now enables us to reproduce the paradox on a new level because of its double character. The relative movement of Achilles and the tortoise is a linear function, as both motions are uniform: $f(x) = ax + b$ (i.e., when Achilles reaches x, the tortoise is at f(x)). The problem: 'At what point does Achilles really catch up with the tortoise?' is now: 'What is the fixed point of f(x)? The fixed point of f can be calculated simply as a function of the constants a and b: $x = f(x) = ax + b$

If we adopt a permanent symbol 'i' for the identity function we can rewrite the above equation as $i(x) = f(x)$ (for certain x). Zeno instead, intending to argue that motion is impossible confuses 'all' and 'exists' by implicitly claiming $i(x) = f(x)$, for all x. Thus Zeno's problem involves a problem of logical types. And, it is, as can be shown (see Otte and Steinbring, 1977) a problem which governs the evolution of the mathematical concept of a continuous function.

Cauchy's definition of a continuous function on the one hand pre-supposes very general and abstract notions of functional correspondence in the sense stated above. On the other hand it provides these notions with mathematical operative meaning by expressing them within the more specific context of an arithmetized version of the notion of continuity. Self-referentiality is just an expression of what above has been called complementarism as an alternative to radical constructivism.

Complementarity of subject and object makes up the starting point of epistemology. Epistemology is not independent of meta-physics, because if we insist on identifying the object with the definition theory gives of it, we also, perhaps unwittingly, tend to define the human subject. This is in opposition to the idea that the essence of man is existential freedom. Mathematics may in part construct its own reality but always in face of the continuum of yet undefined real possibility. Otherwise such a construction loses its subject becoming instead a quasi-mechanized process, as in the case of radical constructivism.

References

BATESON, G. (1973) *Steps to an Ecology of Mind*, St. Albans Frogmore Herts, Paladin.

BIGELOW, J. (1992) *History and Philosophy of Logic*, 13.

BREIDERT, W. (1989) *George Berkeley*, Basel, Birkhauser.

CURRY, H. (1970) *Introduction to Formal Systems*, New York, Dover Press.

HACKING, I. (1984) 'Leibniz und Descartes: Proof and Eternal Truths', in HONDERICH, T. (Ed) *Philosophy Through its Past*, London, Penguin.

LEIBNIZ, G.W. (1956) 'The Leibniz-Clarke correspondence', in ALEXANDER, H.G. (Ed) Manchester, Manchester University Press.

LEIBNIZ, G.W. (1966) Über das Kontinuitätsprinzip in CASSIRER, E. (Ed), *Hauptschriften zur Grundlegung der Philosophie*, 1, Hamburg, Meiner, pp. 84–93.

LOVEJOY, A.O. (1936 and 1964) *The Great Chain of Being*, Cambridge, Massachusetts, Harvard University Press.

LUHMANN, N. (1990) *Soziologische Aufklärung* V, Opladen, Westdeutscher Verlag.

LUHMANN, N. (1992) *Beobachtungen der Moderne*, Opladen, Westdeutscher Verlag.

OTTE, M. (1989) 'The Ideas of Hermann Grassmann in the context of the mathematical and philosophical tradition since Leibniz', *Historia Mathematica*, 16, pp. 1–35.

OTTE, M. (1992) 'Gegenstand und Methode in der Geschichte der Mathematik', *Philosophia naturalis*, 1, pp. 31–68.

OTTE, M. (1993) 'Kontinuitätsprinzip und Prinzip der Identität des Ununterscheidbaren', *Studia Leibnitiana*, XXV, pp. 70–89.

OTTE, M. and STEINBRING, H. (1977) *Probleme der Begriffsentwicklung — zum Stetigkeitsbegriff, Didaktik der Mathematik*, 1, pp. 16–25.

PEIRCE, C.S. (1965) in HARTSHORNE, C. and WEISS, P. (Eds) *Collected Papers*, 8 volumes, Harvard University Press.

PEIRCE, C.S. (1986) *Writings of Charles S. Peirce, A Chronological Edition*, 4, Indiana University Press.

ROTA, G.-C., SHERP, D.H. and SOKOLOWSKI, R. (1988) 'Syntax, Semantics and the problem of the identity of mathematical objects', *Philosophy of Science*, 55, pp. 376–86.

ROSCH, E. and MERVIS, C.B. (1975) 'Family resemblances', *Cognitive Psychology* 7, pp. 573–605.

WEYL, H. (1968) *Gesammelte Werke*, 4 volumes, Heidelberg, Springer.

Chapter 6

Social Constructivism and the Psychology of Mathematics Education

Paul Ernest

Introduction: The Problem

It might be said that the central problem for the psychology of mathematics education is to provide a theory of learning mathematics. That is, to give a theoretical account of learning which facilitates interventions in the processes of its teaching and learning. Piaget's Stage Theory, for example, inspired a substantial body of research on hierarchical theories of conceptual development in the learning of mathematics in the 1970s and 1980s (e.g., in the study of Hart and colleagues, 1981). Piaget's constructivism also led to the currently fashionable radical constructivist theory of learning mathematics, (von Glasersfeld, 1991; Davis *et al.*, 1990). Part of the growth in popularity of radical constructivism is due to its success in accounting for the idiosyncratic construction of meaning by individuals, and thus for systematic errors, misconceptions, and alternative conceptions in the learning of mathematics. It does this in terms of individual cognitive schemas, which it describes as growing and developing to give viable theories of experience by means of Piaget's twin processes of equilibration; those of assimilation and accommodation. Radical constructivism also has the appeal of rejecting absolutism in epistemology, something often associated with behaviourist and cognitivist theories of learning (Ernest, 1991a, b).

It is widely recognized that a variety of different forms of constructivism exist, both radical and otherwise (Ernest, 1991b). However it is the radical version which most strongly prioritizes the individual aspects of learning. It thus regards other aspects, such as the social, to be merely a part of, or reducible to, the individual. A number of authors have criticized this approach for its neglect of the social (Ernest 1991b, 1993d; Goldin 1991; Lerman, 1992, 1994). Thus in claiming to solve one of the problems of the psychology of mathematics education, radical constructivism has raised another: how to account for the social aspects of learning mathematics? This is not a trivial problem, because the social domain includes linguistic factors, cultural factors, interpersonal interactions such as peer interaction, and teaching and the role of the teacher. Thus another of the fundamental problems faced by the psychology of mathematics education is: how to reconcile the private mathematical knowledge, skills, learning, and conceptual development of the individual with the social nature of school mathematics and its context, influences and teaching? In other words: how to reconcile the private and the public, the individual and the collective or social, the psychological and the sociological aspects of the learning (and teaching) of mathematics?

One approach to this problem is to propose a social constructivist theory of learning mathematics. On the face of it, this is a theory which acknowledges that both social processes and individual sense making have central and essential parts to play in the learning of mathematics. Possibly as a consequence of this feature, social constructivism is gaining in popularity. However a problem that needs to be addressed is that of specifying more precisely the nature of this perspective. A number of authors attribute different characteristics to what they term social constructivism. Others are developing theoretical perspectives under other names which might usefully be characterized as social constructivist. Thus there is a lack of consensus about what is meant by the term, and what its underpinning theoretical bases and assumptions are. The aim of this chapter is to clear up some of this confusion by clarifying the origins and nature of social constructivism, and indicating some of the major differences underlying the use of the term. In particular, I shall argue that a major division exists between two types of social constructivism according to whether Piagetian or Vygotskian theories of mind and learning are adopted as underlying assumptions. However to clarify what social constructivism means it is necessary to go back further to its disciplinary origins.

Background Traditions

Although there are few explicit references to social construction in the work of symbolic interactionists and ethno-methodologists such as Mead, Blumer, Wright Mills, Goffman and Garfinkel, their work is centrally concerned with the social construction of persons and with interpersonal relationships. They emphasize conversation and the types of interpersonal negotiation that underpin everyday roles and interactions, such as those of the teacher in the classroom. Mead (1934) in fact offers a conversation-based social theory of mind. Following on from this tradition, a milestone was reached when Berger and Luckmann (1966) published their seminal sociological text 'The social construction of reality'. Drawing on the work of Schütz, Mead, Goffman and others, this book elaborated the theory that our knowledge and perceptions of reality are socially constructed, and that we are socialized in our upbringing to share aspects of that received view. They describe the socialization of an individual as 'an ongoing dialectical process composed of the three moments of externalization, objectivation and internalization . . . [and] the beginning point of this process is internalization.' (Berger and Luckmann, 1966, p. 149)

From the late 1960s or early 1970s, social constructivism became a term applied to the work of sociologists of science and sociologists of knowledge including Barnes, Bloor, Fleck and more recently Knorr-Cetina, Latour, Restivo, and others working in the strong programme in the sociology of knowledge (Bloor, 1976). This tradition drew upon the work of Durkheim, Mannheim, Marx, and others, and its primary object is to account for the social construction of scientific knowledge, including mathematics (Restivo, 1988). Thus the emphasis is on the social institutions and processes that underpin the construction of scientific and mathematical knowledge, and in particular, those that underpin the warranting of knowledge. Recently, there has been further work in this tradition (e.g., by Restivo and Collins) in developing a social theory of mind. This draws on the work of Mead and Vygotsky.

Not long after the development of these sociological traditions, in the 1970s social constructionism became a recognized movement in social psychology through the work of Harré, Gergen, Shotter, Coulter, Secord, and others. These authors have been concerned with a broad range of social psychological issues such as the social construction of the self, personal identity, emotions, gender, and so on (Gergen, 1985). A starting point shared by the social constructionists, but elaborated by different researchers in different ways, is that of Vygotskian theory of mind. This is based on the notion that thought in its higher manifestations is internalized speech or conversation. In consequence, one of the special features of social constructionism in social psychology is the explicit use of conversation as a central metaphor for mind, as well as for interpersonal interaction (e.g., Shotter, 1993).

Within the field of psychology there are other inter-related traditions which build on the work of Vygotsky, and which propose more less well developed social theories of mind. These include Soviet Activity Theorists (Vygotsky, Luria, Leont'ev, Gal'perin, Davydov), and the dialogue theorists, or socio-linguisticians, for want of a better term, including Volosinov, Bakhtin, Lotman, Wertsch, as well as socio-cultural theorists such as Lave, Wenger, Rogoff, Cole and Saxe.

The term 'social constructivism' does not seem to appear in philosophy until the late 1980s, when the growing interdisciplinarity of sociological and social psychological studies, and their terminology, spilled over into philosophy. A social constructivist tradition in spirit, if not in name, can be identified in philosophy, with its basis in the late work of Wittgenstein (1953). However, some scholars, such as Shotter, trace this tradition or at least its anticipations, back to Vico. There are strands in various branches of modern philosophy which might be termed social constructivist. Ordinary language and speech act philosophy, following on from Wittgenstein and Ryle, including the work of Austin, Geach, Grice, Searle and others, makes up one such strand. In the philosophy of science, a mainstream social constructivist strand includes the work of Hanson, Kuhn, Feyerabend, Hesse and others. In continental European philosophy there is an older tradition including Enriques, Bachelard, Canguilhem, Foucault which has explored the formative relations between knowledge, especially scientific knowledge, and its social structure and contexts. Although not known as social constructivist, this strand traces the historical social construction of these traditions and ideas. In social epistemology there is the work of Toulmin, Fuller and others, exploring how scientific knowledge is socially constructed and warranted. Finally, in the philosophy of mathematics there is a tradition including Wittgenstein, Lakatos, Bloor, Davis, Hersh, and Kitcher concerned with the social construction of mathematical knowledge. Ernest (1991a, in press) surveys this tradition, and represents one of the few, perhaps the only philosophical approach to mathematics which explicitly adopts the title of 'social constructivism'.

In the early 1970s the 'social construction (of the knowledge) of reality' thesis became widespread in educational work based on sociological perspectives, such as that of Esland, Young, Bernstein, and others. By the 1980s theories of learning based on Vygotsky were also sometimes termed social constructivist, and although we might now wish to draw distinctions between their positions, researchers such as Andrew Pollard (1987) identified Bruner, Vygotsky, Edwards and Mercer, and Walkerdine as contributing to a social constructivist view of the child and learning.

It appears, however, that the term 'social constructivism' first appeared in mathematics education from two sources. The first is the social constructivist sociology of mathematics of Restivo, which is explicitly related to mathematics education in Restivo (1988). The second is the social constructivist theory of learning mathematics of Weinberg and Gavelek (1987). The latter is based on the theories of both Wittgenstein and Vygotsky, but also mentions the work of Saxe, Bauersfeld and Bishop as important contributions to the area, even though they might not all have called themselves social constructivist. Unfortunately Weinberg and Gavelek never developed their ideas in print. Bishop (1985) made a more powerful impact with his paper on the 'social construction of meaning' in mathematics education, but he did not develop an explicit theory of learning mathematics. Instead he focused more on the social and cultural contexts of the teaching and learning of mathematics. Social constructivism became a more widely recognized position following Ernest (1990, 1991a, 1991b). However, a number of other authors have used and continue to use the word in different ways, such as Bauersfeld (1992) and Bartolini-Bussi (1991). Beyond mere terminology, there are also a number of contributions to mathematics education which might be termed social constructivist, in one sense or other, even though they do not use this title. For example, the approach termed 'socio-constructivism' used by Yackel *et al.* (in press) which is discussed below, can be regarded as social constructivist.

Thus it can be said that social constructivism originates in sociology and philosophy, with additional influences from symbolic interactionism and Soviet psychology. It subsequently influenced modern developments in social psychology and educational studies, before filtering through to mathematics education. Because of these diverse routes of its entry, probably combined with its assimilation into a number of varying paradigms and perspectives in mathematics education, social constructivism is used to refer to widely divergent positions. What they share is the notion that the social domain impacts on the developing individual in some formative way, and that the individual constructs (or appropriates) his or her meanings in response to his or her experiences in social contexts. This description is vague enough to accommodate a range of positions from a slightly socialized version of radical constructivism, through socio-cultural and sociological perspectives, perhaps all the way to fully-fledged post-structuralist views of the subject and of learning.

The *problematique* of social constructivism for mathematics education may be characterized as twofold. It comprises, first, an attempt to answer the question: how to account for the nature of mathematical knowledge as socially constructed? Second, how to give a social constructivist account of the individual's learning and construction of mathematics? Answers to these questions need to accommodate both the personal reconstruction of knowledge, and personal contributions to 'objective' (i.e., socially accepted) mathematical knowledge. An important issue implicated in the second question is that of the centrality of language to knowing and thought. Does language express thought, as Piagetians might view it, or does it form thought, as Vygotsky claims?

Elsewhere I have focused on the first more overtly epistemological question, concerning mathematical knowledge (Ernest, 1991a, 1993b, in-press). However, from the perspective of the psychology of mathematics education, the second question is more immediately important. It is also the source of a major controversy in the mathematics education community. In simplified terms, the key distinction among social constructivist theories of learning mathematics is that between individualistic and cognitively based theories (e.g., Piagetian or radical constructivist theories), on the one hand, and socially based theories (e.g., Vygotskian theories of learning mathematics), on the other.

Although this is a significant distinction, an important feature shared by radical constructivism and the varieties of social constructivism discussed here is a commitment to a fallibilist view of knowledge in general, and to mathematical knowledge in particular. This is discussed elsewhere (Ernest, 1991a, b, in press; Confrey, 1991).

Social Constructivism with a Piagetian Theory of Mind

A number of authors has attempted to develop a form of social constructivism based on what might be termed a Piagetian or neo-Piagetian constructivist theory of mind. Two main strategies have been adopted. First, to start from a radical constructivist position but to add on social aspects of classroom interaction to it. That is, to prioritize the individual aspects of knowledge construction, but to acknowledge the important if secondary place of social interaction. This is apparently the strategy of a number of developments in radical constructivism which seem to fall under this category in all but name (e.g., Richards, 1991). Likewise Confrey (1991) espouses a radical constructivist position which incorporates both social interaction and socially constructed knowledge. Researchers whose positions might be located here commonly prefer to describe their perspectives as constructivist as opposed to social constructivist.

The second strategy is to adopt two complementary and interacting but disparate theoretical frameworks. One framework is intra-individual and concerns the individual construction of meanings and knowledge, following the radical constructivist model. The other is interpersonal, and concerns social interaction and negotiation between persons. This second framework is extended far enough by some theorists to account for cultural items, such as mathematical knowledge. A number of researchers has adopted this complementarist version of social constructivism, including Driver (in press), who accommodates both personal and interpersonal construction of knowledge in science education. Likewise Murray (1992) and her colleagues argue that mathematical knowledge is both an individual and a social construction. Bauersfeld (1992, p. 467) explicitly espouses a social constructivist position based on 'radical constructivist principles . . . and an integrated and compatible elaboration of the role of the social dimension in individual processes of construction as well as the processes of social interaction in the classroom'. Earlier Bauersfeld described this theory as that of the 'triadic nature of human knowledge' 'the subjective structures of knowledge, therefore, are subjective constructions functioning as viable models, which have been formed through adaptations to the resistance of "the world" and through negotiations in social interactions.' (Grouws *et al.*, 1988, p. 39) Most recently Bauersfeld (1994, p. 467) describes his social constructivist perspective as interactionist, sitting between individualist perspectives, such as cognitive psychology, and collectivist perspectives, such as Activity Theory. Thus he explicitly relates it to symbolic interactionism, but he retains a cognitive (radical constructivist) theory of mind complementing his interactionist theory of interpersonal relations.

Another group of researchers is Cobb, Wood and Yackel of the Purdue Project, who describe their position as constructivist, but also emphasize the social negotiation of classroom norms. They have on occasion used the term 'socio-constructivist' for their position (Yackel *et al.*, in press.) Cobb (1989) has explicitly written of the adoption of multiple theoretical perspectives and of their complementarity. These researchers explicitly draw upon the idea of the complementarity of cognitive aspects and acculturation. '[W]hen we talk of students' constructive activities we are emphasising the cognitive aspect of mathematical learning. It then becomes apparent that we need to *complement* the discussion by noting that learning is also a process of acculturation' (Cobb *et al.*, 1992, p. 28, my emphasis). Thus it seems appropriate to identify them as complementarist.

In Ernest (1991a) I proposed a version of social constructivism, which although intended as a philosophy of mathematics, also included a detailed account of subjective knowledge construction. This combined a radical constructivist view of the construction of individual knowledge (with an added special emphasis on the acquisition and use of language) with conventionalism; a fallibilist social theory of mathematics originating with Wittgenstein, Lakatos, Bloor and others.

> The two key features of the account are as follows. First of all, there is the active construction of knowledge, typically concepts and hypotheses, on the basis of experiences and previous knowledge. These provide a basis for understanding and serve the purpose of guiding future actions, Secondly, there is the essential role played by experience and interaction with the physical and social worlds, in both the physical action and speech modes. This experience constitutes the intended use of the knowledge, but it provides the conflicts between intended and perceived outcomes which lead to the restructuring of knowledge, to improve its fit with experience. The shaping effect of experience, to use Quine's metaphor, must not be underestimated. For this is where the full impact of human culture occurs, and where the rules and conventions of language use are constructed by individuals, with the extensive functional outcomes manifested around us in human society. (Ernest, 1991a, p. 72)

> However this conjunction [of social and radical constructivist theories] raises the question as to their mutual consistency. In answer it can be said that they treat different domains, and both involve social negotiation at their boundaries (as Figure 4.1 illustrates [omitted here]). Thus inconsistency seems unlikely, for it could only come about from their straying over the interface of social interaction, into each other's domains . . . there are unifying concepts (or metaphors) which unite the private and social realms, namely construction and negotiation. (Ernest, *ibid.*, pp. 86–7)

In commenting on work that combines a (radical) constructivist perspective with an analysis of classroom interaction and the wider social context, Bartolini-Bussi (1991, p. 3) remarks that 'Coordination between different theoretical frameworks might be considered as a form of *complementarity* as described in Steiner's proposal for TME: the principle of complementarity requires simultaneous use of descriptive models that are theoretically incompatible.' However, some researchers, such as Lerman (1994), argue that there is an inconsistency between the subsumed social theories of knowledge and interaction, and radical constructivism, in this (or any) complementarist version of social constructivism.

In my view, the complementarist forms of social constructivism described above leave intact some of the difficulties associated with radical constructivism. There are first of all many of the problems associated with the assumption of an isolated cognizing subject (Ernest, 1991b). Radical constructivism might be described as being based on the metaphor of an evolving and adapting, but isolated organism, a cognitive alien in hostile environment. Its world-model is that of the cognizing subject's private domain of experience (Ernest 1993c, 1993d). Any form of social constructivism that retains radical constructivism at its core retains these metaphors, at least in some part. Given the separation of the social and individual domain that a complementarist approach assumes, there are also the linked problems of language, semiotic mediation, and the relationship between private and public knowledge. If these are ontologically disparate realms, how can transfer from one to the other take place?

Lerman (1992) proposes that some of the difficulties associated with radical constructivism might be overcome by replacing its Piagetian theory of mind and conceptual development by a Vygotskian theory of mind and language. The outcome might best be seen as a form of social constructivism. More recently, in explicitly taking leave of radical constructivism, Lerman (1994) extends his critique. He argues that any form of social constructivism which retains a radical constructivist account of individual learning of mathematics must fail to account adequately for language and the social dimension. Bartolini-Bussi (1994), however, remains committed to a complementarist approach, and although espousing a Vygotskian position, argues for the value of the co-existence of a Piagetian form of social constructivism, and for the necessity of multiple perspectives. Thus, on the basis of divided opinion and opposing arguments it cannot be legitimately claimed that the Piagetian forms of social constructivism has been refuted or shown to be unviable.

Social Constructivism with a Vygotskian Theory of Mind

In a survey of social constructivist research in the psychology of mathematics education Bartolini-Bussi (1991) distinguishes complementarist work combining constructivist with social perspectives from what she terms social consctructionist work based on a fully integrated social perspective. Some of her attributions of individual projects to these approaches might be questioned. For example, I would locate the diagnostic teaching approach of Alan Bell and his colleagues at Nottingham in a cognitively-based post-Piagetian framework, not one of social constructivism. Nevertheless, the distinction made is important. It supports the definition of a second group of social constructivist perspectives based on a Vygotskian or social theory of mind, as opposed to the constructivist and complementarist approaches described in the previous section.

Weinberg and Gavelek's (1987) proposal falls within this category, since it is a social constructivist theory of learning mathematics explicitly based on Vygotsky's theory of mind. A more fully developed form of social constructivism based on Vygotsky and Activity Theory is that of Bartolini-Bussi (1991, 1994), who emphasizes mind, interaction, conversation, activity and social context as forming an interrelated whole. Going beyond Weinberg and Gavelek's brief but suggestive sketch, Bartolini-Bussi indicates a broad range of classroom and research implications and applications.

In Ernest (1993a, 1993c, 1993d, 1994, in-press) I have been developing a form of social constructivism differing from the earlier version (Ernest, 1990, 1991a) because it similarly draws on Vygotskian roots instead of Piagetian constructivism in accounting for the learning of mathematics. This approach views individual subjects and the realm of the social as indissolubly interconnected, with human subjects formed through their interactions with each other (as well as by their internal processes) in social contexts. These contexts are shared *forms-of-life* and located in them, shared *language-games* (Wittgenstein, 1953). For this version of social constructivism there is no underlying metaphor for the wholly isolated individual mind. It draws instead upon the metaphor of *conversation*, comprising persons in meaningful linguistic and extra-linguistic interaction. (This metaphor for mind is widespread among 'dialogists' e.g., Bakhtin, Wertsch and social constructionists e.g., Harré; Gergen, Shotter.)

Mind is viewed as social and conversational because of the following assumptions. First of all, individual thinking of any complexity originates with, and is formed by, internalized conversation; second, all subsequent individual thinking is structured and natured by this origin; and third, some mental functioning is collective (e.g., group problem-solving). Adopting a Vygotskian perspective means that language and semiotic mediation are accommodated. Through play the basic semiotic fraction of signifier–signified begins to become a powerful factor in the social (and hence personal) construction of meaning or meanings (Vygotsky, 1978).

Conversation also offers a powerful way of accounting for both mind and mathematics. Harré (1979) has elaborated a cyclic Vygotskian theory of the development of mind, personal identity, language acquisition, and the creation and testing of public knowledge. All of these are accommodated in one cyclic pattern of appropriation, transformation, publication, conventionalization. This provides descriptions of both the development of personal knowledge of mathematics in the context of mathematics education (paralleling Berger and Luckmann's socialization cycle), and describes the formative relation between personal and 'objective' mathematical knowledge in the context of academic research mathematics (Ernest, 1991a, 1993b, 1994, in press). Such a theory has the potential to overcome the problems of complementarity discussed above.

Conclusion

It is important to distinguish Vygotskian from radical constructivist varieties of social constructivism, for progress to be made in theoretical aspects of the psychology of mathematics education. The adoption of a Vygotskian version is not, however, a panacea. Piagetian and post-Piagetian work on the cognitive aspects of the psychology of mathematics education remains at a more advanced stage and with a more complete theorization, research methodology and set of practical applications. Nevertheless, Vygotskian versions of social constructivism suggest the importance of a number of fruitful avenues of research, including the following:

- the acquisition of semiotic transformation skills in working with symbolic representations in school mathematics;
- the learning of the accepted rhetorical forms of school mathematical language, both spoken and written;
- the crucial role of the teacher in correcting learner-knowledge productions and warranting learner knowledge; and
- the import of the overall social context of the mathematics classroom as a complex, organized form of life including (a) persons, relationships and roles, (b) material resources, (c) the discourse of school mathematics, including both its content and its modes of communication (Ernest 1993a, in press).

Let me end on a cautionary note. Having delineated a fully social form of social constructivism based on a Vygotskian theory of mind, it is tempting to lump many social and cultural theories together, under this one umbrella. However, significant differences remain between such theories as the sociological theory of mind of Restivo and Collins, Soviet Activity Theory, socio-cultural forms of cognition (Lave, Lave and Wenger), social constructionism in social psychology (Harré, Gergen, Shotter), and Post-structuralist theories of mind (Henriques *et al.*, 1984; Walkerdine, 1988). These theories may have more in common with the Vygotskian form of social constructivism than with Piagetian forms. But to lump these diverse perspectives together would lead to far more confusion and inconsistency than that which I have been trying to dispel in this chapter with my analysis of social constructivism.

References

Bauersfeld, H. (1992) 'Classroom cultures from a social constructivist's perspective', *Educational Studies in Mathematics*, 23, pp. 467–81.

Bauersfeld, H. (1994) 'Theoretical perspectives on interaction in the mathematics classroom', in Biehler, R., Scholz, R.W., Straesser, R. and Winkelmann, (Eds) *The Didactics of Mathematics as a Scientific Discipline*, Dordrecht, Kluwer, pp. 133–46.

Bartolini-Bussi, M.G. (1991) 'Social interaction and mathematical knowledge', in *Proceedings of PME-15*, Assisi, Italy, 1, pp. 1–16.

Bartolini-Bussi, M.G. (1994) 'Theoretical and empirical approaches to classroom interaction', in Biehler, R., Scholz, R.W., Straesser, R. and Winkelmann (Eds) *The Didactics of Mathematics as a Scientific Discipline*, Dordrecht, Kluwer, pp. 121–32.

Berger, P. and Luckmann, T. (1966) *The Social Construction of Reality*, London, Penguin Books.

Bishop, A.J. (1985) 'The social construction of meaning: A significant development for mathematics education?', *For the Learning of Mathematics*, 5, 1, pp. 24–8.

Bloor, D. (1976) *Knowledge and Social Imagery*, London, Routledge and Kegan Paul.

Cobb, P. (1989) 'Experiential, cognitive, and anthropological perspectives in mathematics education', *For the Learning of Mathematics*, 9, 2, pp. 32–42.

Cobb, P., Wood, T. and Yackel, E. (1992) 'A constructivist alternative to the representational view of mind in mathematics education', *Journal for Research in Mathematics Education*, 23, 1, pp. 2–33.

CONFREY, J. (1991) 'Learning to listen: A student's understanding of powers of ten', in GLASERSFELD, E. (1991), pp. 111–38.

DAVIS, R.B., MAHER, C.A. and NODDINGS, N. (Eds) (1990) *Constructivist Views on the Teaching and Learning of Mathematics*, Reston, Virginia, National Council of Teachers of Mathematics.

DRIVER, R. (in press) 'Constructivist approaches to Science teaching', in STEFFE, L.P. (Ed) *Alternative Epistemologies in Education*, London, Erlbaum.

ERNEST, P. (1990) 'Social constructivism as a philosophy of mathematics: Radical constructivism rehabilitated?', Poster paper, *PME 14 Conference*, Mexico, July.

ERNEST, P. (1991a) *The Philosophy of Mathematics Education*, London, The Falmer Press.

ERNEST, P. (1991b) 'Constructivism, the psychology of learning, and the nature of mathematics: Some critical issues', in *Proceedings of PME-15* (Italy), 2, pp. 25–32. (Reprinted in *Science and Education*, 2, 2, 1993, pp. 87–93.)

ERNEST, P. (1993a) 'Mathematical activity and rhetoric: Towards a social constructivist account', *Proceedings of PME 17 Conference*, Tsukuba, Japan, 2, pp. 238–45.

ERNEST, P. (1993b) 'Dialectics in Mathematics', Paper presented at 19th International Conference on the History of Science, Zaragoza, Spain, August. (A revised version is: 'The dialogical nature of mathematics', in ERNEST, P. (Ed) (1994) *Mathematics, Education and Philosophy: An International Perspective*, London, The Falmer Press.

ERNEST, P. (1993c) 'Metaphors for mind and world', *Chreods*, 6, pp. 3–10.

ERNEST, P. (1993d) 'Constructivism and the problem of the social', in JULIE, C., ANGELIS, D. and DAVIS, Z. (Eds) *Political Dimensions of Mathematics Education: Curriculum Reconstruction for Society in Transition*, Johannesburg, South Africa, Maskew, Miller and Longman, pp. 121–30.

ERNEST, P. (1994) 'Conversation as a Metaphor for Mathematics and Learning', *Proceedings of BSRLM Annual Conference*, Manchester, November, 1993, pp. 58–63.

ERNEST, P. (in press) *Social Constructivism as a Philosophy of Mathematics*, Albany, New York, SUNY Press.

GERGEN, K. (1985) 'The social constructionist movement in modern psychology', *American Psychologist*, 40, pp. 266–75.

GOLDIN, G.A. (1991) 'Epistemology, constructivism and discovery learning mathematics', in DAVIS, R.B., MAHER, C.A. and NODDINGS, N. (Eds) (1990) *Constructivist Views on the Teaching and Learning of Mathematics*, Reston, Virginia, National Council of Teachers of Mathematics.

GROUWS, D.A., COONEY, T.J. and JONES, D. (Eds) (1988) *Effective Mathematics, Teaching*, Reston, Virginia, National Council of Teachers of Mathematics.

HARRÉ, R. (1979) *Personal Being*, Oxford, Basil Blackwell.

HART, K. (Ed) (1981) *Children's Understanding of Mathematics: 11–16*, London, John Murray.

HENRIQUES, J., HOLLOWAY, W., URWIN, C., VENN, C. and WALKERDINE, V. (1984) *Changing the Subject*, London, Methuen.

LERMAN, S. (1992) 'The function of language in radical constructivism: A vygotskian perspective', *Proceedings of PME-16 Conference*, Durham, New Hampshire, 2, pp. 40–7.

LERMAN, S. (1994) 'Can we talk about constructivism?', *Proceedings of BSRLM Conference*, Manchester, November, 1993, pp. 20–3.

MEAD, G.H. (1934) *Mind, Self and Society*, Chicago, University of Chicago Press.

MURRAY, H. (1992) 'Learning Mathematics Through Social Interaction', Paper presented to Working Group 4, *ICME 7 Conference*, Québec, August.

POLLARD, A. (Ed) (1987) *Children and their Primary Schools*, Lewes, East Sussex, The Falmer Press.

RESTIVO, S. (1988) 'The social construction of mathematics', *ZDM*, 20, 1, pp. 15–19.

RICHARDS, J. (1991) 'Mathematical Discussions', in VON GLASERSFELD, E. (1991) pp. 13–51.

SHOTTER, J. (1993) *Conversational Realities: Constructing Life Through Language*, London, Sage.

VON GLASERSFELD, E. (Ed) (1991) *Radical Constructivism in Mathematics Education*, Dordrecht, Kluwer.

VYGOTSKY, L. (1978) *Mind in Society*, Cambridge, Massachusetts, Harvard University Press.

WALKERDINE, V. (1988) *The Mastery of Reason*, London, Routledge.

WEINBERG, D. and GAVELEK, J. (1987) 'A Social Constructivist Theory of Instruction and the Development of Mathematical Cognition', *Proceedings of PME 11*, Montréal, Québec, 3, pp. 346–52.

WITTGENSTEIN, L. (1953) *Philosophical Investigations* (translation G.E.M. Anscombe), Oxford, Basil Blackwell.

YACKEL, E., COBB, P. and WOOD, T. (in press) 'Instructional development and assessment from a socioconstructivist perspective', in LEDER, G. (Ed) *Curriculum + Planning = Learning*.

Chapter 7

Mathematics, Computers and People: Individual and Social Perspectives

Erick Smith

In an interview on the PBS radio programme, 'All Things Considered' (13 December, 1991) D. Ford Bailey Jr., son of the first black member of the Grand Old Opry was asked what made his father's harmonica playing so different:

Bailey: He played his ideas.
ATC: What do you mean by 'he played his ideas'?
Bailey: You couldn't write them down. He just played the ideas in his head.

Introduction

In a recent paper on the role of computers in education, Gavriel Salamon states that there has been 'a shift of attention from Piaget to Vygotsky' (Salamon, 1992, p. 25). Although Salamon made this remark in connection with the importance of understanding the social changes in the classroom that accompany the introduction of computers, he also captured the essence of an important debate which is beginning to emerge in education. In this debate a challenge is being issued to what might be called conventional constructivist views based on the work of Piaget. This Piagetian view is often characterized as seeing constructive activity as an individual process isolated from both cultural artifacts and more knowledgeable others ((Scott, Cole and Engel, 1992, p. 191). The emerging challenge takes two forms. One places primary emphasis on the social and linguistic basis of knowledge (Ernest, 1991; Gergen, 1992; Restivo, 1993), while the second emphasizes the importance of social and cultural factors in individual learning (Cole, 1985; van Oers, 1992; Wertsch, 1979). Often these two forms are lumped together under the general rubric 'social constructivism', a genre of constructivism which emphasizes social/cultural factors and which is currently dominated by those working in the legacy of Vygotsky. Thus the dispute is often seen as one between the traditional (or Piagetian) constructivists and the social (or Vygotskian) constructivists. Although in this chapter, I will argue for making a distinction between two constructivist approaches, I will argue first that the Piagetian/Vygotskian criteria may not be the most useful and second that the two approaches to constructivism I suggest are better viewed as complementary than as competitive. Ultimately, however, I argue that the essence of what we call human mathematical activity requires what I call an individual constructivist perspective.

The stage for this discussion is set by examining the role of computers in mathematics instruction, specifically by asking the question: Can computers think? This question has both literal importance, in relation to the role of the computer in the classroom, and metaphorical importance in emphasizing that the heart of the educational enterprise lies not in simply inducting students into the social world, but in educating the whole person. This becomes even more true in mathematics, where the history of formal symbolic (linguistic) trappings can too easily lull us into forgetting or ignoring the individual as sense-maker in her experiential world.

The Computer as Social Participant: Can Computers Think?

In his book *Artificial Experts, Social Knowledge and Intelligent Machines* Harry Collins describes a test originally proposed by Alfred Turing for answering the question: 'Can machines think?' (1990, p. 181). In Collins' version of Turing's test, a person and a computer are placed in separate rooms. An interrogator in a third room does not know in which room each is located, but attempts to find out through written interactions. The question is whether the interrogator will make a distinction between the person and the computer. Turing himself, while believing that the literal question of whether a computer could think was 'meaningless', also stated 'that at the end of the century the use of words and general educated opinion will have altered so much that one will be able to speak of machines thinking without expecting to be contradicted' (quoted in Collins, 1990, p. 184). Thus turing seems to frame the issue in terms of social relativism. The question of whether the interrogator will make a distinction has more to do with how she has constructed her concept of computer and of thinking than with how the computer is physically constructed or the particular software in use. To the extent that social (cultural, linguistic) distinctions between computer activity and thinking diminish, we will become less likely, as interrogators, to make this distinction.

Collins makes several additional observations: First Turing's test is designed to 'mimic the interactions of a human being'. Thus it is a test that equates thinking with acceptable social interaction, or 'a test of our abilities to make an artificial human that will fit into a social organism as opposed to a test of a machine's ability to mimic a brain' (p. 183). Second he raises the question of the attitude and experience of the interrogator. Is the interrogator actively attempting to make a distinction or simply interacting with a more social intention? The act of distinguishing between humans and computers is closely related to the intention to make such a distinction. As Collins states: 'These different features of Turing Test protocols must make a difference to what we think of as intelligence and what we think of as the test' (p. 188).

Collins points out that Turing did not think of the test as a laboratory experiment under controlled conditions run by an expert on computers, rather as a 'broad statement about likely trends in the development of computers and their impact upon the general public' (p. 184). The issue, according to Collins 'has little to do with the development of computers and much more to do with the way we think about them' (p. 184). In this sense, the real importance of the Turing Test is not so much whether we *can* distinguish between computers and people as whether we *do* make that distinction. Collins makes a metaphorical comparison to artificial heart implants and the ability of the body to detect, and thus reject them. Computers are the artificial implants of the social world, a 'social prosthesis' (p. 14). He continues:

> Luckily for transplant patients the sensitivity of the immune system is not fixed; it can be reduced by drugs. According to the state of the immune system, the same prosthesis might be treated as an alien invasion or as a familiar part of the body. In the same way the social organism can be more or less sensitive to artifacts in our midst; one might say that it is a matter of the alertness of our social immune system. . . . One of the things I will try to do . . . is to make our immune system more sensitive to mechanical strangers, for just as reducing the sensitivity of the physiological immune system carries with it enormous costs for the body, reducing sensitivity to mechanical invasion has costs too. (Collins, 1990, p. 15)

A question I want to raise revolves around what those costs may be in a mathematical classroom. I would suggest that in a certain sense the distinctions we make between people and computers are equivalent to the distinctions we make between educating and training. The acceptance of computers as 'social prosthesis' increases the difficulty of distinguishing between educating and training. As computers become more integrated into our social world, how will we distinguish educating from learning to act like a computer? In the balance of this chapter I will first suggest that the uncritical acceptance of computers in classrooms, even when introduced in conjunction with tool and simulation software, may diminish our awareness of this social prosthesis. Like Collins, I will suggest that realizing the potential of computers in education may be closely related to our ability to 'make our immune system more sensitive to mechanical strangers'.

Student Examples

In a review of computers in education, Gaviel Salamon describes three historical approaches to educational computing: computer-aided instruction/intelligent tutoring systems (CAI/ITS), programming, and tools. By tools, he means 'cognitive tools' and includes such software as "model builders, data sets affording manipulation, conceptual map-makers, simulations and similarly partly or wholly open-ended instruments for the student to operate and manipulate'" (1992, pp. 11–12). Salamon claims that neither CAI/ITS nor programming have shown promise and that the development and use of appropriate software tools provides the most hopeful route to successful use of computers in schools. Although one can agree with his conclusion, it is certainly also the case that even the best tools can be used poorly, that is tool software itself is not the sole requirement for positive change. Thus a question we face is what it means to use the tools 'well'. The example below has been chosen to illustrate these issues, particularly in relation to the role of computer as social participant in the classroom. In this sense, I present it to push at the edges of what it means to learn in a computer environment with the intention of raising issues that challenge us to better define our pedagogical commitments.

Can Computers do Mathematics?

Three students are engaged as a group with one computer and the software tool, Function Probe© (FP, Confrey, 1992a) in a mathematics classroom. They are using the table window in FP to build a table of values for a contextual problem which can be modelled linearly (distance as a function of years). They have filled one column in the table window with years by using the 'fill' command and are attempting to enter the values in the second column. They have already decided that the initial distance is 241.5 inches and that this distance decreases by 2.75 inches each year. Since one option when using the 'fill' command in FP is to fill a column by declaring an initial value, a final value, and an iterative arithmetic operation (+, –,*, or ÷) by a specified value, one student, Dan, suggests that they use the 'fill' command to enter values iteratively in the distance column by starting at 241.5 and subtracting 2.75 each row until 0 is reached. He attempts this, but neglects to change the iterative operation from its default action (plus) to the action he needs (minus). Thus FP attempts to fill the column by starting at 241.5 and ending at 0 by adding 2.75. Only one value is entered in the column, 241.5, and the 'fill' stops. Dan assumes that he has made an operational mistake in the programme. Another student, Tess, objects:

> You have to make an equation don't you understand? Because we want the years to correspond with the inches.

Since the goal is to create a relationship between years and distance, Tess' argument seems to be that this correspondence can only be created through an equation.[1] However, Dan insists that a 'fill' should work and the third student, Lila, seems to agree. They try several more times to fill the column, failing each time for the same reason.[2] As they do so, Tess becomes increasingly upset, attributing their failure to their unwillingness to listen to her point. Although Dan still expresses certainty about his method, Lila begins to waiver. Tess states:

> I know but . . . you have to have an equation otherwise you have to do each one separately . . .

To which Lila responds:

> I think Tess is right. We have to make an equation, otherwise *it won't know what to do*. (my emphasis)

Reluctantly, Dan goes along and the group eventually finds an equation and finishes the problem.

Had the teacher reached this group at an appropriate time, he would very likely have shown them how to use minus in the 'fill' operation. However, an interesting issue would be how a teacher could have responded to Tess and Lila, had he taken seriously the reasons they attributed to the failure of 'fill' for the group. How could he have responded to their suggestion that the computer in some sense ***knew*** that years and distance must correspond and thus was unable to carry out an operation which, in their eyes, would fail to make this correspondence? Although the students were fairly new to FP, they had worked with the table enough to know that it could algorithmically carry out the indicated operations. Thus it would appear that they were attributing, in some sense, a very human quality to the computer, an ability to create a 'problematic' (Confrey, 1991) within the problem, which one might describe as an ability for the computer to 'do mathematics'.

Seeking an Appropriate Role

The intention in providing this example is to point out how computers can become an unrecognized 'social prosthesis' and how this can serve to undermine the kind of exploratory constructive processes necessary for meaningful learning. If students do not distinguish between 'meaningful learning' (Novak and Gowin, 1984) and learning to do what the computer does, then we have lost a significant part of the educational process. On the other hand, the programme used in the above example is representative of just that kind of tool software which Salamon sees as the hope of educational computing. Thus we must find an appropriate role for these tools which will allow them to reach their potential. They are valuable as a 'social prosthesis', as long as they are recognized as 'mechanical strangers' and as long as we build this understanding into our instructional practices. Thus we might ask how our classroom practices allow students to understand what it is in their own actions that is different from that which a computer does. This issue has particular relevance in mathematics instruction, particularly in how it is that we view our subject matter, mathematics.

Turing's dismissal of the possibility of a computer being able to think seems quite straightforward in many areas. As Wittgenstein points out, Computers lack that 'rest of behaviour'. (1953, par. 344). This is not something that will be overcome as computers become more sophisticated, for that 'rest of behaviour' is that which makes us distinctly human. If we could imagine a computer which had the level of caring about others that people do, whose 'mental operation' was intrinsically tied to its participation in a social network, which had a sense of humour, and which worried about being unplugged the same way we worry about dying, we might be able to imagine a computer which could exhibit the 'rest of behaviour'.

But perhaps mathematics is different. It has recently been seriously proposed that computers can do mathematics. Following the publication of a computer algorithm that theoretically allows a computer to check any formal proof, one of its authors states that since the creation of *Principia Mathematica*:

> mathematicians have rested assured that all their ingenious proofs could, in principal, be transformed into a dull-string of symbols which could be verified mechanically. (Babai, 1992, p. 1)

Figure 7.1: Individual Constructivist Perspective

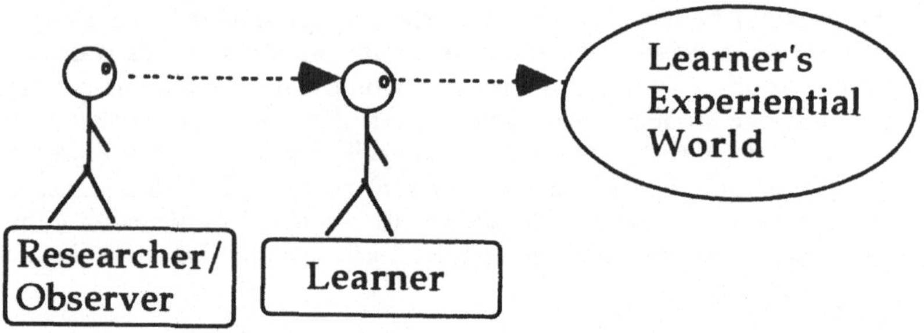

Despite the optimism of the mathematician, one might ask him whether, in verifying these mathematical proofs, the computer would come to *know* the mathematics. Is verification simply a matter of formal manipulation or does it require one to answer the question 'verified for whom?' Is there a distinction between what a computer can do and mathematics? Mathematically, how important is the 'rest of behaviour'?

Individual Perspective Versus Social Perspective

In making a distinction between individual and social constructivists' perspectives, I will argue that there is not a single answer to the questions raised above, rather that different answers may be more or less viable depending on the perspective that one takes — that is on how one situates oneself as an observer. From the claim that: 'Every distinction is made by an observer' (Maturana and Varela, 1987). I would further emphasize that the distinctions made are made as a result of an observer who has chosen a perspective. Thus distinctions are not there for the observing, but must be constructed by the observer. In this sense we can learn something about constructivist researchers by initially asking ourselves where it is that we see a researcher placing herself as observer. I have found two general observer categories which I believe are useful in discussing educational issues, particularly the kinds of issues raised in the above discussion. I have chosen to use individual perspective (or individual constructivist) and social perspective (or social constructivist) to distinguish these two categories. From an individual perspective one imagines oneself as observer from the perspective of the mind of the individual as sense-maker in his experiential world. This is not a claim that the researcher can ever actually know what is in the mind of another, rather a claim that the questions they wish to answer can best be facilitated if the models one creates are made as if they represent the learner in relation to his experiential world. Although social and cultural factors are often seen as intrinsic to learning by individual constructivists, they tend to emphasize the individual constructive activity regulated through a personal sense of 'viability' (von Glasersfeld, 1988). This model is illustrated in Figure 7.1, where the researcher/observer attempts to model the experiential world of the learner 'through the learner's eyes'. In this model, the social and cultural worlds are included in the learner's experiential world.

Figure 7.2: Social Constructivist Perspective

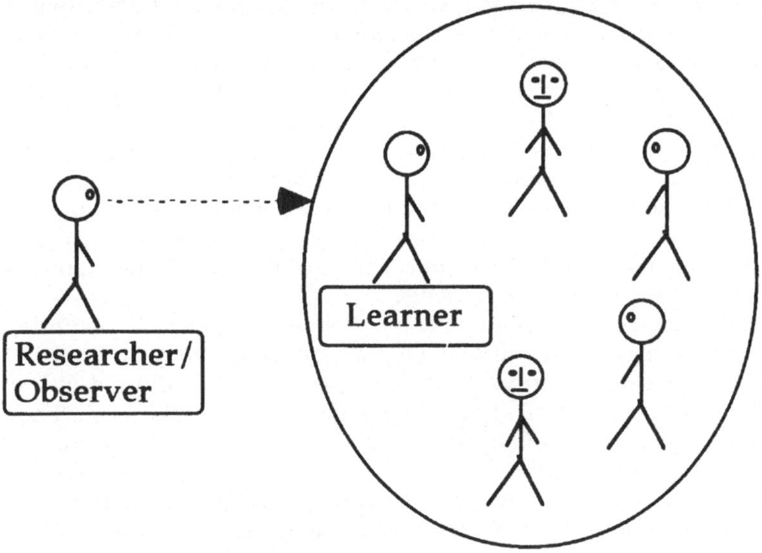

A second perspective that an observer might take is to place himself within a social setting with a focus on language and other kinds of interaction and viewing the individual in relation to his or her role as social participant. Educational researchers who take this perspective focus on issues best understood by modelling the interactions that take place between people. In Figure 7.2, the researcher/observer places himself or herself outside a circle which encompasses the social/cultural milieu, that is as observer of that milieu. This does not necessarily mean that they will not focus on the individual, but the focus will be on individual as participant in the social milieu, rather than on individual as sense-maker in an experiential world. I will refer to such researchers as social constructivists. Although social constructivists do not necessarily reject the individual constructive actions of the learner, they tend to give language and the social/cultural context primary importance.

By these criteria, it would seem clear that traditional Piagetian researchers take an individual perspective while others take a distinctly social perspective (e.g., Ernest, 1991[3]; Restivo, 1993; Gergen, 1992; Lave and Wenger, 1991). This distinction may not appear so obvious for either social interactionists or Vygotskians. However, if one focuses on how these researchers focus their observations, it seems possible to make distinctions. For example, from a social interactionist perspective, Yackel, Cobb, and Wood have, in recent years, provided an explanatory framework for social interactions within elementary classrooms but have done so from an individual perspective, that is imagining what it is like for the individual in the social setting of the classroom (1990). Likewise I would place the work of both Bauersfeld (1991) and Voigt (1992) in the individual constructivist category. Although it is possible to use concepts developed by Vygotsky and his students to elucidate an individual perspective (Smith and Confrey, 1990), when one looks at current research being done in Vygotskian or socio-cultural framework, one tends to see the social constructivist perspective dominating. For example, in summarizing some socio-cultural work in relation to mathematics education, Renshaw states: '. . . a *particular* meaning is never acquired in the sense that all possible meanings can be reduced to or replaced by a single cultural meaning. Rather, meaning is socially situated in an ongoing dialogue between the frameworks that a child can bring to an activity, or that can be introduced into an activity by other participants such as another peer or teacher' (1992, p. 24, his emphasis). Although Renshaw is acknowledging the possibility of different meanings among individuals, he is doing so from a social perspective, that is viewing these meanings through the social interactions of the participants, rather than the individual's construction of knowledge.

This might seem to bring up the question of whether these two perspectives are necessarily separate or whether they might productively be bridged. I have argued elsewhere that attempts to do so have been unsuccessful and such attempts have often led to a linguistic 'knowledge slide', the use of the word 'knowledge' without discrimination to indicate both social and individual constructs (Smith, 1993). In that work examples were provided with the suggestion that those who try hardest to bridge these two perspectives, primarily the social interactionists and the Vygotskians, are most likely to fall into the use of a knowledge slide. Rather than attempting to bridge the two perspectives, it is important to recognize their complementarity. Each allows for a possible way of viewing and understanding that is difficult or impossible from the other.

As a convenient metaphor for distinguishing these approaches we might say that for social constructivists: 'Individual constructivists don't see the forest for the trees'. Whereas for an individual constructivists: 'Social constructivists don't see the trees for the forest.' Just as the biologist and ecologist ask different questions from different perspectives, individual and social constructivists may do the same, and it may be neither possible nor desirable to reduce their viewpoints to a common perspective. However, results from one framework may (and should) continually provide a rich source of new problems from the perspective of the other (Smith, in progress). Using these two perspectives as a framework, we can re-examine the questions regarding computers and mathematics described above. To begin, it will be helpful to describe a recently published social constructivist view of mathematics (Ernest, 1991).

Social Constructivist Mathematics

In his book *The Philosophy of Mathematics Education* (1991) Paul Ernest argues for a social constructivist view of mathematics. This mathematics has three descriptive characteristics:

1. It is descriptive rather than prescriptive.
2. It is based on quasi-empiricism, conventionalism, and radical constructivism.
3. It consists of both subjective and objective knowledge.

Objective knowledge consists of socially accepted forms of linguistic expressions which evolve over time (thus are fallibilistic) through processes similar to those described as quasi-empiricist by Lakatos (1976). In mathematics, this objective knowledge concerns the linguistic rules, theorems, formulas, and algorithms which have become accepted within a particular culture, for example school mathematics within the United States. Ernest attributes the certainty we feel for this 'objective knowledge' to linguistic conventionalism.

Since the linguistic rules for 'objective mathematics' are fairly well-defined and change slowly, one can imagine that a 'mathematical computer' might have an easier time with the Turing test than a more conversational one. As computers become more sophisticated, the linguistic traps, proposed by Collins may be harder to find in a world of 'objective mathematics' and may be invisible to unaware participants. Thus if we are to detect these mechanical strangers in Ernest's social constructivist world, we may need a way to account for the 'rest of behaviour'. Ernest does attempt to bring an individual perspective to his work by introducing individual or 'subjective knowledge' which he ties to radical constructivism. This, however, poses another problem.

Subjective Knowledge and Radical Constructivism

Once one places knowledge within a linguistic domain, it becomes difficult to escape these bounds when discussing individual knowledge. Individual knowledge is subject referential, that is it implicitly assumes a knower. To place such knowledge in language requires some means of explaining how it is that individuals are knowers of that knowledge. For Ernest, all mathematical knowledge (both objective and subjective) has a linguistic basis: '. . . the foundation of mathematical knowledge, both genetic and justificatory, is acquired with language' (1991, p. 75). Because Ernest depends upon the common linguistic basis to bridge his 'objective knowledge' to individual ('subjective') knowledge, however, he has difficulty tying this knowledge to individual (radical) constructivism with its emphasis on individual constructive actions. In fact, in arguing his case, I believe Ernest comes perilously close to the use of a knowledge slide. From a radical constructivist perspective, operations are based on actions (Piaget, 1970; Steffe *et al.*, 1983; Confrey, 1991) and there is no direct relationship to linguistic structures. For example, Confrey's recent work on splitting (1994) suggests that there are at least two fundamental kinds of actions underlying the operation of multiplication which young children develop despite the fact that our language provides little basis for this distinction. This conflict is also apparent when looking at the issue of fallibility. Recently, in separate discussions on the same day, Paul Ernest stated that the mathematical statement $2 + 2 = 4$ is fallible (personal communication, 1992) and Ernest von Glasersfeld said that the same statement was true with certainty (personal communication, 1992). Given Ernest's social constructivist perspective, especially with its ties to conventionalism, he must allow that this statement, like all mathematical statements is fallible, simply because language use changes and is both culturally and historically dependent. For von Glasersfeld, however, to claim certainty for this mathematical statement is to claim that one has abstracted the mental operations of isolating two sets of two units each and has combined these sets into one set which numbers four. For von Glasersfeld, the mental operations create a certainty for which the language $2 + 2 = 4$ is arbitrary and merely descriptive. If one's use of language changes over time or situation, it will not alter this certainty. Thus in articulating a radical constructivist view, von Glasersfeld would not place the basis for mathematical knowledge in language. Henderson carries this argument further emphasizing that diversity of background engenders diversity of mathematics. A viable proof must 'answer my why question and relate my meanings of the concepts involved. The proof that satisfies someone else may not satisfy me because their meanings and why-questions are different from mine. . . . Persons who are different from me (for example, in terms of cultural background, race, gender) are most likely to have different meanings and thus have different why-questions and different proofs' (Henderson, 1992, pp. 3–4). These concerns about reconciling social constructivist mathematics with an individual constructivist approach are similar to those raised by Confrey (1992b). She also sees the role of the observer as central to the two perspectives:

> This concern for the observer's status poses a fundamental dilemma for the socio-constructivist. If the intermental exchanges are what creates and determines the intramental thinking, then what would lead one to expect diversity in students' interpretation of a statement, argument or justification? Is it simply accorded to the diversity of social experiences children have undergone? Giving voice, recognition and encouragement to these constructive inventions is a primary goal of the constructivist. If classroom interactional descriptions . . . ascribe a uniformity to students' experience, then the opportunities to build on the richness and diversity of student approaches will evaporate. . . . The pivotal question as I see it is how do they [socio-constructivists] bring such ideas and experience into coordination (and, to some extent only, reconciliation or also, potentially, confrontation) with the public arena of debate and negotiation. (Confrey, 1992b, p. 27)

Confrey's concern is directly related to the particular context of education. Without an individual perspective, one loses access to the richness of individual constructions that are at the core of the learning process. Thus the 'viability of the child's model from the child's perspective' receives little attention (p. 19) and many 'researchers in this tradition focus so heavily on the social interactions that the rich world of individual constructs remains relatively unarticulated and discussed in their writings' (p. 24). If one accepts the validity of separating these two perspectives as I have done, then a question to social constructivists would be: Given that all eduction is situated within social and cultural settings, is that enough to argue that a social constructivist perspective is sufficient by itself to account for educational activity? The problem which I see, and which seems to be central to Confrey's concern, is that with their focus on language and social interaction, social constructivists may, like computers, leave out the 'rest of behaviour', or in this case, the 'rest of mathematics'.

Computers in a Social Constructivist Classroom

The concerns expressed about relying on a social constructivist perspective in mathematics education carry over to models of computers in education. In particular, how does one, from a social constructivist perspective, recognize the 'mechanical stranger' if it is quite good at speaking the language of 'objective' mathematics? The two social constructivist accounts discussed below provide one basis for a addressing this question

Computer as Intellectual Partner

Pea and Salamon have recently proposed the idea of computer as intellectual partner, 'a sharing of cognitive load, the off loading of a cognitive burden onto a tool which partakes in the handling of the intellectual process' (Salamon, 1992, p. 17). What makes computers unique according to Salamon is that they have intelligence. Computers can not only handle lower-level tasks like calculating, drawing, and so on, but allow a new conception of partnership, where cognitions become 'distributed' in the sense that computer and partner *think jointly* (Pea, in press, from Salamon, 1992, p. 17). In this situation, Salamon argues for what he calls 'distributed intelligence', where instead of focusing on individual accomplishment, the emphasis is on the 'joint product'. The computer becomes a 'performance-oriented tool' (p. 18). This seems to be a strong socially situated view of knowledge. As Salamon states:

> The common assumption is that intelligence resides in people's heads and that situations can invite, afford the employment of, or constrain the exercise of this intelligence. The intellectual partnership with computers, particularly that kind of partnership which is joint-performance-oriented, suggests the possibility that the resources that enable and shape activity do not reside in one or another agent but are distributed between persons, situations, and tools. In this sense, 'intelligence is *accomplished* rather than possessed' [Pea, in press]; it is brought about, or realized, when humans and tools pool their intelligences and jointly tackle a problem or task. (Salamon, 1992, p. 18)

There are two questions that are not directly addressed in this chapter. First, one wonders if there is, in their view, no longer a distinction between the roles of computers and individuals, or if so, how that distinction would be made in the classroom. Second, if intelligence is 'accomplished', what is it that we are trying to do in schools. Do we offer no possibility that students obtain 'something' that they can carry away with them? If problem-solving is nothing more than solving a particular problem in a particular setting, then the 'accomplishment' of solving that problem does seem to be paramount and perhaps it makes little difference how the task was divided between human and computer. The value of the experience lies only in the ability of the participants to be able to repeat under similar conditions. Given the preponderance of evidence, however, that humans do have the capability to creatively remake old knowledge in new situations, one must wonder whether the distinctions between the roles of human and computer should be so diminished.

These questions point to the focus on a social perspective in their work. From this perspective, Pea and Salamon do offer valuable insights on the role of computers. As computers become integrated into school settings, the nature of the learning activity does change. Tools are not simply amplifiers of an independently existing subject matter (Pea, 1987), rather tool and problem (or subject matter) are inherently connected. Thus we can reject the idea that representations (including computer interfaces) can be 'invisible' and begin to seek how our understanding of what it means to learn mathematics changes when using computers.

Computer as Cultural Tool

In a move Vygotskian approach, Scott, Cole and Engel (1992) describe what they call 'cultural constructivism' which they characterize, in contrast with Piagetian constructivism, as assuming 'not only an active child but an equally active and usually more powerful adult in interaction . . . Moreover, cultural constructivism emphasizes that all human activity is mediated by cultural artifacts, which themselves have been constructed over the course of human history' (p. 191). For them computers play the role of a cultural tool which serves to mediate human activity:

> From this perspective, the historically conditioned forms of activity mediated through computers must be studied for the qualitatively distinctive forms of interaction that these artifacts afford and the social arrangements that they help constitute. Moreover, one is encouraged to seek explanation of current uses of computers in terms of the history of the technology and the social practices that the technology mediates; one needs to consider the 'effects' of interacting in this medium not only as they are refracted through transfer tests or in local activity systems (such as classroom lessons) but also in the entire system of social relations of which they are a part. (Scott *et al.*, 1992, p. 192)

There is much in the Scott *et al.* description that facilitates our understanding of computers in the classroom. The idea that computer-as-tool mediates the activity of learning supports the idea, also made by Pea (1987), that computers do not simply 'amplify' our ability to learn but alter the ways in which we learn and, thus, the nature of what is learned. This suggests that the question of 'what' is learned may be a more important question than simply 'how much', for Scott *et al.* as well as Pea seem to be arguing that the question of 'what' is learned cannot be answered without also knowing 'how' that learning takes place. In addition, Scott *et al.* also point out the importance of developing extended learning contexts which become embedded in computer cultures, noting several research centres which they describe as having 'an interest in sustaining learning contexts over time and beyond the confines of their own experimentation and an interest in creating a research agenda (and therefore an educational practice) in contrast to the technicist tradition in educational computing' (p. 242).

There is a broad agreement among researchers of many persuasions in the importance of developing sustained learning environments in which computers become embedded in the learning culture. If this does not happen, there seems to be little chance that computers will have a widespread significant influence in education. However, as suggested in Harry Collin's work above, this can be a double-edged sword. For just as the benefit of computers may be dependent upon their becoming culturally embedded, this same process may foster other concerns.

Although Scott *et al.* provide little detail about how they would view students' interactions with computers within a subject-matter context, they do refer to Wertsch's interpretation of Vygotsky (1985). The idea of a tool 'mediating' experience, in our case mathematical experience, brings forth the image of an individual and the mathematical subject matter with the tool somehow sitting between, 'mediating'. Mathematical content is separated from the doing of mathematics. This seems to be consistent with the general Vygotskian view of the individual and physical reality with culture mediating their interactions. As Wertsch states, a tool 'serves as a conductor of humans' influence on the object of their activity . . . [It] changes nothing in the object, it is a means for psychologically influencing behaviour . . .' (p. 78). In this case, the 'object' is mathematics. One need not postulate the independent existence of mathematics in this view, for it would seem that Ernest's 'objective mathematics', a socially constructed view of mathematics, would be quite consistent with the idea of mediation. However, this also leads one to the same problem which faced Ernest, reconciling this view with individual goal-oriented experience. For example, Confrey describes three possible methods which students might have when working on a specific problem involving multiplication, each a valid approach to the problem. She states: 'I see little in Vygotsky to help a more knowledgeable other to choose between the options, for each option would be legitimate to a more knowledgeable other . . . Vygotsky seems to argue that the more sophisticated method would be the one that relied on semiotic mediation of psychological tools to accomplish more complex operations' (1992, p. 18). Looking back at the student example in this chapter, if one takes the view that finding the equation is the 'more sophisticated method', then we must ask if, from a social constructivist perspective, there would be any pedagogical concern that Dan was never able to successfully fill the distance column, since the group did end up finding a correct equation. How could one place a value in allowing Dan to explore this problem from his own perspective, when his inability to do so ultimately led to the 'more sophisticated' solution? One might then ask why the software programme should have the capability to fill a column iteratively if the same result can be obtained with an equation, or, perhaps, why the computer cannot become the cultural conduit or 'more knowledgeable other'? Pushing this point a bit further, we might remember that when Turing originally posed his test, he was not asking whether the interrogator *could* distinguish the computer from the person, but questioning whether, in the course of normal interactions, the interrogator *would* make that distinction. From a social constructivist perspective, one must wonder whether there is any reason why students should make that distinction, if we eventually reach the point where, in the 'normal' course of classroom interactions they do not do so. Where are the 'enormous costs' that Collins was so concerned with?

Individual Constructivism and the Computer

Both of the above social constructivist models seem to want to 'pull the computer in' to make it, in a certain sense, part of the way one thinks and acts on problems which exist 'out there' in the social and physical world. From the social constructivist perspective, the computer is part of the same social milieu as the individuals, part of the interactive communicative processes that are being observed. Thus Figure 7.3 represents an extension of Figure 7.2 which includes the computer as part of those communicative processes. Individual constructivists, on the other hand, see the individual as making sense of an experiential world which includes both computers and mathematical problems. From an individual

Figure 7.3: Computer from a Social Constructivist Perspective

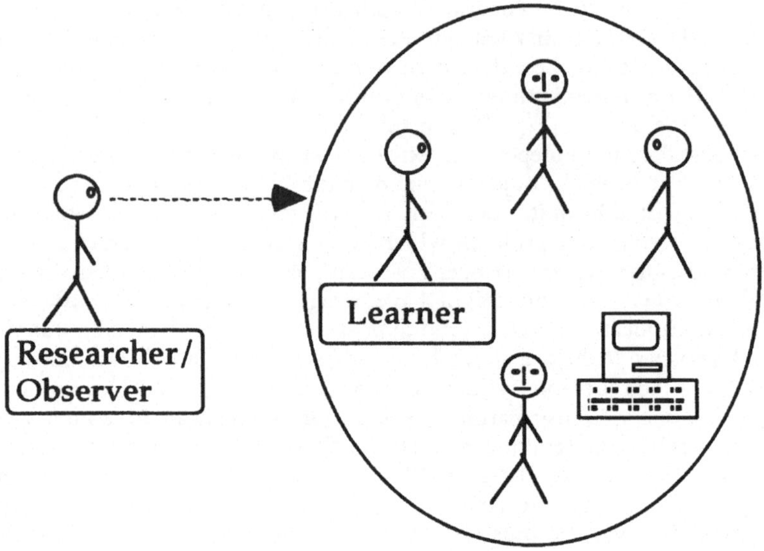

perspective the computer is not separable from the experiential world — it is part of the world from which the individual is constructing his or her mathematics, and the computer as constructed by the individual in conjunction with the problem. Both, are 'out there' in the experiential world of the individual. This difference is illustrated in Figure 7.4 where the computer is part of the experiential world of the learner. In terms of the previous distinction between social and individual constructivism, one sees the social constructivist observer as observer of 'culturally situated mathematics', observing the interaction between individual and computer. The individual constructivist observer, on the other hand, would be attempting to model the individual's possibilities for acting upon an experiential world and seeing mathematics being constructed in relation to actions which the individual sees as viable within his or her experiential world which includes a computer.

Thus for the individual constructivist, instead of viewing computers as a tool that changes the ways I do an independently existing (perhaps socially constructed) mathematics, an individual constructivist tends to see the computer as intrinsically related to one's understanding of mathematics itself. In the previously mentioned example from von Glasersfeld describing the possible mental actions one might take in constructing the certainty that $2 + 2 = 4$, it is clear that he places mathematics in the mental actions one takes. Those mental actions are created as tentative viable responses to problematical situations within one's experiential world. Within the experiential world, 'mathematical situations' and cultural tools are seen as inseparable, that is one creates problematical situations in relation to how one views possible ways to resolve those problems. Likewise, objects are created and defined through our actions. Thus to say that tools do not affect the objects, but only change the way we act is too weak a role. Statements such as: 'I made

Figure 7.4: Computer from an Individual Constructivist Perspective

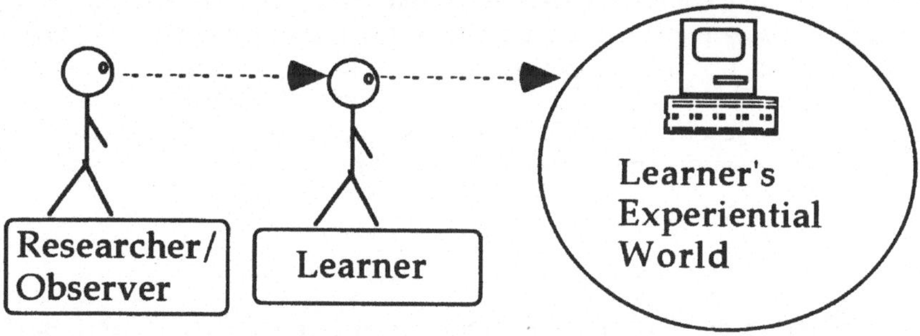

a graph', 'I filled a table column', or 'I divided two numbers' become statements about a different mathematics when one has access to tool software versus paper and pencil, just as they did when we started using writing tools. Computers are not simply a tool with which we act upon the world, but become part of the world upon which we act.

There is significant evidence in the literature illustrating how an individual constructivist perspective allows one to see a change in mathematics that evolves as students use computers (and this change grows as the 'computer culture' (Scott *et al.*, 1992) becomes more firmly established). For example, Rizzuti (1991) documented the power of a 'covariation' understanding of functions by students working with the fill command in tables; Afamasaga-Fuatai (1992) described students' construction of quadratic functions in contextual problems using a process of evaluating net accumulations over time in the table window of Function Probe; Borba (1993) has documented a fundamentally different understanding of functional transformations when working with students who began their explorations using visually dynamic mouse-driven actions directly on computer graphs; and finally, my own work suggests that students' mathematical language when problem-solving in small groups may change significantly when their efforts are centered around a problem-solving tool (Smith, 1993). In addition, much of the research on the use of Logo supports the idea that technological tools lead to changes in the how students think of mathematics (Hoyles and Noss, 1992; Papert, 1980). In all of these cases, the power of the computer became evident to the extent that it allowed students to instantiate their actions in relation to their own understanding of the problem. This work along with that of others (Kaput, 1986; Rubin, 1990) has been part of the development of an 'epistemology of multiple representations' based on the idea that 'mathematical concepts' are neither singular nor embedded in individual representations, but are constructed as a process of interweaving multiple representational forms with the actions we make within problematical situations. Thus actions, representations and concepts are inseparable. Students create their own problematics in relation to posed problems where the problematic they construct is inseparable from their understanding of the actions which were possible using the available tools (Confrey and Smith, 1991; 1992).

The strongest argument for the importance of the individual constructivist approach may be that its goals are directly aimed at understanding the 'rest of behaviour', that part of people which makes us human and which can never be part of a computer. From an individual constructivist perspective, the question of whether a computer can do mathematics is as meaningless as the question of whether computers could think was to Alfred Turing, for one has chosen a perspective that places the computer outside of the thinking individual. This is the point from which one starts, from which one looks out, and from which one tries to make sense of the world. Failing to pay attention to the individual constructivist perspective may result, in Collin's terms, in the failure of our social immune system to recognize the 'mechanical stranger' in our midst.

Conclusion

There has been a big push in recent years to increase the use of computers in school mathematics. However, since computers do much of what has traditionally been taught and valued (i.e., carry out computation), this push has been accompanied by some fairly hazy descriptions of what students are supposed to learn in these environments, especially at the elementary level. The arguments for the importance of an individual perspective underlie the importance of helping teachers and students develop an understanding of what people do and what computers do not do. Articulating these arguments should go a long way towards breaking down popular views of mathematics as a technical subject and support the movement towards a more humane view of our subject.

However, I am not arguing for the ultimate value of one perspective over another, rather to describe a legitimate tension between social and individual constructivists approaches to learning, using computers as a vehicle for describing that debate. The concerns that underlie the social constructivist programmes have led to essential insights into teaching and the social organization of the classroom, many of which have been neglected or even denied in many individual constructivist research findings. For example there has sometimes been a tendency to assume that results from individual studies of students working on disconnected problems in isolated situations can be generalized to broad learning environments. In this sense, bringing in the concepts of cultural tool, mediation, situated learning, etc. have provided an essential framework in which to evaluate constructivist research programmes. However, the concern expressed in this chapter is that these social concerns should not prevent us from still trying to understand how individuals make sense of their world. How do social constructivists allow individuals to create the kinds of culturally rich knowledge that contributes to the continual evolution of our culture if they do not, in some sense, allow those students to stand separate from the cultural milieu?

The computer is a convenient medium to explore this issue because many of our societal fears of 'technicizing' education fit well with many individual constructivists' fears of 'culturalizing' education, making education nothing more than a cultural (or linguistic) exercise. Looking back at the opening quote in this chapter, I would not deny the historical/cultural basis for the music of D. Ford Bailey, but would suggest that, had the Grand Old Opry restricted its membership to those who could 'write down' their music, Mr Bailey may never have had the opportunity to 'play the ideas in his head' on that stage. The question of how we create educational settings which allow all students to 'play the ideas in their heads' cannot be ignored in the process of educational reform.

Notes

1. The two approaches to creating this function suggested by D and T have been described as the covariational and correspondence approach respectively (Rizzuti, 1991; Confrey and Smith, 1991).
2. It should be noted that this is only the second week that these students have been using FP. They later become adept at using the 'fill' command in a variety of situations.
3. In this case I am specifically refering to Ernest's account of 'objective mathematics'.

References

AFAMASAGA-FUATAI, K. (1992) *Students' Strategies for Solving Contextual Problems on Quadratic Functions*, Doctoral dissertation, Cornell University, Ithaca New York.

BABAI, L. (1992) 'Transparent proofs', *Focus*, 12, 3, pp. 1–2.

BAUERSFELD, H. (1991) 'The structuring of the structures,' Paper presented at the *Visiting Lecture Series*, University of Georgia, Athens, Georgia.

BORBA, M. (1993) *Students' Understanding of Transformations of Functions using Multi-Representational Software*, Doctoral dissertation, Cornell University, Ithaca New York.

COLE, M. (1985) 'The zone of proximal development: Where culture and cognition create each other', in WERTSCH, J.V. (Ed) *Culture, Communication, and Cognition: Vygotskian Perspectives*, Cambridge, Cambridge University Press, pp. 146–61.

COLLINS, H.M. (1990) *Artificial Experts, Social Knowledge and Intelligent Machines*, Cambridge, Massachusetts, MIT Press.

CONFREY, J. (1991) 'The concept of exponential functions: A student's perspective', in STEFFE, L. (Ed) *Epistemological Foundations of Mathematical Experience*, New York, Springer-Verlag, pp. 124–59.

CONFREY, J. (1992a) *Function Probe© v.2.3.5.*, Santa Barbara, California, Intellimation Library for the Macintosh.

CONFREY, J. (1992b) 'The relationship between radical constructivism and social constructivism', Paper presented at the *State of the Art Conference on Alternative Epistemologies in Education with Special Reference to Mathematics and Teacher Education*, Athens, Georgia.

CONFREY, J. (1994) 'Splitting, similarity, and rate of change: New approaches to multiplication and exponential functions', in HAREL, G. and CONFREY, J. (Eds) *The Development of Multiplicative Reasoning in the Learning of Mathematics* Albany, State University of New York Press.

CONFREY, J. and SMITH, E. (1991) 'A framework for functions: Prototypes, multiple representations, and transformations', in UNDERHILL, R. (Ed) *Proceedings of the 13th Annual Meeting of PME-NA*.

CONFREY, J. and SMITH, E. (1992) 'Exponential functions, rates of change, and the multiplicative unit', Paper presented at the *Annual Meeting of the American Educational Research Association*, San Francisco.

ERNEST, P. (1991) *The Philosophy of Mathematics Education*, London, The Falmer Press.

GERGEN, K. (1992) 'From construction in context to reconstruction in education', Paper presented at the *State of the Art Conference on Alternative Epistemologies in Education*, Athens, Georgia, 19–23 February.

HENDERSON, D. (1992) 'Proof as a convincing argument that answers — Why?', Paper presented at the *Seventh International Congress of Mathematics Education*, Québec, 17–21 August.

HOYLES, C. and NOSS, R. (1992) *Learning Mathematics and Logo*, Cambridge, Massachusetts, MIT Press.

KAPUT, J. (1986) 'Information technology and mathematics: Opening new representational windows', *Journal of Mathematical Behaviour*, 5, pp. 187–207.

LAKATOS, I. (1976) *Proofs and Refutations: The Logic of Mathematical Discovery*, Cambridge, Cambridge University Press.

LAVE, J. and WENGER, E. (1991) *Situated Learning: Legitimate Peripheral Participation*, Cambridge, Cambridge University Press.

MATURANA, H. and VARELA, F. (1987) *The Tree of Knowledge*, Boston, Shambhala Publications.

NOVAK, J. and GOWIN, R. (1984) *Learning How to Learn*, Cambridge, Cambridge University Press.

PAPERT, S. (1980) *Mindstorms*, New York, Basic Books.

PEA, R. (1987) 'Socializing the knowledge transfer problem,' *International Journal of Educational Research*, 11, 6, pp. 639–63.

PIAGET, J. (1970) *Genetic Epistemology*, New York, Norton and Norton.

RENSHAW, P. (1992) 'Synthesizing the indiviudal and the social: Sociocultural theory applied to the mathematics education of young children', Paper presented at the *Seventh International Conference on Mathematics Education*, Québec, August.

RESTIVO, S. (1993) 'The social life of mathematics,' in RESTIVO, S. (Ed) *Math Worlds*, Albany, New York, SUNY Press, pp. 247–78.

RIZZUTI, J. (1991) *Students' Conceptualizations of Mathematical Functions: The Effects of a Pedagogical Approach Involving Multiple Representations*, Doctoral dissertation, Cornell University, Ithaca, New York.

RUBIN, A. (1990) 'Concreteness in the abstract: Dimensions of multiple representations in data analysis', in BOWEN, B. (Ed) *Design for Learning*, Cupertino, California External Research Division, Apple Computer, Inc., pp. 45–51.

SALAMON, G. (1992) 'Computer's first decade: Golem, Camelot, or the Promised Land?', Paper presented at the *Annual Meeting of the American Educational Research Association*, San Francisco, 19–24 April.

SCOTT, T., COLE, M. and ENGEL, M. (1992) 'Computers and education: A cultural constructivist perspective', in GRANT, G. (Ed) Review of Research in Education, Washington, DC, AERA, pp. 191–251.

SMITH, E. (1993) *Practice in a Radical Constructivist Setting: The Role of Virtues and Activities in Mathematical Knowing*, Doctoral Thesis, Cornell University, Ithaca, New York.

SMITH, E. (in press) 'A radical constructivist critique of radical constructivism'.

SMITH, E. and CONFREY, J. (1990) 'The social construction of mathematical knowledge: Issues and results from a classroom experience', Paper presented at the *Annual Meeting of the American Educational Research Association*, Boston, 16–20 April.

STEFFE, L., VON GLASERSFELD, E., RICHARDS, J. and COBB, P. (1983) *Children's Counting Types: Philosophy, Theory, and Application*, New York, Praeger.

VAN OERS, B. (1992) 'Learning mathematics as meaningful activity', Paper presented at the *Seventh International Congress of Mathematics Education*, Québec, August.

VOIGT, J. (1992) 'Negotiation of mathematical meaning in classroom processes', Paper presented at the *Seventh International Congress of Mathematics Education*, Québec, August.

VON GLASERSFELD, E. (1988) 'Environment and Communication', Paper presented at the *ICME-6*, Budapest, 27 July–3 August.

WERTSCH, J. (1979) 'From social interaction to higher psychological processes', *Human Development*, 22, pp. 1–22.

WERTSCH, J. (1985) *Culture, Communication, and Cognition*, London, Cambridge University Press.

WITTGENSTEIN, L. (1953) *Philosophical Investigations* (G.E.M. Anscombe Translation), Oxford, Basil Blackwell.

YACKEL, E., COBB, P. and WOOD, T. (1990) 'Learning through interaction: A case study', Paper presented at the *American Educational Research Association*, Boston, 16–20 April.

Chapter 8

The Context of Cognition: The Challenge of Technology

Katherine Crawford

Do not all charms fly
At the touch of cold philosophy?
There was an aweful rainbow once in heaven:
We know her woof, her texture; she is given
In the dull catalogue of common things,
Philosophy will clip an Angel's wings,
Conquer all mysteries by rule and line,
Empty the haunted air, and gnomed mine-
Unweave a rainbow. . . . (Keats)[1]

Introduction

The context for learners has changed. Most obviously, the advent of information technology has changed physical aspects of the learning environment through the increasing presence of computers and the new possibilities that are offered by educational software, multi-media and electronic communication. Less tangibly, but possibly more powerfully and irrevocably, the new technologies appear to occupy a different niche in the system of human activities and discourses and to have changed the relationships between culture, science and technologies — the patterns of their respective evolutionary paths. They present a challenge to traditional conceptions of cognition, of epistemology and the very purpose of education. Nowhere is this challenge more evident than in the field of mathematics.

The impact of modern science and technologies on cultures has been largely mediated through military and industrial activities and the acquisition of scientific and technological capability has become an intensely political process. Perhaps because of this, in many countries, the state has assumed more direct responsibility for the development of science and technology. This responsibility and concern is reflected in the priorities of state funding for schooling, technological development and research in scientific fields. A lot has happened since the time of Aristotle when learning about technology was excluded from educational activity, *a priori*, as degrading.

The significance of technology and its relationship to human society has changed in ways that present a real challenge to educational institutions. Formerly, technology was essentially an aggregate of practical know-how, devoid of any real theoretical justification. Technical evolution was slow and scientific rationalization typically followed technical developments. In general, technology was about injecting energy into systems.

In many ways, current educational practice in schools and universities is still widely viewed in terms of older forms of technology by politicians, administrators and the wider community. Their view may be justified. The educational community has been slow to provide a strong theoretical justification for the 'techniques' of schooling. Much teacher education involves acculturation into 'good' practices and 'techniques' for classroom management without a strong theoretical rationale. In addition, educational institutions have been slow to review the central role of paper and pencil as the appropriate technology for the development of knowledge in all fields.

Because of the state of the 'technology' of traditional schooling, education systems have been vulnerable, particularly in the United Kingdom, to new demands for increased conscious control, systematic planning and evaluation. In part, the demands reflect changed ideas in western cultures about the nature of knowledge, its relationship to science and technology and the responsibilities of educators. However, the strategies used to control and plan in education also reflect older views and measurement techniques associated with reproductive curricula and transmission models of learning. They also provide a 'scientific' rationalization for the new possibilities in social organization, record keeping, system analysis and 'quality control' of institutionalized educational processes that have become possible with the advent of computer technologies. Science, technology and mathematics are widely viewed as closely connected. According to Ladriere (1977) science and technology are more closely intertwined but still distinguishable.

The object of scientific investigations is the acquisition of more knowledge about the world. Mathematics is widely used as part of the process of investigation. In the process of scientific and mathematical activity information about existing system organization is transformed into abstract mental representations. Techniques may be used to further manipulate or investigate abstract representations to develop further knowledge and a changed view of reality. Romanyshyn (1989) describes the ways in which the distanced and objective scientific view changes our experience. He suggests that 'not only does the self create its own vision, but it creates the world in the image of that vision'.

He notes that our modern mathematized notions of reality have provided the basis for powerful technological intervention because:

> . . . it is only into a universe of purely mathematical relations that we could launch ourselves as astronauts, because prior to this emptying of the universe of all but mathematical quantities the heavens were either filled with gods or the home of one God and his angels. Angels first had to be erased from the heavens by mathematising the sky before we could launch ourselves into space. (Romanyshyn, 1989, pp. 79–80)

In contrast, the object of modern technology is to intervene to change reality. Information, as well as energy, is injected into systems in order to change events or create new objects. The effect of modern technologies is to transform abstract mental representations into information in the form of objective organization.

This process places a strong emphasis on the self as a creator. No longer is man merely an actor in the natural universe but a creator of his environment — his consciousness.

In modern information technologies knowledge and action appear fused. No longer is the abstract mental representation of the knower separate from the technology and its objective impact on reality. No longer are new scientific or mathematical understandings something within the minds of the elite. Now they are often derived from, and also embedded in, new technologies — in the virtual reality of urban life.

It is now widely believed that it is possible to plan for the future. In a planned future, collaboration, utility and productivity are highly valued. There is also value attached to detachment from a particular point of view. However, an increased sense that reality can be manipulated — knowledge can be used to construct new processes and objects as well as to utilize nature — has lead to a new form of political contest. To the extent that mathematics and education are viewed as sciences or technologies, the politics of the change is evident in educational management, planning and curriculum development. The political process is usually evident in four stages. These are:

1. An awareness of the ethical responsibilities. In a planned future in which man intervenes the responsibilities are considerable.
2. Political action in the form of competing lobbies for limited research and development funding from the state. Information becomes increasingly commodified and thus valued for its economic advantages. (e.g., statistics and probability in mathematics)
3. Anti-science/technology movements. These call in question the very foundations of modern rationalism. Contact, communication, communion, intuition, affectivity etc. are highly valued.
4. Emerging awareness of the *limits* of science and technology — the limits of the value of their separateness from mankind and nature — the limits and dangers of their autonomous progress without reference to human activities, human purposes and meanings.

There is mounting concern about the practical implications of the changes for human development and survival. Carter (1991, p. 278) suggests that there is a need to 'resituate' science and technology in a social context. He also notes that the impacts of technological developments on humanity have often been 'written out of discussion'. De Hunt Hurd (1991) advocates a future oriented curriculum in science and technology. He suggests that:

> The new curriculum is to be future oriented, not in the sense of predicting the future, but rather in providing students with knowledge and habits of thought for shaping and managing a favourable future for human survival and a desired quality of life. (De Hunt Hurd, 1991, p. 258)

Technology presents a challenge to society because it changes the environment in ways that may not always be beneficial for human survival and development. There is a growing awareness of the limits of former beliefs about cognition — habits of thinking and ways of knowing. In the past these were seen as natural and unchangeable. Now, it is widely believed that human cognitive activity can be changed and new ways of thinking and a new human reality can be created.

Changed Views of Cognition: What Is Cognition?

Information technologies are changing the conscious reality of humans in western society. They are changing the context of human activity, including learning, in increasingly rapid, more systematic and consciously controlled ways. Reality, even the reality of human capability and development, has become something to change — to improve. It is hardly surprising that such a changeable context for education, cognition and human development has received a closer scrutiny and considerable critique.

In the not too distant past cognitive processes were considered as processes carried out in the minds of individuals. Cognition was considered to be related to sensory perception but separate from physical activity. Many researchers and psychologists still measure cognitive 'abilities' as though they operate without reference to the contexts in which they are used (e.g., WISC and Binet tests). Further, in western psychology, notions of absolute consciousness have prevailed until recently. Sensory perceptions were thought to be equally available — or evident — to all. Cattell's (1971) 'culture fair' test of visual abilities exemplifies the position. Because visual perception was assumed to be objective Cattell (1971) was able to postulate that:

> Probably in the normal human range, differences in sensory experience opportunities are trivial and rare, so that the differences we measure in Pv (*visual abilities*) will turn out to be largely those of neural endowment. (Cattell, 1971, pp. 299–300)

Abilities were seen as (biological) characteristics of individuals and unrelated to the context or experience. In mathematics education researchers (such as Fennema and Sherman, 1977; Guay and McDaniel, 1977) explored the cognitive abilities that were associated with various tasks without reference to the context of the activities.

Piaget's (1977) epistemological theory was highly influential in mathematics education. His important observation of the developmental changes in the ways people process information has been a cornerstone of curriculum development in mathematics. However, although Piaget recognized that cognitive development and sensory information processing were influenced by experience, his theory proposed logico-genetic laws that limit the extent and rate of intellectual growth. More importantly, his theory has been largely interpreted in schools in terms of an age stage developmental sequence in which individual differences in 'readiness' for a particular stage were the result of inherent biological rather than experiential variations.

More recently, two basic assumptions, underlying earlier cognitive research in western countries, have been challenged. The notions of absolute consciousness have been questioned by constructivists and the theories of Vygotsky (1978) and his colleagues. The cartesian duality between mind and body, man and his environment have also been questioned. A more holistic and systemic view of humans acting, intellectually or physically, in a context has emerged.

First, the absolute views of consciousness that have prevailed in western psychology have been contested by the neo-piagetian constructivist theorists. The work of von Glasersfeld (1984) and others has been highly influential in mathematics education. It is now generally accepted by researchers (Crawford, 1986, 1992a; Cobb, 1988; Resnick, 1987; Cobb, Yackel and Wood, 1992) that humans learn through their experience and that the nature of the experience influences the nature of the resulting knowledge. Coming to know is an adaptive process that organizes one's experiential world; one does not discover an independent, pre-existing world outside the mind of the knower. Constructivism is an epistemological position that stresses the relationships between the experience of the knower and the quality of the resulting knowledge. It implies a relationship between the context of learning and the learning process. However, it tells us little about the cognitive processes involved in learning.

Non-absolute views of consciousness have lead to an interest in investigating levels of awareness of the context — individual and social interpretation of the constraints and supports of a setting. There has also been an interest in the meta-cognitive processes, self-awareness and self-regulation, associated with learning in a social context mediated by cultural artefacts. For example, researchers at the tertiary level (Marton, 1988), Volet and Lawrence, 1989), Crawford, Gordon, Nicholas and Prosser, 1993)) have explored in detail the relationships between students' personal conceptions of a task and its context, the nature of, and extent of, awareness of their own learning strategies, and learning outcomes.

The context of cognition and the relationships between cognition and context have received attention. The work of Rogoff and Lave (1984) and Lave (1988) have explored this relationship in detail. Lave's exploration of a 'setting', which she defines as the relationship between the actor and the arena in which he or she acts, is a seminal example of the less bounded view of cognition that is evolving among researchers.

There is a growing interest in the more holistic view, of Vygotsky (1978), Davidov and Markova (1981) and other Russian psychologists, of thinking acting and feeling in the world within a community consciousness that is constantly evolving through social activity. This view perhaps represents the least bounded view of human cognition and the strongest contest to the assumptions of cartesian duality that underlie other paradigms including the constructivist approach. A fundamental tenet of Vygotsky's theory is the integration of the internal and the external — of the self and other, through activity in context.

The activity theorists (Leont'ev, 1981; Davidov and Markova, 1983) exemplify the fusion of knowledge, thinking, feeling and action as a basis for cognitive development. In their interactive and systemic model of cognition, 'activity', either physical, emotional or intellectual is stimulated by subjective perceptions of needs in any context and subordinated to subjective 'images' of the goal of any action. In turn activity has objective results which form the context for later actions. For the activity theorists, activity occurs first in a social context and is later internalized to create new cognitive structures that will form the basis of future consciousness and activity. Within any activity, Leont'ev *et al.* make a distinction between 'actions' which occur in cases of conscious decision-making, planning and problem-solving and the unconscious 'operations' that are the automated routine procedures that are tools that usually form part of any activity. They note that operations require little intellectual effort, are not usually available for change and review. Engestrom (1990) uses the term 'activity system' to describe the processes occurring when people act in a group in an institutional setting such as a school.

The notion of Vygotsky and his colleagues that human cognitive capabilities are formed through action in context is now accepted by a number of researchers. Not only do we consider the responsibilities of a planned future in terms of the environment, social organization or technological development but also in terms of human cognition. The four stages of the politics of science and technology are exemplified in the development of cognitive science.

Life was much simpler when talent was God given and the context for human development was natural and unquestioned. The possibilities and responsibilities associated with changing human consciousness, the cognitive processes and the context in which learning occurs are all now open for discussion. For example Ornstein (1991) suggests that biological evolution of human consciousness is too slow and steps must be taken to speed things up. He suggests that knowledge of cognitive science must be applied to the problem as:

> This is a time when the need for conscious evolution is becoming a necessity for all humanity, not just a few individual isolates fasting and praying in a hilltop monastery. (Ornstein, 1991, p. 273)

Newman, Griffin and Cole (1989, p. 147) describe their research in working for cognitive change in school. They conclude that where they 'resonate most strongly with current formulations of cognitive science is in the notion of creating artificial systems . . . Research on education is also a science of the artificial: the study of how educational interactions work can never be far removed from the task of engineering them to work better.' Their approach exemplifies the responsibility that comes with knowledge that can be used to create something new — knowledge that can be injected into a system to create objective results. For them knowledge and action are integrated if not actually fused. The lobbies for educational research funding and the impact of politics on the scope and form of research are well known. At present, 'scientific' research with a technological base is still at a premium in most countries.

The task questioning of the dominant culture of research and interpretation of reality has been taken up with vigour by the post-structuralists and postmodernists. For example, Foucault (1977, p. 199) describes the 'discursive practises' through which all institutions, disciplines, practices and technologies have been humanly, historically, socially constructed. Such writers have demanded a radical reflection on interpretive frames and the motives and means of a humanly constructed reality. In arguing forcibly that people create their own worlds they have attempted to debunk the myths about the universality, progress and autonomy of science and technology. Lather (1991) comments on the effects of this process as follows:

> Given the postmodern foregrounding of the ways we create our worlds via language, perhaps for the first time the complex question of political commitment and its relation to scholarly enquiry can be seriously

> addressed. While the concept of advocacy research remains as oxymoron to many who take scholarly objectivity as both a possible and desirable goal in the human sciences, the mantle of objectivity which largely shielded the sciences from such questions has been irreparably rent. (Lather, 1991, p. 14)

She goes on to suggest that within the vast array of ways of knowing the questions have changed about data or information from 'Is the data biased?' to 'In whose interest is the bias?'. This somewhat cynical position in relation to the information base of the information age is reminiscent of a much earlier comment by Vygotsky (1939/62) about interpreting language. He said:

> To understand another's speech, it is not sufficient to understand his words — we must understand his thought. But even that is not enough — we must know its motivation. (Vygotsky, 1962, p. 51)

In a systemic approach to human cognition which explores the thinking, feeling and acting of humans in an increasingly man-made context, computers and other programmable forms of technology add an extra dimension to the complexity. They exemplify the objective results of technological knowledge. They also add another explicitly interactive relationship to the existing interpersonal interaction choices in the socio-cultural environment.

Public perceptions of the line between computers and human cognition is blurred in ways that are confusing. The confusion seems inevitable in view of the function of computers to organize information (previously the sole domain of conscious human intelligence) and the metaphors that have always been used to describe them. We talk glibly of 'artificial intelligence' and 'memory' and 'machine learning'. The computer in my office goes into 'sleep' mode if I leave it unattended. Computers are widely perceived as intimately connected to our identity and function as human beings (Winograd and Flores, 1986).

Computers do function to carry out many of the routine cognitive tasks that were once only a function of humans. As Leont'ev (1981, 64) states, it is usually the function of cognitive operations to become 'technicalized' as the function of machines. For example, when a person uses a computer to solve a mathematical problem, the machine carries out the computation as part of the human action in solving the problem. The human activity is not broken by the 'extra-cerebral link'.

Mathematical techniques have dominated most curricula in schools and universities. The advent of computers, and their capacity to carry out routine procedures fast and to repeat them many times, has changed the nature of mathematical knowledge and the ways in which people might think mathematically in at least four ways.

First, computers have made mathematical ideas routinely accessible in ways that were not possible before. For example the notions of chaos and fractal geometry have captured the public imagination and rapidly revolutionized the ways in which many people think about reality. Barnoly (1989) captures the essence of the wonder of the new images and insights as follows:

> The beauty and freshness of fractal geometry suggests that once again we are at the start of science and mathematics . . . women and men will look back on this era much as we look back at the early Greeks. (Barnoly, 1989, p. 5)

Second, the use of computers to carry out routine tasks has the potential to shift the focus of learners', and teachers', attention away from the memorization of techniques towards other cognitive capabilities such as interpreting the mathematical information in contexts, posing problems, devising mathematical models of systems, and interpreting and evaluating the results of computation. In other words, if a computer carries out the routine techniques the human part of the mathematical activities in a mathematics course, (and the quality of the mathematical knowledge of graduates) will change and might expand. Some universities are already using software systems, such, as 'Maple' or 'lcetl', as a computational base for mathematics courses. Abelson and DiSessa (1981) have described in detail the possibilities of 'turtle geometry' using Logo-style graphic representations of geomety. Such changes to the context of learning mathematics involve a re-evaluation of the cognitive demands of the mathematical tasks, the quality of learning that might be expected or possible, and the forms of assessment that might be appropriate.

Third, computers, and other complex technologies, have changed the context of mathematical activity in society and the relationship between mathematics and other spheres of human endeavour. Mathematical knowledge is now applied to technological systems in most fields in order to create objective results. Mathematical knowledge is also embedded in the very structure and design of technological systems. In information technology, mathematical knowledge and action are fused. Fujita (1993) suggests that with the extensive use of computers in modern industry, the nature of mathematical literacy has qualitatively changed. In particular, he stresses the importance of skills in mathematical modelling and interpretation of non-routine problems as capabilities in applied mathematics that are now needed by graduates in most professions where computers are used.

Finally, the new technologies present fundamentally new possibilities for the design and adaptation of interactive learning environments. As Kaput (1992) points out the impact of computers as a new feature of the learning environment for school mathematics has been very recent. In addition the rapidly changing technical possibilities mean that there must be caution about extrapolating from even the recent past. Once separate technologies are now merging. For example, as the digitalization of video progresses, computers and television are merging. Supercalculators now have the functions of computers. To date the impact of computers on educational practice has been marginal (Crawford, Groundwater-Smith and Milan (1990). As might be expected, much computer use in education represents alternative ways of following traditional practices. However, a number of software development projects (e.g., Cabri Geometry) are designed to be used as tools to facilitate the change in the nature of human mathematical activity that is implied by the presence of machines to carry out routine procedures. Less structured computational media such as 'Logo' or 'Boxer' also offer the possibility of more tailored 'microworlds' for exploratory, active, independent learning in mathematics.

The introduction of the new technologies into an institutional learning environment is a complex process. They represent but the newest element in a complex system of human activity which has an already well established set of assumptions and practices. The three examples outlined below illustrate some aspects of the change.

Drawing on Television

Children begin their experience of institutionalized education with a complex knowledge base, expectations based on their experience, and well developed language abilities. Thus, when a simplified form of 'Logo' was introduced in a preschool setting the 4-year-old children labelled the activity 'drawing on television'. The turtle made lines, which seemed like drawing on paper, and the monitor looked just like the familiar television set at home.

The naming of the activity immediately embedded it in their previous experience and shaped the ways that the children chose to explore the possibilities of the new activity. Their attention was riveted by the wrap-around effect. This was not like paper and pencil! Chubby little hands pressed and repressed the forward key and watched the emerging pattern of lines on the screen. With a limited set single key commands (including F — (FD 10); B — (BK 10); L — (LT 30); R — (RT 30); U — PU, D — PD, H —HT, and S — ST) they cheerfully set about exploring the possibilities of playing with the new 'toy'. Adult interference in their exploration was kept to a minimum as earlier sessions indicated that teachers tended to take over the decision-making and instruct them about what to do. In particular, in the presence of an adult the girls exhibited dependent learning behaviour. All children were encouraged to talk about what they were doing — to point to the left and right, to describe the patterns they were making to each other and the teachers.

Exploration of the new environment generally followed the following pattern:

1. Random experimentation with the commands.
2. Attempts to predict where the turtle was after using the hide command.
3. Experimentation with the wrap around effect using the forward command.
4. Attempts to control the direction of the turtle resulting in attempts to create horizontal and vertical lines. The terms 'horizontal' 'pattern of development' . . . 'horizontal vertical'.
5. Planned filling of available space with shapes.

By the end of a twelve-week period, most children were able to competently discuss the activity in terms of distance, direction, horizontal–vertical, left–right, and planned methods to produce different shapes. Many were exploring the possibility of repeating closed shapes in different positions. Some were expressing frustration at the clumsiness of the L and R commands which were set to give a turn of 30 degrees (For a more detailed account see Crawford, 1988).

Experience experimenting with the new learning environment focused children's attention on concepts that are not usually developed until rather later than 4 years of age. However, in the emerging 'drawing on television' culture they were familiar and empowering ideas and thus accessible. The project provided ample evidence of the powerful impact of experiences of the computer-based environment on children's learning. Three aspects of the situation appear particularly important. These were:

1. The children were not yet socialized into school learning. They expected to make sense of the new activity and played independently.
2. The mathematical nature of the 'Logo' environment was evident to the children and intrinsically interesting.
3. The activity allowed the children to start with their own conceptions about 'drawing' and 'television' and through the use of commands convert their ideas into objective results. In turn, the results provided the basis for new concepts and ideas.

Designing Something New

The following scenario describes a typical scene in a project investigating learning in mathematics and science using Lego-logo materials at a girls primary school (Crawford, 1992a). Three girls are sitting together trying to build a Lego model in first session of the school computer club. Later they will be expected to use the Lego-logo software to programme their model. The girls attend an exclusive, traditionally oriented school. They are neatly dressed and well organized. The lego-logo activities are part of the computer club as it was feared that they might interfere with the usual curriculum. In fact it soon becomes apparent that the usual curriculum of the school has all but paralyzed the group. 'Is this correct?' one girl asks the tentatively. The other two look perplexed. The engine is connected directly to the steering wheel so that it spins furiously when the engine is switched on. Another girl approaches the group to look at the model car she is holding. The model is snatched away with cries of 'Don't copy!' Eventually order and security is restored during the first session by providing the whole group with a simple task of building a model by copying an illustration. The teacher was most puzzled about the lack of confidence and creative initiative. She remarked that the girls had 'done' three dimensional shapes in mathematics lessons.

The above description illustrates the tensions that emerge as part of the challenge of technology to culture and in particular to the cultural practice of schooling. The girls' cultural experience outside school was rich in activities such as ballet lessons, piano lessons and excursions to the theatre. However, they were very inexperienced in using materials such as the Lego blocks and they had paid very little serious attention to how cars worked. As a result, despite their enthusiasm for the computer-club activities, they were culturally impoverished in terms of access to a conceptual basis for the building and modelling activities. They had 'covered' three dimensional shapes in the mathematics curriculum but had difficulty building three dimensional models with the Lego blocks. In addition, their socialization in the traditional school context and knowledge of 'appropriate' behaviour as learners, although powerful in focusing their attention and motivation with regard to teacher-centred activities involving imitation and memorization, had left them intellectually dependent and with the notion that collaboration was a form of cheating.

Educational institutions have been traditionally structured in ways that are appropriate for 'transmission' of knowledge from expert to novice and for sorting students on the basis of the cognitive abilities that are demanded in such contexts. From a student perspective, the curriculum is imposed, the ideas and knowledge are acquired, memorized and demonstrated in assessments of one kind or another. The challenge of information technology to most fields of human activity is that the development of capabilities in representing knowledge as objective organization appears to require a different 'setting' (Lave, 1988) for learning and knowledge of a different kind.

Information Technology and Teaching

The definitions of teaching are also challenged. A computer, potentially at least, provides an alternative source of authority and information in an educational setting. There is evidence of concern among teachers that Educational Intelligent Systems (EISs) may supplant their role in schools and also evidence that computers have been largely marginalized in the mainstream curriculum (Crawford, Groundwater-Smith and Milan, 1990). Many teachers are ambivalent about computers and wary about using tutoring systems and teacher-support software to carry out routine tasks.

However, the more profound challenge to teaching practice stems from the epistemological changes that are implicit when new technologies now carry out routine procedures and provide accessible declarative knowledge. That is, reproduction has now been largely mechanized and this presents a challenge for traditional teaching practice and assessment which has demanded reproduction of knowledge and techniques from learners. Educating people to carry out the functions of machines rather than to use them creatively appears anachronistic at the very least.

The results of a study of computer-mediated teaching and learning, using Boxer, exemplify the challenges presented by the new technology (Crawford and Lambert, 1993). It was found that, for teachers, the process of designing and constructing a computer-based environment required extensive relearning about the task domain and theories of learning as well as learning about programming in 'Boxer'. That is, knowledge about learning theories and mathematical techniques was not sufficient basis for the design of a computer-based learning environment in mathematics. However, re-learning occurred through the process of converting knowledge about learning and about the domain into objective organization as the task of designing, making and modifying a suitable computer-based environment was carried out. In the 'Boxer' environment this effect was tangible and unavoidable. Many tacit beliefs and assumptions of a teacher must be communicated overtly in order to organize the computer context. The results of initial research support claims about the educational benefits of 'bricolage', 'constructionism' and design (Turkle and Papert 1990; Harel, 1991; Pratt, 1991) for younger children. The inadequacy of previously learned knowledge and routines was apparent. Similar effects have been noted for teachers designing exploratory learning environments in mathematics (Crawford, 1992b; Ball, 1987). That is, the traditional modes of education that have produced the 'expertise' of teachers in discipline-based knowledge and educational theory do not necessarily equip them for tasks involving the design and responsive modification of new settings for learning.

However, the process of designing contexts for learning provides important opportunities for teachers to gain a deeper understanding of domain knowledge and of educational theory. Working within the constraints and supports of the 'Boxer' system also leads to a deeper understanding of the possibilities of the computational medium.

The students, that later used the 'Boxer' learning environment, learned relatively less than the teacher who constructed it about either the knowledge domain or the 'Boxer' interface. They asked for constant support in carrying out tasks and expected adults present to assist them at each point of difficulty. Their notion of the expert teacher was someone who helps when difficulties arise. The results support Pratt's (1992) claim, in relation to Logo microworlds, that the educational power of new technologies lies in the opportunities to gain experience of converting ideas and knowledge into objective organization through the creation of 'microworlds' rather than in the interaction with microworlds that have been developed by others.

Conclusion

The new technologies have evolved, in part at least, from a mathematized view of the world. Certainly the application of mathematics in technological systems has been a generative process in creating new artefacts and contexts for new learning. The new context for human evolution and survival raises questions about the processes involved in 'educating' humans to create new artefacts using available technologies and the objective organization of knowledge. In particular, it seems that there is a need for isomorphism between the conditions and processes used in learning and the cognitive demands of later application of knowledge. This isomorphism was generally present when public-education systems were first set up with structures which facilitated the 'transmission' of knowledge. Learning with the intent to reproduce knowledge and procedures was appropriate for most of the population when the rates of cultural change were slower, book keeping and other clerical tasks were done by hand, and the ability to allude to, and quote from, literature was more highly valued by 'men of letters'.

The new role of machines in organizing information and carrying out complex routine tasks, and the impact of the new technologies on human development present a challenge for technologically based cultures. First, now that the impact of technological artefacts on human activities and development is recognized, there is a need for a review of the relationships between technological and scientific activities and development, on the one hand, and other human values and purposes, on the other. Second, with a higher priority given to creative and adaptive human capabilities there is a need for a review of the social organization of educational contexts. Implicit prerequisites to the creative use of knowledge as a basis for objective organization are more independent and creative learning experiences for students. Understanding the logic and purpose of new knowledge is one thing, being able to apply it to transform a new context is another.

The most beneficial possibilities of a technologically based culture will only be realized through a process whereby recent knowledge about learning and human development, and new technology-based learning contexts, are applied, as a means of objective organization, in educational institutions. The process seems likely to be slow and to involve the resolution of complex tensions. There are substantial vested interests in the prevailing organization and the recent push for competency based curricula, in some countries, seems likely to reinforce reproductive, teacher-centred curriculum design and implementation. A change towards more creative and self-directed learning experiences for students in technological contexts will place considerable demand on teachers to relearn and rethink their knowledge of mathematics and technology, and their knowledge of the learning–teaching process. The challenge presented by technology is different from the challenge of science. Science has changed the way we look at the natural environment — a process of unweaving rainbows. Technology in constructing a new reality presents the challenge of deconstruction and reconstruction of the environment, and also socially constructed institutional organizations, and the responsibility of more planned human development in a humanly created context.

Note

1. Jack Stillinger (Ed) (1978) *The Poems of John Keats*, The Belknap Press, Cambridge, pp. 472–3.

References

Abelson, H. and DiSessa, A (1981) *Turtle Geometry: The computer as a Medium for Exploring Mathematics*, Cambridge, Massachusetts, MIT Press.

Ball, D. (1987) 'Unlearning to teach mathematics', *For the Learning of Mathematics*, 8, 1, pp. 40–7.

Barnoly, M. (1989) *The Desktop Fractal Design Handbook*, New York, Academic Press.

Carter, C. (1991) 'Science-technology-society and access to scientific knowledge', *Theory into Practice*, 31, 4, pp. 273–9.

Cattell, R.B. (1971) *Abilities, Their Structure, Growth and Action*, Boston, Houghton Mifflin.

Cobb, P. (1988) 'The tension between theories of learning and instruction in mathematics', *Educational Psychologist*, 23, 2, pp. 87–104.

Cobb, P., Yackel, E. and Wood, T. (1992) 'A constructivist alternative to the representational view of mind in mathematics education,' *Journal of Research in Mathematics Education*, 23, 1, pp. 2–23.

Crawford, K.P. (1986) 'Cognitive and social factors in problem solving behaviour', in the *Proceedings of the Tenth Conference of the International Group for the Psychology of Mathematics Education*, London, July, pp. 412–15.

Crawford, K.P. (1988) 'New contexts for learning mathematics', in the *Proceedings of the 11th Annual Conference of the International Group for the Psychology in Mathematics Education*, Vesprem, Hungary, August, pp. 239–46.

Crawford, K.P. (1992a) 'Playing with Lego/Logo: School definitions of work and their influence on learning behaviour', in Nevile, L. (Ed) *Proceedings of the Logo and Mathematics Education Conference*, LME5 Lake Tineroo, Queensland, Australia, April 1991, pp. 45–55.

Crawford, K.P. (1992b) 'Applying theory in teacher education: Changing practice in mathematics education', in Geeslin, W. and Graham, K. (Eds) *The Proceedings of the Sixteenth PME Conference*, University of New Hampshire, Durham, New Hampshire, August pp. 6–11, pp. 161–7.

CRAWFORD, K. and LAMBERT, B. (1993) 'Study of Computer Mediated Teaching and Learning, using Boxer', Unpublished Paper, University of Sydney.

CRAWFORD, K.P., GORDON, S., NICHOLAS, J. and PROSSER, M. (1993) 'Learning mathematics at university level', in ATWEH, W. (Ed) *Contexts in Mathematics Education, The Proceedings of the 16th Annual conference of the Mathematics Education Research Group of Australasia*, Brisbane, July, pp. 209–14.

CRAWFORD, K.P., GROUNDWATER-SMITH, S. and MILAN, M. (1990) *Gender and the Evolution of Computer Literacy*, Revised research report to the NSW Ministry of Education, published by the Government Printing Office, 1990.

DAVIDOV, V. and MARKOVA, K. (1983) 'A concept of educational activity for school children', *Soviet Psychology*, 21, 2, pp. 50–77.

DE HUNT HURD, P. (1991) 'Closing the educational gaps between science technology and society', *Theory into Practice*, 31, 4, pp. 251–9.

ENGESTROM, Y. (1990) 'Developing thinking at the changing workplace: towards a redefinition of expertise', *Technical Report CHIP 130* of the Centre for Human Information Processing, University of California, La Jolla.

FENNEMA, E. and SHERMAN, J. (1977) 'Sex related differences in mathematics achievement, visualisation and affective factors', *American Educational Research Journal*, 14, 1, pp. 51–71.

FOUCAULT, M. (1977) 'History and systems of thought', in BOUCHARD, D.F. (Ed) *Language Counter-memory, Practice: Selected Essays and Interviews with Michael Foucault*, Ithaca, New York, Cornell Press, pp. 199–204.

FUGITA, H. (1993) 'Principles in Organising University and Pre-University Mathematics Curricula For Scientists and Engineers', Paper presented at the South East Asian Congress on Mathematics Education, 7–11 June, Surabaya, Indonesia.

GUAY, R. and MCDANIEL, E. (1977) 'The relationships between mathematics achievement and spatial abilities among elementary school children', *Journal of Research in Mathematics Education*, 8, 3, pp. 211–15.

HAREL, I. (1991) *Children Designers*, Norwood, New Jersey, Ablex.

KAPUT, J. (1992) 'Technology and mathematics education', in GROUWS, D. (Ed) *A Handbook of Research in Mathematics Learning and Teaching*, New York, NCTM/Macmillan.

LADRIERE, J. (1977) 'The Challenge Presented to Cultures by Science and Technology', Paris, UNESCO.

LATHER, P. (1991) *Getting Smart*, New York, Routledge.

LAVE, J. (1988) *Cognition in Practice*, Cambridge, Cambridge University Press.

LEONT'EV, A.N. (1981) 'The problem of activity in psychology', in WERTSCH, J. (Ed) *The Concept of Activity in Soviet Psychology*, USA, M.E. Sharpe.

MARTON, F. (1988) 'Describing and improving learning', in SCHMECK, R. (Ed) *Learning Strategies and Learning Styles*, New York, Plenum Press, pp. 53–82.

NEWMAN, D., GRIFFIN, P. and COLE, M. (1989) *The Construction Zone: Working for Cognitive Change in School*, Cambridge, Massachusetts, Cambridge University Press.

ORNSTEIN, R. (1991) *The Evolution of Consciousness*, New York, Prentice Hall.

PIAGET, J. (1977) *The Principles of Genetic Epistemology*, London, Routledge and Kegan Paul.

PRATT, D. (1991) 'The design of microworlds', in NEVILE, L. (Ed) *The Proceedings of the Logo and Mathematics Education Conference* LME5, Lake Tineroo, Queensland, July, pp. 25–43.

RESNICK, L.B. (1987) *Education and Learning to Think*, Washington, DC, National Academy Press.

ROGOFF, B. and LAVE, J. (1984) (Eds) *Everyday Cognition: Its Development in a Social Context*, Cambridge, Massachusetts, Harvard University Press.

ROMANYSHYN, R.D. (1989) *Technology as Symptom and Dream*, New York, Routledge.

TURKLE, S. and PAPERT, S. (1990) 'Epistemological pluralism: Styles and voices within the computer culture', *Signs*, 16, 1, pp. 128–57.

VOLET, S. and LAWRENCE, J. (1989) 'Goals in adaptive learning of university students', *Learning and Instruction: European Research in an International Context*, 2, 3, pp. 297–516.

VON GLASERSFELD, E. (1984) 'An introduction to radical constructivism', in WATZLAWICK, P. (Ed) *The Invented Reality*, New York, Norton, pp. 17–40.

VYGOTSKY, L.S. (1962) *Thought and Language*, Cambridge, Massachusetts, MIT Press.

VYGOTSKY, L.S. (1978) 'Mind in Society', in COLE, M. *et al.* (Eds), Cambridge, Massachusetts, Harvard University Press.

WINOGRAD, T. and FLORES, F. (1986) *Understanding Computers and Cognition*, Norwood, New Jersey, Ablex.

Part 2

Psychology, Epistemology and Hermeneutics

> Hermes is cunning, and occasionally violent: a trickster a robber. So it is not surprising that he is also the patron saint of interpreters. (Kermode, 1979, p. 1)

One of the central themes of constructivism is that of individual interpretation. This concerns the reading that persons make of their environment, experiences and indeed of texts, in the course of learning and living. Interpretation is of course not a concern exclusive to constructivism, and there are other traditions for which interpretation is central. These include, for a start, psychoanalysis, hermeneutics, phenomenology, ethno-methodology, anthropology, literary criticism, post-structuralism and post-modernism. Thus, for example, in Derrida's 'deconstructive' approach to post-modernism, the issue of interpretation and reading is central. His position is that there is no privileged interpretation of any text, irrespective of the intentions of the author, as the following commentary makes clear.

> In writing, the text is set free from the writer. It is released to the public who find meaning in it as they read it. These readings are the product of circumstance. The same holds true even for philosophy. There can be no way of fixing readings . . . (Anderson *et al.*, 1986, p. 124)

Hermeneutics does not go so far as to postulate a universe of free-floating texts and interpretations. Instead it anchors itself in the persons making an interpretation, and their relationship with the text. Indeed, Ricoeur emphasizes the transformative power of the text on the reader through his or her engagement with it.

> The theory of appropriation . . . follows from the displacement undergone by the whole problematic of interpretation . . . To understand is not to project oneself into the text; it is to receive an enlarged self from the apprehension of proposed worlds. (Ricoeur, 1981, p. 182)

Such a perspective locates hermeneutics closer to (but distinct from) constructivism. It also raises some of the concerns of psychoanalysis, namely the transformation of the person through their interpretations of events, dreams, etc. One of Freud's key contributions is the recognition of the symbolic nature of mind (Henriques *et al.*, 1984). How in its productions and interpretations the unconscious mind condenses meanings into certain symbols, which successful psychoanalysis can interpret. Of course such meanings can also be displaced. This gives rise to pathologies such as phobias and fetishes. But it may be that these very functions are essential to the formation of the human mind as we know and understand it (to the extent that we do!). Indeed Lacan argues that by acquiring language we submit to a symbolic order, one that structures our unconscious like a language. In this homology, condensation and displacement mirror linguistic metaphor and metonymy.

Thus a consideration of interpretation takes us from the individual making sense of events and texts, via a free-floating realm of texts, to events and texts structuring the individual. Perhaps the post-modernist claim that all is text is not so far-fetched at all? The realm of interpretation, text and culture flows around and within the individual, both structuring the individual in its multiplicity and changing, following the individual's actions. This dissolves the boundary between the individual and the world and perhaps creates a realm of 'potential space' where the mysterious objects of mathematics are to be found (Winnicott, 1971).

Looking at the architecture of the mind raises the issue of the impact of other persons, which is of course central in Freud, Lacan and other post-Freudians, But it is also central in the theories of Vygotsky, the Soviet activity theorists, and in Mead and the social theories of mind to which his work gave rise. These theorists also stress language and conversation as progenitors of mind.

Ultimately interpretation becomes a vast portmanteau concept, encompassing all relations between knower and known, between subject and object, and between speaker and spoken. It is the shared idea in theories of knowledge that make room for the knower. It also requires an attitude of humility, once we realize that what we say to learners and each other will be received in a different spirit than the one in which it was intended. There is no essence of meaning we can transfer to each other, to guarantee a fixed interpretation. There is no essence of meaning in these words I am writing now. I am in an imagined meaning world of my own, and I have my intentions as I write this, but I will doubtless look at these words in a different light when they come off the press. What you the reader will make of them, is yet another matter! But having 'popped up' my presence in the text, let me pop it down again to relate these matters to education.

In mathematics education the issue of interpretation is especially important. Teachers direct, structure and control mathematics learning conversation both to present mathematical knowledge to learners directly or indirectly, and to participate in the dialectical process of criticism and warranting of mathematical knowledge claims (i.e., assessment). The primary aim of such conversation is that of ensuring that the learner is appropriating collective mathematical knowledge and competencies, and not some partial or distorted version. Appropriated mathematical knowledge is potentially unique and idiosyncratic, because of the inevitability of human creativity in interpretation and sense-making. This possibility is exaggerated because school mathematical knowledge is not something that emerges out of the shared meaning and purpose of a pre-given form of life. Instead it is a set of artificially contrived symbolic practices whose meaning is not already given, but a significant part of it is deferred until the future. Interpretation, and its offspring understanding, are central and familiar issues in the context of schooling.

This section includes a number of forward looking and provocative chapters on some of these themes. I believe that the kinds of explorations in this section make up an important theoretical growth point for research in mathematics education, and that we will be hearing more from these perspectives in the future. For example, ten years ago the idea of a mathematics education journal publishing articles on psychoanalysis and mathematics education, let alone devoting the whole issue to it, was unthinkable. Yet a recently we saw just that.[1]

Note

1. *For the Learning of Mathematics*, February 1993, Volume 13, Number 1.

References

ANDERSON, R.J., HUGHES, J.A. and SHARROCK, W.W. (1986) *Philosophy and the Human Science*, Beckenham, Croom Helm.

KERMODE, F. (1979) *The Genesis of Secrecy*, Cambridge, Massachusetts, Harvard University Press.

HENRIQUES, J., HOLLOWAY, W., URWIN, C., VENN, C. and WALKERDINE, V. (1984) *Changing the Subject*, London, Methuen.

RICOEUR, P. (1981) *Hermeneutics and the Human Sciences*, Cambridge, Cambridge University Press.

WINNICOTT, D.W. (1971) *Playing and Reality*, London, Tavistock.

Chapter 9

Another Psychology of Mathematics Education

David Pimm

> Psychoanalysis is part of psychology. It is not medical psychology in its traditional sense, nor the psychology of pathological processes. It is psychology proper; certainly not all of psychology, but its substratum, possibly its very foundation. (Freud, 1955)

> Mathematical science . . . is the language of unseen relations between things. But to use and apply that language, we must be able fully to appreciate, to feel, to seize, the unseen, the unconscious. (Ada Lovelace)

> Method, Method, what do you want from me? You know that I have eaten of the fruit of the subconscious. (Jules Laforgue)

> Mathematics is a disembodied version of the human personality. (Winnicott, 1990)

For much of this century (from at least the time of Piaget and Thorndike), the development of mathematical concepts, particularly early mathematical concepts, has apparently held a fascination for psychologists. Increasingly, this preoccupation has grown to the point where some psychologists seem to consider this to be their natural domain. Moreover, it is on occasion rationality itself that is ostensibly to be studied by means of an examination of thinking and the use of mathematical concepts (see, for example, Sutherland, 1992 also Paulos, 1993 and Pimm, 1993a). Walkerdine (1988, 1990), a psychologist herself by training, has in some sense continued this tradition while at the same time sharply criticized its presumptions.

Relatively recent developments in social aspects of cognition, in particular following the rehabilitation in the West of the work of Vygotsky, have led outwards, into the social world, into the learning community (see, for instance, Lave, 1988). In philosophy of mathematics too, discussion of the isolated, supremely rational, individual, idealized mathematician has given way to a consideration of communities of mathematicians (see, for instance, Kitcher (1983), but also Rotman's (1988) fascinating analysis of the language and multiple agency of proof, explained in his 1993 book *Ad Infinitum*), as well as to wider, cultural embodiments of patterned thought (see, for example, D'Ambrosio, 1991), often unhelpfully termed 'ethno-mathematics'.[1]

My intention here is relatively straightforward. I choose to turn my attention in the opposite direction — inwards, not outwards — not because I think the social to be misplaced as a locus of exploration, but because I firmly believe it is not the only one of interest. A mathematician, David Henderson, once commented to me, 'I do mathematics to find out about myself.' And, in an enticing passage, George Spencer Brown (1977) writes of his sense of congruence between mathematical and psychoanalytical activity.

> In arriving at proofs, I have often been struck by the apparent alignment of mathematics with psycho-analytic theory. In each discipline we attempt to find out, by a mixture of contemplation, symbolic representation, communion, and communication, what it is we already know. In mathematics, as in other forms of self-analysis, we do not have to go exploring the physical world to find what we are looking for. (Brown, 1977, p. xix)

So, for now, I turn away from both the social and rationality and towards a brief exploration of individual 'irrationality' in the context of mathematics — and I use that term deliberately, with both its mathematical and illogical connotations resonating. My focus for discussion will be language. Paul Ricoeur (1970, p. 4) reminds us: 'I contend that the psychoanalyst is a leading participant in any general discussion about language.' In particular, psychoanalysis may prove insightfully relevant in that important subset of language, the interplay of symbols and meanings. This is probably the richest arena we have for particular discussion about the teaching and learning of mathematics.

One of the most often cited but seldom explored quotations in mathematics education is the following from the mathematician René Thom (1973) offered in a plenary lecture at ICME 2.

> *The real problem which confronts mathematics is not that of rigour, but the problem of the development of 'meaning', of the 'existence' of mathematical objects.* (Thom, 1973, p. 202)

His belief that meaning and existence comprise mathematics education's central problem is reflected in the fact that this is the only one of two italicized sentences in the entire piece.[2] What is seldom pointed out is that the central part of his talk, the one which culminates in the above summary sentence, concerns the role of 'unconscious activity' in mathematics, while the etymology of the word 'existence' invokes something standing (or being placed?) *outside.*[3]

Meaning is, on occasion, primarily about reference alone; but can also be about much more. Meaning is often also about associations of all sorts (including verbal similarities). Young children seem particularly open to the playful aspects of the *sound* similarities of language, and the surprising connections that can sometimes be made through this version of 'moving along the metonymic chain'. In particular, I believe meaning is partly about unaware associations, about subterranean roots than are no longer visible even to oneself, but are nonetheless active and functioning.[4]

Michel Serres (1982) writing of the origins of mathematics, claims:[5]

> The history of mathematical sciences, in its global continuity or its sudden fits and starts, slowly resolves the question of origin without ever exhausting it. It is constantly providing an answer to and freeing itself from this question. The tale of inauguration is that interminable discourse that we have untiringly repeated since our own dawn. What is, in fact, an interminable discourse? That which speaks of an absent object, of an object that absents itself, inaccessibly. (Serres, 1982, p. 97)

How are we implicated in making ourselves (and others) unaware of certain connections? Are our referential meanings in mathematics absenting themselves, with or without our express permission? What processes may be at work?

Forms of Words

> At the bottom of each word
> I'm a spectator at my birth.
> (Alain Bosquet)

There is a terminological difficulty in referring to the potential area of study I have been alluding to. Freud's term 'psychoanalysis', as a general theoretical term, seems to require adherence to a Freudian interpretation of phenomena; 'analytical psychology' to a Jungian one; 'psychotherapy', though more theory-neutral, to a more pragmatic activity. Other possibilities include 'psychodynamics' or even 'psychosynthesis'.[6] So while I sometimes employ and enjoy the coded term 'unconscious-raising' when asked what I am currently about, I am still looking for a helpful term which signals 'attention to unconscious elements'.

Even the very word 'unconscious' has difficulties. First, it is negative: it tells us what it is not, rather than what it is. (This is a phenomenon not unknown in mathematics, with terms such as 'irrational' or 'infinity'.[7]) The equivalent German term *Das Unbewusste*, first coined in the psychoanalytic sense by Freud 100 years ago, shares this first difficulty of being negative, but has greater resonance. Shattering the word in different places provides:

un — a negative prefix;
bewusst — an adjective meaning 'aware';
wusste — the first person singular, imperfect tense of the verb *wissen*, meaning 'to know'.

Putting these together produces a compilation meaning of:

> [that which I am] not [now] aware [that] I used to know.

Thus, the sense of 'I' is much stronger in the German word than in the corresponding English one.[8]

There is a second difficulty in pointing to that which is of interest to me: invisibility. The image of an eclipse can perhaps help. One thing that eclipses can do is to blot out the blatant, the obvious, and allow things to be seen that normally are invisible to the naked eye. How to bring about classroom eclipses deliberately is a deeply problematic undertaking, but occasionally fleeting glimpses into the psychic lives of teachers and students of mathematics can be obtained — merely by being attuned to the possibility *and* being in the right place at the right time.[9]

What is on the list of 'windows into the unconscious' that Freud came up with? It includes: slips of the tongue, dreams (see Early, 1992 in relation to mathematics), things you know well but forget or are unable to recall (e.g., names of people), and 'faulty achievements' (where I achieve something faultlessly, but it wasn't 'really' what I intended to do). Are there any particularly mathematical ones? Sherry Turkle at the end of her book on psychoanalyst Jacques Lacan ('The French Freud'), writes:

> For Lacan, mathematics is not disembodied knowledge. It is constantly in touch with its roots in the unconscious. This contact has two consequences: first, that mathematical creativity draws on the unconscious, and second, that mathematics repays its debt by giving us a window back to the unconscious. [. . .] so that doing mathematics, like dreaming, can, if properly understood, give us access to what is normally hidden from us. (Turkle, 1992, p. 240)

One possibility for such direct access may be through working with geometric forms themselves, those strange self-referentialities for which the symbol and the referent are one: 'the symbol for a circle is a circle.' In a book with the evocative title *The presence of the past,* Rupert Sheldrake (1989) talks of 'morphic resonance', an expression which captures exactly what I am concerned with. The power of forms to evoke has been long exploited by sculptors. For instance, Barbara Hepworth has recalled: 'All my early memories are of forms and shapes and textures' and her sometime fellow art-school student Henry Moore observed: 'There are universal shapes to which everyone is subconsciously conditioned and to which they can respond.'

As a teenager at a single-sex school, I can recall on different occasions being both excitedly part of, and embarrassedly outside of, groups of adolescent boys giggling over the resemblance of certain quartic graphs to a pair of breasts, and still remember the charge that being asked to mark in the turning points on many curves generated in me. Thus geometric diagrams at certain times can become 'psycho diagrams' as well (see Davie, 1992, p. 2).

I also remember as an adult watching a geometric film intended for high-school pupils (Schattschneider, 1990) about a stellated figure (a stellar octangula) comprised of two interpenetrating tetrahedra. The lingering, almost balletic construction of 'Stellar' (and despite associations with the female name) put me strongly in mind of sexual union, and due to the fact that both 'participants' were the same and pointed, of buggery.

More significantly and fundamentally, dynamic Nicolet films can on occasion put us in touch once again with our forming former selves that are still with us (those that used to know how we should become), as well as recontacting those forms from which we became.[10]

Mathematical Analysis

Occasionally in the writing of psychoanalysts, mention of mathematics arises. Sometimes it is autobiographical material, such as the following account. Carl Jung writes of his bewilderment and fear at school with mathematics in general, and with algebra in particular.

> The teacher pretended that algebra was a perfectly natural affair, to be taken for granted, whereas I didn't even know what numbers were. They were not flowers, not animals, not fossils; they were nothing that could be imagined, mere quantities that resulted from counting. [. . .] No one could tell me what numbers were, and I was unable even to formulate the question. To my horror, I found that no one understood my difficulty. [. . .] Whenever it was a question of an equivalence, then it was said that a = a, b = b, and so on. This I could accept, whereas a = b seemed to me a downright lie or fraud. I was equally outraged when the teacher stated in the teeth of his own definition of parallel lines that they meet at infinity. [. . .] My intellectual morality fought against these whimsical inconsistencies, which have forever debarred me from understanding mathematics. [. . .] All my life it remained a puzzle to me why it was that I never managed to get my bearings in mathematics when there was no doubt whatever that I could calculate properly. Least of all did I understand my own *moral* doubts concerning mathematics. (Jung, 1965, pp. 27–8)

Sometimes, mathematics appears in accounts of material which emerged from work with patients. The child psychoanalyst Melanie Klein (1923) provided a number of brief accounts of mathematical interest, but the excerpts I offer below illustrate some of the acute difficulties which lie in this area of 'pointing' at the phenomena of interest by means of a written account and then offering an account of what has been pointed to. She writes of connections between schooling and the sexual interest of children: 'The extremely important *rôle* played by the school is in general based upon the fact that school and learning are from the first *libidinally* determined for everyone, since by its demands school compels a child to sublimate his libidinal instinctual energies' (p. 59). In particular, she details a number of arithmetic/sexual links that she came across in working with young patients. Klein's work here offers unchallengeably theory-laden reports ('unchallengeably' in the sense that the only access we have is through her account), with the implicit assumption being perhaps: 'Trust me. I know. I have analysed this child. She is my patient.'

For instance, Klein (1923; 1975) writes of 17-year-old Lisa:[11]

> Lisa considered the number '3' insupportable because 'a third person is of course always superfluous' and 'two can run races with one another' — the goal being a flag — but the third has no business there. Lisa, who had a taste for mathematics, but was very inhibited where it was concerned, told me that actually she only thoroughly understood the idea of *addition*; she could grasp 'that one joins with another when both are the same', but how were they added up when they were different? This idea was con-

> ditioned by her castration complex, it concerned the difference between male and female genitals. The idea of 'addition' proved to be determined for her by parental *coitus*. She could well understand on the other hand, that in multiplication different things were used, and that then, too, the result was different. The 'result' is the child. (Klein, 1923; 1975, pp. 67–8)

And she also cites her work with 6-year-old Ernst:

> He showed me too what he meant and wrote the sum '1 + 1 = 2' in the little boxes. The first little box in which he wrote the '1' was larger than the others. Thereupon he said: 'What is coming next has a smaller box'. 'It is mummy's *popöchen* [penis],' he added, 'and (pointing to the first '1') that is father's *popöpchen*, and between them the 'and' (+) is me'. He further explained that the horizontal stroke of the + (which he made very small too) didn't concern him at all, he and his *popöchen* were the *straight* stroke. Addition for him, too, is parental *coitus*. (*ibid.*, p. 68)

It may not have escaped your notice that this description of human procreation also fits the mathematical idea of group. The early notion of group (Galois actually wrote of *groupements*; that is, groupings; as Piaget was to in the next century) is of two elements [of a collection of substitutions] combining to produce a third: historically, that seems to have been the first explicit awareness. Fifty years later, in 1880, there was still extensive discussion as to whether or not identity elements or inverses need to be present. Sophus Lie explicitly added to the specification of a transformation group (so it is no longer the same as a group of transformations) that the transformations can be ordered into pairs of inverse elements. In 1918, George Miller commented: 'These difficulties [with regard to 'the most desirable definitions of the technical term *group*'] cannot be regarded as solved at the present time' (p. 383). I find faint echoes of the struggles of Lisa and Ernst in these mathematical uncertainties and confusions over the desirable. Some contemporary definitions of group completely subsume (sublimate?) the primary notion of the union of elements in the notion of a binary operation: the only explicit mention in the axiom list being of those (historically) subsequent requirements (identity, associativity, inverses).

Fidelity

> And how reliable can any truth be that is got
> By observing oneself and then just inserting a Not?
> (W.H. Auden)

One general difficulty I have encountered in writing this article is offering convincing descriptions of (let alone accounts *for*) purportedly unconsciously-influenced phenomena: another is the intensely personal nature of much of this material. However, having acknowledged such reservations, here is an attempted description and attendant discussion of an incident arising from a videotaped classroom lesson which I witnessed.

A secondary teacher (T) started a lesson with a class of 13-year-olds on addition and subtraction of negative numbers by invoking the assistance of an image of 'The Linesman'.

T:	I want to remind you of a little cartoon character that we've been looking at over the last few weeks — the Linesman — our little stick man. And he's been helping us do some calculations using positive numbers and the new ones we've been looking at — the negative numbers. Let me show you a picture of the Linesman — here he is. [He puts up an overhead projector slide.] There, right, remember him? And we've drawn various pictures of him. I want to remind you of how he was able to help us do some calculations about positive and negative numbers. Can anyone remember where his number line came from? Where did he get his number line?
P:	His suitcase.
T:	In his suitcase, yes. How long was it? How long was his suitcase? Lorna?
L:	Fidel [stumbles], fidelity.
T:	Fidelity? Fidelity? [He speaks with rising tones of surprise and disbelief.]
L:	No. [She laughs, very embarrassedly, completely hiding her face.]
T:	How long was the Linesman's number line? David?
D:	As long as you want it to be.
T:	As long as you want it to be. OK. We do decide just to take a piece of it, don't we. But if we wanted to take all of his number line, how long would it be. Gary?
G:	Infinity.
T:	Infinity. [To Lorna] Is that the word you're looking for? What does that mean, 'infinity'?
Pupil:	It never ends.
T:	It never ends, it never ends. Right, let's have a look . . .

For me, this brief classroom excerpt offers a plausible example of unconscious processes interfering with conscious language. The sexual *double entendres* of the whole dialogue are rampant, as is the wishful thinking: the 'going on forever', the 'being as long as you want'.[12] Although the words 'infinity' and 'fidelity' both have four syllables and the last two of both are the same, why did Lorna did not say '*in*fidelity', surely a more direct metonymic association with 'infinity'?

Freud writes:

> The subject matter of a repressed image or thought can make its way into consciousness on condition that it is *denied*. Negation is a way of taking account of what is repressed; indeed it is actually a removal of the repression, though not, of course, an acceptance of what is repressed. It is to be seen how the intellectual function is here distinct from the affective process. The result is a kind of intellectual acceptance of what is repressed, though in all essentials the repression persists. (Freud, 1955, p. 235)

Lemaire (1986, p. 75) glosses this as: 'The repressed signifier is always present in the negation, but, in another sense, it retains the repression through the "not".'[13]

My account for this incident must obviously remain highly speculative, partly because so little is above ground. I remain struck by the fact that she did not offer 'infidelity'. Assuming part of her attention (unawarely, I expect) was taken by the sexual interpretation of the above dialogue, and also perhaps that the charged term 'infidelity' was playing a role in her home life (even suitcases, perhaps), then the above described process of negation could have turned 'infidelity' into the more acceptable 'fidelity' (two negatives making a positive here at least), even though I doubt she has ever heard the parallel neologism 'finity'. Her stumbling over the first uttering of 'fidelity' could indicate either the emotional force of what was at work, or some half-aware realization that this was not quite right.

Dick Tahta has offered me an alternative account. It concerns the fact that there are strong resonances and connections between a *meaning* invested in 'infinity' and that of 'fidelity', through the notion of 'going on forever'. He offers the thought that there is a strong adolescent investment in fidelity and a corresponding unease about infidelity. Teenage magazines are filled with storied discussions and explorations of this idea.

It is clearly impossible to ascertain what was the case here. But the two accounts given illustrate two general routes for accounting for such occurrences. The first posits a preoccupation that is being kept down, but a connection is made between this concern and the mathematical topic at hand. The preoccupation finds a way to return to consciousness, but is only allowed to do so in a negated form. The links are primarily metonymic: from 'infinity' to 'infidelity' to 'fidelity', which is what finally surfaces. In the second account, the links are primarily semantic, metaphoric. She cannot produce the word 'infinity' for some reason, but the closest word with a strongly related meaning she can get is 'fidelity'. But unlike with mathematics' excluding, exclusive use of 'or', the one account need not exclude the other. Metaphor and metonymy are offered as *axes* rather than categories by linguist Roman Jakobson, and hence elements of both can, indeed *must* be present at the same time.

The Resonance of Words

> There seems to be no mathematical idea of any importance or profundity that is not mirrored, with an almost uncanny accuracy, in the common use of words. (George Spencer Brown)

Mathematicians 'borrow' many everyday words to describe mathematical phenomena of interest. They also use the same words over and over in order to reflect perceived structural or functional similarities, resulting in condensations of experience around terms such as 'normal', 'similar', 'multiplication', 'number'. Curiously, they seem to expect a conscious, overt denial of links to be sufficient to render such connections no longer heard. (I recall asking in a graduate theoretical course on tensor analysis what the curvature tensor has to do with curvature and receiving the answer 'Nothing'.)

But such words can and do evoke connections and links, particularly I believe in cases where the mathematical content does not generate powerful images and feelings of its own. The looseness and gap between symbol and referent, regularly exploited for mathematical ends, also permits such slippage to a far greater extent than in other disciplines. 'Circumscribed' is very 'close' to 'circumcised' and the connection is not arbitrary. Teenage girls working on the period of a function can and on occasion do make overt connections with menstrual periods. Adolescents can become preoccupied with freedom and constraints upon themselves, and geometry can offer them the possibility of working with the same material as well as the same terms.

Tahta (1993) comments:

> Mathematics teachers do not normally expect to make connections between the *relations* of mathematics and the *relationships* of family life. Indeed, many emphatically deny that there is any connection at all; for them, equality of algebraic expressions would be something quite different from, say, equality of esteem. (Tahta, 1993, p. 48)

What do these allusive stories have in common? The first is the resonant quality of language, both in the forms of words and in the topology of those forms, reflected in the belief that nearby symbols have nearby meanings. Thus, metonymic links or 'slips' have semantic consequences by presumed association.

The second commonality raises for me the whole question of 'meaning' in mathematics, in particular some of its unaware contributory components. One reason for teaching mathematics may be so that our students can develop this means of finding out about themselves, in addition to our offering them access to a shared inheritance of mathematical images and ideas, language and symbolism, and the uses for mathematics which humans have so far developed. To the extent that students are enabled to think like mathematicians, this possibility is made available for them. I would like to express this possibility in terms of mathematics deriving from both inner and outer experiences, and meaning as being generated in the overlapping, transitional space between these two powerful and sometimes competing arenas.

Postscript: And Katie Makes Three?

> . . . a mathematics in which persons are not added together nor separated from one another; but in which there is an original whole, from which, through all the transformations of experience, one is never estranged. (M.C. Richards)

Katie is 3. She, her mother and I are having breakfast in the kitchen. I am telling Barbara who the ten were at the table for dinner in the Chinese restaurant the night before. I list nine names and struggle for a while to name the tenth. Then it 'comes' to me: *I* am the tenth. The number 'ten' refers to the group as a whole with no distinguished counter, a fact I have filed away. In retelling it here, *I* am the counter, generating names from my point of view when visualizing the table. It still takes a while for me to name myself. When I am counting people where I am to be one of the counted as well as the counter, I seem to arrange things so the counting ends with me (where counter and counted are one), partly so I know I am done.

A little while after, Katie counts, pointing to each of us turn: I am to be 'one', her mother is 'two', and she is 'two' too. She does it again: 'One, two, two.' She is insistent on how we are to be counted: 'You are one', she declares firmly, looking at me. She adds unprompted: 'I know it is "one, two, three", but I am pretending.' I think only at the time that perhaps she is identifying with her mother, with being female. Perhaps there are only two sorts to count: those like herself, and others. We move on to talk of other things. I think about this later in the day. Why is she 'two' and not 'three'? A few months earlier, Katie had her third birthday, so there are at least two separate senses (the other being the conventional arithmetic of the count) in which she could legitimately number herself 'three'.

I then remember that, moments before this reckoning, Katie had remarked to me 'Daddy dead'. Her father was killed not long after she was 1. I also remember too the night before this breakfast when she arrives home from childcare and I am already in the house: she seems surprised into unaccustomed silence. And after a very short while she says to me through the banisters: 'I thought you were daddy.' (She has mentioned this to me before: sometimes it is 'I nearly called you daddy.' I am only an occasional visitor to their now two-fold household, but bear some physical resemblance to her father.) It comes to me that 'three' was also what they were as a family before becoming two, and Katie was the third: 'Three, the simplest possible family number.' (Winnicott, 1990. p. 55) Was she avoiding saying 'three' so as to avoid thinking about what she once had but no longer has? Is she denying that the 'three' before her, despite all the similarities, was '*the* three'? Or had she managed things so that her desired answer was reached?

Notes

1. See Chevallard (1990) for a trenchant and penetrating analysis of some of the difficulties inherent in the notion of 'culture' in mathematics education, and in particular his distinction between 'ethno-mathematics' and 'proto-mathematics', the latter offering fertile sources for mathematization without itself *being* mathematics.

   For an illuminating account of the reasons why Garfinkel coined the term 'ethno-methodology', see Garfinkel (1968). There he indicates that his initial exploration in 1945 was of the function of being a juror, by means of the question 'What makes them jurors?'. He tells how he was going through files in the Yale library looking for a term to describe the methodological concerns and preoccupations of these people with 'being jurors'. He found many tags like 'ethno-botany', 'ethno-physiology', and 'ethno-physics'.

   > 'Ethno' seemed to refer, somehow or other, to the availability to a member of common-sense knowledge of his society as common-sense knowledge of the 'whatever'. If it were 'ethnobotany', then it had to do somehow or other with his knowledge of and his grasp of what were for members adequate methods for dealing with botanical matters. (Garfinkel, 1968, p. 7)

Garfinkel goes on to detail the ways in which ethno-methodology has become a shibboleth and comments: 'I think the term [ethno-methodology] may, in fact, be a mistake.'

My concern with some uses of 'ethno-mathematics' is not the studies themselves of particular cultural practices, but the absence of any sense that the members of the particular culture believe they are doing mathematics or are concerned with mathematics when working on, for instance, sand drawings or making baskets. The knowledge they have may just as well be said to be about the activity itself (and hence is better seen as 'proto-mathematical').

2. In the other, Thom criticizes the assumption (of mathematical 'modernists') that:

> *By making the implicit mechanisms, or techniques, of thought conscious and explicit, one makes these techniques easier.* (Thom, 1973, p. 197)

This is the complete reverse of my belief that automation of functioning and the liberation that this can bring results from the making unconscious of conscious control mechanisms. Thom adds:

> Certainly, this detachment [of the thinker from his thought] is a necessary step in the process of mathematical advancement: but the inverse operation, which is the reabsorption of the explicit into the implicit, is no less important, no less necessary. (*ibid.*, p. 199)

3. Hadamard (1954), too, devotes two chapters to general issues of the unconscious in the context of mathematical creativity.
4. There is a difficulty relating to the word 'associations', as it can seem to suggest a conscious, intellectual process, and so 'unaware associations' may seem an oxymoron. Bettelheim (1989) writes of the problem of translating Freud's term *Das Einfall* as 'free association'.

> Also, using the technical term 'free association' to describe a procedure entails the *a priori* assumption that two or more seemingly entirely disconnected events are indeed fairly closely connected. *Webster's* definition of the verb 'associate' 'to join together, connect makes this amply clear. In the translator's use of 'free association', what ought to be two separate processes letting something come spontaneously to one's mind, and examining how it may be connected to some immediately preceding stimulus are merged into one, and it is predicted what the result of this investigation will be. (Bettelheim, 1989, pp. 94–5)

5. For an insightful account of some of the mathematical writings of Serres, see Wheeler, 1993.
6. Tahta edited a special issue, 13, 1, of the journal *For the learning of mathematics*, on the theme of the psychodynamics of mathematics education. The English pronunciation of the word 'psychoanalysis' lays stress on 'analysis' and hence brings out possible links with the powerful Greek mathematical distinction between the methods of 'analysis' and 'synthesis'.
7. For Cantor, A is *finite* if it is equipollent to {1, 2, . . . , n}, otherwise it is infinite. For Dedekind, A is *infinite* if it is equipollent to a proper subset of itself, otherwise it is finite. Thus, what for Dedekind was a definition, for Cantor was a property; a switch of field and ground. But the 'negation' is still attached to the word 'infinity'.
8. Given my remarks in the opening paragraphs about the social in relation to

mathematics education, Jung's notion of the 'collective unconscious', with mathematical forms as archetypes, might offer a possible synthesis of these two foci of interest. Jung wrote:

> The collective unconscious is part of the psyche which can be negatively distinguished from a personal unconscious by the fact that it does not, like the latter, own its existence to personal experience and consequently is not a personal acquisition. While the personal unconscious is made up eventually of contents which have at one time been conscious but have disappeared from consciousness through having been forgotten or repressed, the contents of the collective unconscious have never been in consciousness and therefore have never been individually acquired, but owe their existence exclusively to heredity. Whereas the personal unconscious consists for the most part of complexes, the content of the collective unconscious is made up essentially of archetypes. (Jung, 1959, p. 42)

This certainly casts the Socratic view of 'learning as remembering' in quite a different light. However, it also indicates that the term 'collective unconscious' is another potential oxymoron, in German at least, due to the presence of the tacit claim of '*ich wusste*' in *Das Unbewusste.*

9. There is also the decision to be made between exploring events arising in everyday settings versus interviews or even some form of mathematics-focused, individual or group analysis. (See, for example, the work of Claudine Blanchard-Laville, 1991, 1992.)
10. A schoolteacher working on such a geometric film with students in late adolescence can be seen in Love, 1987.
11. The play *Mrs Klein* by Nicholas Wright identifies 'Lisa' as being Klein's own daughter, whom she analysed as an adolescent.
12. I have no doubt this occurred completely unawarely on the teacher's part, and it certainly was on mine, despite my being present at the filming. I had remembered the 'fidelity' remark as a complete singularity — and my memory was that she had actually said 'infidelity'.
13. There is also a possible link between proof by contradiction and the psychological process of negation (see Pimm, 1993b). Both produce negated statements with the (tacit) intent of asserting the opposite. In proof by contradiction, the assertion is made that not-P is true, but the underlying reading that is required is that P is to be believed.

References

BETTELHEIM, B. (1989) *Freud and Man's Soul*, Penguin, Harmondsworth.

BLANCHARD-LAVILLE, C. (1991) 'La dimension du travail psychique dans la formation continue des enseignant(e)s de mathématiques', *Proceedings of the Fifteenth PME Conference*, Assisi, Italy, 1, pp. 152–9.

BLANCHARD-LAVILLE, C. (1992) 'Applications of psychoanalysis to the in-service training of mathematics teachers', *For the learning of mathematics*, 12, 3, pp. 45–51.

CHEVALLARD, Y, (1990) 'On mathematics education and culture: Critical afterthoughts', *Educational Studies in Mathematics*, 21, 1, pp. 3–27.

D'AMBROSIO, U. (1991) 'Ethnomathematics and its place in the history and pedagogy of mathematics', in HARRIS, M. (Ed) *Schools, Mathematics and Work*, London, The Falmer Press, pp. 15–25.

Davie, A. (1992) *The Quest for the Miraculous*, Tucker, M. (Ed), London, Lund Humphries.

Early, R. (1992) 'The alchemy of mathematical experience', *For the Learning of Mathematics*, 12, 1, pp. 15–20.

Freud, S. (1955) *Complete Works of Sigmund Freud*, Standard Edition, 19, London, Hogarth Press.

Garfinkel, H. (1968) 'On the origins of the term "ethnomethodology" ', in Hill, R. and Stones Crittenden, K. (Eds) *Proceedings of the Purdue Symposium on Ethnomethodology*, Institute Monograph Series No 1, Institute for the Study of Social Change, Purdue University, pp. 5–11.

Hadamard, J. (1954) *The Psychology of Invention in the Mathematical Field*, Dover, New York.

Jung, C. (1959) *The Archetypes and the Collective Unconscious*, London, Routledge and Kegan Paul.

Jung, C. (1965) *Memories, Dreams, Reflections*, Harmondsworth, Penguin.

Kitcher, P. (1983) *The Nature of Mathematical Knowledge*, New York, Oxford University Press.

Klein, M. (1923; 1975) 'The *rôle* of the school in the libidinal development of the child', in Klein, M. *Love, Guilt and Reparation and Other Works 1921–1945*, London, Hogarth Press, pp. 59–76.

Lave, J. (1988) *Cognition in Practice: Mind, Mathematics and Culture in Everyday Life*, Cambridge, Cambridge University Press.

Lemaire, A. (1986) *Jacques Lacan*, London, Routledge and Kegan Paul.

Love, E. (1987) *Working Mathematically on Film with Sixth Formers* (PM647B videotape), Milton Keynes, Open University.

Miller, G. (1918) 'Mathematical encyclopaedic dictionary', *American Mathematical Monthly*, 25, pp. 383–7.

Paulos, A. (1993) 'Think about it', *London Review of Books*, 11 March, pp. 9–10.

Pimm, D. (1993a) 'Down with decision theory', *London Review of Books*, 8 April, pp. 4–5.

Pimm, D. (1993b) 'The silence of the body', *For the Learning of Mathematics*, 13, 1, pp. 35–8.

Ricoeur, P. (1970) *Freud and Philosophy: An Essay on Interpretation*, New Haven, Connecticut, Yale University Press.

Rotman, B. (1988) 'Towards a semiotics of mathematics', *Semiotica*, 72, 1/2, pp. 1–35.

Rotman, B. (1993) *Ad Infinitum: The Ghost in Turing's Machine*, Stanford, California Stanford University Press.

Schattschneider, D. (1990) *Stellar Octangula*, Berkeley, California, Key Curriculum Press.

Serres, M. (1982) *Hermes: Literature, Science, Philosophy*, London, The Johns Hopkins University Press.

Sheldrake, R. (1989) *The Presence of the Past: Morphic Resonance and the Habits of Nature*, New York, Vintage Books.

Spencer Brown, G. (1977) *Laws of Form*, New York, The Julian Press.

Sutherland, S. (1992) *Irrationality: The Enemy Within*, London, Constable.

Tahta, D. (1993) 'Victoire sur les Maths', *For the Learning of Mathematics*, 13, 1, pp. 47–8.

Thom, R. (1973) 'Modern mathematics: Does it exist?', in Howson, A.G. (Ed), *Developments in Mathematical Education*, Cambridge, Cambridge University Press, pp. 194–209.

Turkle, S. (1992) *Psychoanalytic Politics: Jacques Lacan and Freud's French Revolution*, (2nd edition), London, Free Association Books.

Walkerdine, V. (1988) *The Mastery of Reason*, London, Routledge.

WALKERDINE, V. (1990) 'Difference, cognition and mathematics education', *For the Learning of Mathematics*, 10, 3, pp. 51–6.

WHEELER, D. (1993) 'Knowledge at the crossroads', *For the Learning of Mathematics*, 13, 1, pp. 53–5.

WINNICOTT, D. (1990) *Home is Where We Start From: Essays by a Psychoanalyst*, Harmondsworth, Penguin.

Chapter 10

On Interpretation

Dick Tahta

Plato's dialogue, the *Theatetus*, contains a nested sequence of accounts — interpretations of interpretations — that lead back to a purported discussion on the nature of knowledge between Socrates and two mathematicians, Theodorus and his young student Theatetus. This discussion is of interest to historians of mathematics because, in passing, it describes Theodorus as having established the irrationality of square roots of non-square numbers up to 17 — 'where for some reason he got stuck' — and it indicates that Theatetus may have generalized the result. It is not clear why Theodorus stopped at 17 and many plausible and implausible interpretations have been proposed. But what is more important in the context of the Platonic dialogue is that Theatetus invokes the binary classification of numbers in terms of rationality as a possible metaphor for what he imagines are two sorts of knowledge, namely the sciences on the one hand as opposed to something else on the other, that he does not yet quite understand, but which is implicit in Socrates' method of inquiry.

Further issues about the nature of knowledge may be found in the multiple resonances that Plato establishes in his writing. Plato starts by presenting a conversation between two philosophers, Euclid of Megara (i.e., not the mathematician) and Terpsion. This is said to take place on the day in 369BC that Theatetus is dying from dysentery and wounds incurred in the defence of Corinth. Euclid recalls that many years previously Socrates had described a discussion that he had had with the then young Theatetus and his teacher, Theodorus, Euclid recounts that he himself had taken notes during Socrates' spoken account and had then written up a version in direct speech, which he claims he had afterwards carefully checked over with Socrates. Terpsion suggests this might be a good time to go through the written version. The two philosophers then settle down to hear a reading of the piece by a boy servant, and Plato provides the script for us. It is not until the very end that Terpsion — and the reader of Plato's book — realizes that the discussion about knowledge must be supposed to have taken place in 399BC on the same day that Socrates was charged with corrupting Athenian youth. Thirty years later, the philosophers are reminded of the death of Socrates at the very time that they are aware of the dying Theatetus.

Plato presents some arguments about the nature of knowledge in the form of an account (his) of an account (Euclid's) of an account (Socrates') of a conversation between three people. And the present reader is now reminded that what is being read is my account of someone's translation of this sequence of accounts. Which of these most truly reflects what is supposed to have happened? Can there be knowledge without some sort of representation, without some sort of account?

Or, to put the problem in another context, suppose someone gives you an account of a transcript of a film of some classroom dialogue: how confident are you in your understanding of the original event? Consider, for example, the account given by David Pimm (see Chapter 9 of this book) of a transcript of a sequence from an edited videotape of a classroom episode: what do you feel you know about the original event? Would you have felt more confident if you had been there yourself when the episode was being filmed? Some knowledge may be immediate; but much of what we know comes to us mediated through representation, through preconception, through theory. Socratic ignorance may be preferable where the only other choice seems to lie — in the words of Seth Bernadete, a recent translator of the *Theatetus* — 'between an immediacy that is not available and a mediacy that is uneliminable.' (Bernadete, 1986a, p. 88).

But another alternative might be to embrace mediacy and make it serve one's ends. Socrates seeks to persuade Theatetus to discard an understanding of knowledge as 'true opinion'. He is critical of the view — ascribed to the philosopher Protagoras and advanced here by the two mathematicians — that knowledge is based on perception and that man is the measure of all things. For then knowledge becomes relative, and this was unacceptable, at any rate for Plato. But it is more commonly accepted nowadays that truth is a problematic notion and that what we can demand of opinion is that it be fruitful. And according to Goethe, *was fruchtbar ist, allein ist wahr* (only that which is fruitful is true). This is certainly the approach taken by psychoanalysts, who are of course very centrally concerned with interpretation in their work. Analysts invoke a distinction between narrative truth and historical truth: interpretation is understood to be a creative construction rather than a supposedly accurate historical re-construction.

> Interpretations are persuasive . . . not because of their evidential value but because of their rhetorical appeal; conviction emerges because the fit is good, not because we have necessarily made contact with the past. (Spence, 1982, p. 32)

So, in considering classroom accounts, it may be helpful to consider various interpretations, judging them not for some supposed veracity but in terms of their fruitfulness for the matter in hand, which may be supposed ultimately to be the improvement of our understanding and practice of the teaching and learning of mathematics. It may also be helpful to borrow some of the notions invoked by psychoanalysts when considering our own interpretations of classroom events, whether mediated or not. Thus, I hope to show that it could be fruitful to invoke the unconscious processes of displacement and condensation. These have also been characterized by linguists as associated respectively with the grammatical notions of metonymy and metaphor, and I have described elsewhere how interpretations of classroom practice, and of mathematical history, might be made in these terms (see Tahta, 1991, p. 229). Here, I want to try to illustrate the working of displacement and condensation in our understanding, or interpretation, of a mathematical theorem, and then in some further comments on Pimm's account of a classroom episode.

A Classical Theorem

It seems appropriate to take as an example the irrationality of the square root of 2, the classical result which Plato describes Theodorus and Theatetus as having generalized. It is not known for certain how it might originally have been proved that the diagonal of a square was not commensurable with its side: there are some plausible, geometric reconstructions. The well-known, dramatically elegant, arithmetic proof — ascribed by later Greek writers to Pythagoras — is a typical (and often claimed to be the first) mathematical example of proof by so-called *reductio ad absurdum*, a logical argument used over and over again by Socrates in Plato's dialogues. The proof is by analysis (as opposed to synthesis) in the sense that it starts with what is required, only in this case with its negative: you pretend (liar!) that you know the square root of 2 to be rational. In modern algebraic terms, the pretence is that $\sqrt{2}$ is a fraction a/b where a and b may be supposed to have no common factor. A sequence of transformations yields in the first place the result that a must be even — in which case b, having no common factor with a, much be odd; but further transformation then yields that b must be even. It seems that b must be both odd and even. Not accepting this, you backtrack to find the mistake in your argument, to find (surprise, surprise!) that it can only have been with your initial pretence.

The sequence of steps from one algebraic equation to the next can be associated with the psychoanalytic notion of *displacement*, a process in which links are established through a chain of connections some of which may not always be conscious. A typical example would be the slip of the tongue, in which a word gets substituted by another word. An example — from *The Merchant of Venice* — is Portia's hint to Bassanio that she is wholly his: 'One half of me is yours, the other half is yours — mine own, I would say.' A sequence of such shifts often occurs in dreams or in the activity of free association of words. Algebraic transformation of the equation $\sqrt{2} = a/b$ also involves a chain of substitutions. Seymour Papert has suggested that many students faced with equations of this sort engage — 'almost as if they have read Freud' — in a process of mathematical free association (Papert, 1980, p. 198). And just as the psychoanalyst occasionally interrupts the flow of associations, the algebraic transformations are punctuated by an occasional conscious interpretation — for instance, when $a^2 = 2b^2$ is understood to mean that a^2, and so also a, must be even.

In the course of the proof, the square root sign is got rid of, but the 2 reappears, not in its original form but as a sign of evenness. Papert describes this feature as a sort of mathematical pun and associates this with the psychoanalytic notion of *condensation*, a process in which a single idea or symbol is at an intersection of several associative chains and coagulates a cluster of meanings. Dreams, for example, are shorter and more compressed than their verbal descriptions and eventual interpretations; a play on words — a pun — operates in the same way. Mathematics is itself a powerful condensation of experience; its 'abstractions' are derived from a number of different 'concrete' examples.

Moreover, as mathematical theorems are transmitted across generations, they condense the accumulated experience of past interpretations. This yields another sequence of accounts of accounts; but, as in the game of 'Chinese whispers', the final account may be quite unlike the initial one. Irrationality is not quite the same notion as incommensurability: in my end is *not* my beginning. John Mason has suggested that many named theorems would not now be recognized by their originators:

> Cayley's theorem, a fundamental theorem in group theory, doesn't exist in his writings. Unless you dig and dig and dig and recognise — because of your current awareness of what is a significant result — a sort of little kernel, a little something or other which you want to hold on to and say, 'that is a theorem'. It seems to me that in mathematics we transform everytime we present a theorem. (Mason, 1991, p. 16)

There is a further sense in which the proof of irrationality condenses a number of very dramatic, *external* associations that do not illuminate the mathematics as such, but have an expressive, mythical quality of their own. Such are the stories — and they may well be fictions — about the supposed consternation of the Pythagoreans at the discovery of irrationality. Some historians of mathematics would now deny that these are historical truths, and they would not wish to multiply narrative truths. Others — and I declare my own sympathy with these — are prepared to weave as rich a tapestry of associations as possible. For example, Michel Serres reads the supposed crisis at the dawn of Greek mathematics in terms of three deaths: not only the legendary death of the shipwrecked Hippasus who was held to have divulged the discovery of irrationality, and the historical death of Theatetus who developed a classification of irrational numbers, but also the turning away from some of the teachings of the revered Parmenides, which Plato somewhat startlingly called a form of parricide (Serres, 1982, p. 130).

The third death — the parricide — arises in the following way. Parmenides had formulated the law of contradiction invoked in proofs by *reductio ad absurdum*. But he was emphatic that Being (that which is) is One; its opposite, non-Being (that which is not) cannot be. Whatever exists can be thought of, and conversely, everything that can be thought of exists. But, he claimed, you cannot think of what is not, you could not know it, you could not even say it. However, in another Platonic dialogue, the *Sophist*, Theodorus brings along a philosopher, referred to as a Stranger, who tells Theatetus,

> Don't take me to be, as it were, a kind of parricide . . . It will be necessary . . . to put the speech of our father Parmenides to the torture and force it to say that 'that which is not' *is* in some respects, and again, in turn, 'that which is' is not. (Bernadete, 1986b, p. 33)

The Stranger is prepared — despite Parmenides — to take opposites into account; the supposed parricide may be seen as a return to the dualism of the early Pythagoreans. (Theatetus, the mathematician for whom everything is countable, is certainly already prepared to take opposites into a count!) 'That which is not' can be conceived: for 'thinking makes it so', as Hamlet claimed, illustrating this with a pair of opposites, namely good and bad. How would the early Pythagoreans have conceived of that which is not commensurable? John Fauvel has pointed out that incommensurability is not in itself plausible without some sort of proof: its first proof must have constituted its discovery (Fauvel, 1987, p. 18). Proving irrationality made it so!

The proof establishes the contradiction that a number can be both even and odd. By a series of intriguing displacements, Serres suggests that because '*even* means equal, united, flat, *same*; while *odd* means bizarre, unmatched, extra, left over, unequal, in short, *other*', the contradictory result may be described as asserting that 'Same is Other'. This cannot be so; but because that which is not rational has been proved to exist as well as what is rational, then numbers may just as well be irrational as rational. This dualism — masterfully explored by Theatetus — means that irrationality, or what Parmenides would have called non-Being, is in a sense on the same footing as rationality, or Being. So, according to Serres, Same is indeed Other 'after a fashion'; and this is the parricide of Parmenides.

'Legend, myth, history, philosophy, and pure science have common borders over which a unitary schema builds bridges', writes Serres, stretching the condensations even further to include a play on the name, Metapontum, of the birthplace of he who was shipwrecked. The 'unitary schema' reasserts the oneness of Being with a vengeance; but the crossing of boundaries brings disparate and sometimes contradictory things together. Following Parmenides, we usually assume that contradictions cannot — must not — be simultaneously entertained: either something is or it is not. But condensations demand otherwise. David Pimm has suggested that this is why metaphor can be so disquieting in mathematics:

> for the very essence of metaphor . . . is to be able to claim *at one and the same time* that 'it is *and* it is not'! I assert 'a function is a machine' (and yet I also know it is not one) — the strength of the metaphoric assertion comes through the use of the verb *to be* — yet it carries with it implicitly its own negation. (Pimm, 1995, Chapter 10)

Narrative Truths

The ambivalence of simultaneous assertion and negation is familiar in psychoanalysis. It is perhaps this accommodation of the contradiction inherent in metaphor that makes some psychoanalytic narratives seem, at first sight, implausible. David Pimm quotes some startling examples from the work of Melanie Klein, including that of a 17-year-old patient, known as Lisa, who had understood addition when the things being added were the same, but not when they were different. Klein suggests, in effect, that addition was a metaphor that condensed various meanings for Lisa, including her difficulty in entertaining the idea of parental coitus, where different genitals are brought together. These remarks occur in a very comprehensive survey — based on some of her cases — of the role of the school in the libidinal development of children. Lisa is mentioned quite often in this survey; for example, it is reported that she recalled always finding it difficult to divide a large number by a smaller one. She associated this difficulty with a dream involving mutilation of a horse, and her going shopping for an orange and a candle. This — and its inevitable (Kleinian) interpretation in terms of castration — may seem far-fetched to some people, but it is precisely the way of the unconscious to be far-fetched. Another analyst might have made a less severe interpretation, but the issue must always be how fruitful this particular interpretation had been in the course of Lisa's analysis, and this, of course, we do not know.

For Klein, 'the tendency to overcome [the fear of castration] seems in general to form one of the roots from which counting and arithmetic have evolved' (Klein, 1950, p. 80). That arithmetical operations might symbolize such matters is supported by evidence from many other analysts. Klein assembles a lot of further detail about Lisa's — and other children's — unconscious thinking in her account. For instance, Lisa disliked the number 3 'because a third person is of course always superfluous'. Similar oedipal conflicts have been reported recently by Lusiane Weyl-Keiley, a therapist and teacher, who has shown how some typical problems encountered in remedial work in mathematics may be interpreted psychoanalytically. For example, 3 may be associated with the family triple — mother, father, child — and Weyl-Keiley describes a depressed adolescent for whom 5–2 was always 2. Asked to display this by folding fingers of one hand, he was unable to sustain the display of three fingers and had to fold down another. 'You see it makes two', he announced, keeping himself, it is suggested, out of the family conflict (Weyl-Keiley, 1985, p. 38).

Another example of the possible psychic significance — for some individuals — of a mathematical topic may be found in Klein's report that Lisa recalled never understanding an equation with more than one unknown. The exercise of interpreting that in psychoanalytic terms is, as they say, left to the reader. But another sort of condensed meaning may be found in the reminder that in the early development of algebra there must also have been something particularly difficult about the notion of two unknowns. For Diophantos, who tackled various problems with apparently two or more unknowns, seemed unable — or unwilling — to symbolize more than one. A second unknown was always arbitrarily given a particular numerical value, the problem then being expressed in terms of one variable. When this caused some inconsistency or infelicity, the value of the second unknown was modified in order to satisfy all the conditions of the problem.

For example, Diophantos sought a Pythagorean triple such that the hypotenuse less each side is a cube (see van der Waerden, 1988, p. 288). His method is here more conveniently described in modern notation. Pythagorean triples can be expressed in terms of two parameters; calling one of the associated parameters s, Diophantos arbitrarily took the other to be 3. The base, height and hypoteneuse are then $s^2 - 9$, $6s$ and $s^2 + 9$, respectively. Hypotenuse less base is 18, which is not a cube. To make it one, you need to have assigned the second parameter a number whose square, doubled, is a cube. So try 2! The sides are then $s^2 - 4$, $4s$, $s^2 + 4$. Hypotenuse less height is now the square of $s - 2$, which has to be a cube. So set $s = 10$! The sides are now as required. This was also the approach to such problems taken later — and independently — by the Arab mathematician, Al-Khwarizmi; it re-appears once again in the mediaeval 'rule of false position'.

We admire the crucial step that Diophantos made — the awareness that he could name the as-yet-unknown by a special symbol (it looked like an *s* and may have been an abbreviation of the Greek word for number, *arithmos*). We have no idea why he was unable, or unwilling, to invoke a second symbol at the same time. But it was clearly a difficult issue. The point here is, of course, not to interpret Diophantos as having some kind of unconscious block, but to establish that entertaining two unknowns may be problematic in itself, and to suggest that in Lisa's case, where other relevant evidence is also available, this might be seen as having come also to symbolize unresolved issues about her oedipal conflicts. It is not, of course, that two unknowns will always represent two parents for everyone, but rather that they may attract such unconscious condensations for some individuals at some time of their lives.

Listening to, and trying to make sense of, other people's accounts of their experience can be difficult. Interpretation is a fragile instrument which is often made to bear too much ontological weight. Socrates explains to Theodorus that he is afraid that they will never be able to understand the thought of Parmenides, and that in their discussion of the nature of knowledge they can only multiply interpretations which will never allow them to reach a satisfactory conclusion. All that Socrates, whose mother was a midwife, can do is to use his 'maieutic art' — or intellectual obstetrics — to help Theatetus deliver his own understanding.

> So I'm afraid that we'll fail as much to understand what he [Parmenides] was saying as we'll fall far short of what he thought when he spoke, and — this is the greatest thing — that for whose sake the speech has started out, about knowledge, whatever it is, that that will prove to be unexamined under the press of the speeches that are bursting in like revellers, if anyone will obey them. And this is all the more the case now, since the speech we now awaken makes it impossible to handle by its immensity, regardless of what one will do. For if one will examine it incidentally, it would undergo what it does not deserve, and if one will do it adequately, it will by its lengthening wipe out the issue of knowledge. We must do neither, but we must try by means of the maieutic art to deliver Theaetetus from whatever he's pregnant with in regard to knowledge. (Bernadete, 1986a, p. 51)

Finding the Truth in the Mistake

Introducing his own discussion of a transcript of a videotape of a classroom dialogue, Pimm emphasizes how difficult it is to present convincing descriptions of supposed unconscious processes. His interpretations of the quoted extract in terms of the linguistic dimensions of metonymy and metaphor seem to be both apt and convincing. His account could also be seen — more or less equivalently — in terms of the associated psychoanalytical processes of displacement and condensation. In the first place, when the teacher asks Lorna to recall the word 'infinity', her answer 'fidelity' may be interpreted in terms of displacement: the associated, but possibly disturbing, word 'infidelity' is swiftly disavowed and replaced by 'fidelity'. On the other hand, taking infinity to be a powerful condensation of meanings, her answer may be interpreted as directly relating to an associated meaning, that of 'lasting for ever'; unable to recall the actual word required she responds with the word 'fidelity' that has a similar, and for adolescents a particularly potent, meaning. Both interpretations can be taken into account at the same time, for the unconscious accommodates contradictions and converses.

The example underlines how irrelevant the quest for historical truth may be in such cases. But why, someone may ask, would anyone be interested in some narrative truth, let alone a proliferation of such truths? One possible answer is clear from a further reading of the transcript, which like so many classroom reports, is a record of someone's public success (here, significantly, that of the two boys, David and Gary) at the expense of someone's humiliating public failure.

The teacher is described as repeating Lorna's word 'fidelity' with 'rising tones of surprise and disbelief'. The rest of the class can be left in no doubt that Lorna has *got it wrong*. This is common enough in the contemporary classroom. But, to paraphrase an oft-quoted sentence of Jules Henry's, 'To a Zuni, Hopi or Dakota Indian, David and Gary's performance would seem cruel beyond belief, for competition, the wringing of success from somebody's failure, is a form of torture foreign to these non-competitive cultures.'

The point about the possible interpretations of Lorna's mistake is that they suggest ways in which she may, in fact, have got it right. They support a classroom attitude which encourages everyone — teacher and fellow students — *to find the truth in the mistake*. Whatever our view of what psychoanalysts say, we know, at least, that they *listen*, and this is something that we should all do more of in our classrooms. We can also strive to contain our premature interpretations and let our students deliver their own.

> Andrew, aged 5, is enormously energetic and egocentric. The world revolves round him. He has an exuberant preoccupation with words — talks incessantly, rhyming when he can and following up word-associations with unconcealed glee and self-delight. He pushes people and things, jumps up and down, shouts, tears and throws paper, stamps his feet. His aggressive masculinity is at once charming and intolerable. Andrew and friends are measuring. Ian solemnly records that he is '3 recorders and 1 pen tall'. Andrew has been measuring the length of the table using a block of wood. He asks me to write out an appropriate sentence for him. 'The table is . . . pieces of wood long.' Gripping the pencil like a dagger, he inserts a 7. He then stabs the paper a few times and produces what looks like a row of 7's under the sentence. I say nothing, but I feel puzzled. Is Andrew being exuberant and generous with his numerals in the way he is with spoken words? After a brief, shared silence, Andrew explains to me: 'you write it seven times because there were seven of them.' (Tahta, 1975, p. 10)

In a moving postscript to his chapter, David Pimm relates what he heard when he listened to 3-year-old Katie. She counts, pointing a finger in turn at him, her mother and herself: 'One, two, two'. She says she knows it should be one, two, three, but adds, 'I am pretending'. The previous evening she had told him, 'I thought you were daddy' and earlier that morning she had announced, 'daddy dead'. He offers some sensitive possible interpretations of her apparent avoidance of 'three'. It seems overwhelmingly obvious that feelings about the missing father are being processed in all this. This confirms the authenticity of Klein's account of Lisa's version of the (not always homely!) maxim, 'two's company, three's a crowd', and of the other examples of number avoidance mentioned by Lusiane Weyl-Keiley among others. I cannot resist adding a further possible interpretation that links with Andrew's explanation of his seven sevens: in counting the two two times because there were two of them in her family, perhaps Katie was also unobtrusively telling David that she knew that he was not her father and *that was all right*. The unconscious can be a creative as well as a destructive force, and young children often tell it like it is. Moreover, alternative interpretations can be simultaneously entertained, even where they may seem to be contradictory. For that is the way of the unconscious: same is other, after a fashion.[1]

Note

1. I am particularly grateful to John Mason, David Pimm and David Wheeler for their perceptive comments on earlier drafts of this chapter.

References

BERNADETE, S. (1986a) *Plato's Theatetus*, University of Chicago Press.
BERNADETE, S. (1986b) *Plato's Sophist*, University of Chicago Press.
FAUVEL, J. (1987) *Mathematics in the Greek world*, MA290-unit 2, Milton Keynes, Open University.
KLEIN, M. (1950) *Contributions to Psychoanalysis*, Hogarth Press.
MASON, J. (1991) in TAHTA, D. (Ed) *The history of mathematics in terms of awareness*, transcript of a seminar with Caleb Gattegno, Reading, Educational Solutions.
PAPERT, S. (1980) *Mindstorms*, Basic Books.
PIMM, D. (1995) *Symbols and Meanings in School Mathematics*, London, Routledge.
SERRES, M. (1982) *Hermes*, John Hopkins University Press.
SPENCE, D. (1982) *Narrative Truth and Historical Truth*, Norton.
TAHTA, D. (1975) 'Seven times', *Recognitions I*, Derby, Association of Teachers of Mathematics.
TAHTA, D. (1991) 'Understanding and desire', in PIMM, D. and LOVE, E. (Eds) *Teaching and Learning School Mathematics*, Hodder and Stoughton.
VAN DER WAERDEN, B. (1988) *Science Awakening*, Amsterdam, Kluwer.
WEYL-KEILEY, L. (1985) *Victoires sur les maths*, Paris, Robert Laffont.

Chapter 11

Potential Space and Mathematical Reality

Philip Maher

The term space, as used *within* mathematics, has a wide variety of meanings: e.g., Hilbert space, metric space, topological space. Here, in writing *about* mathematics the term 'potential space' has a psychological, indeed psychoanalytical, meaning.

Hardy, in his famous little book *about* mathematics. *A Mathematician's Apology* posits the existence of what he calls mathematical reality. I quote:

> For me, and I suppose for most mathematicians, there is another reality, which I will call 'mathematical reality'; and there is no sort of agreement about the nature of mathematical reality amongst either mathematicians or philosophers. Some hold that it is 'mental' and that in some sense we construct it, others that it is outside and independent of us . . . I will state my own position dogmatically . . . I believe that mathematical reality lies outside us, that our function is to discover or *observe* it, and that the theorems which we prove, and which we describe grandiloquently as our creations, are simply our notes of our observations. This view has been held in one form or another by many philosophers . . . from Plato onwards . . . (Hardy, 1941, pp. 63–4)

Although Hardy, in the above passage, is careful to allow the possibility that mathematical reality is 'mental' in that we 'in some sense construct it' his very usage of the term 'mathematical reality' implies a Platonic view; and when later he says 'A mathematician . . . is working with his own mathematical reality' (*ibid.*, pp. 69–70) he is careful to maintain what he calls a 'realistic' view of mathematical reality: mathematical reality is independent of how our minds are made.

Now whilst many of Hardy's views have been questioned (his dichotomy of the ugly and useful versus the beautiful and useless mathematics, for instance) and whilst *A Mathematician's Apology* may seem to us now to resonate with the cultural ideals of the Bloomsbury group, Hardy's notion of mathematical reality has retained its spell over pure, and applied, mathematicians alike. Thus, for example, Penrose (who began his career as a pure mathematician and is now a theoretical physicist) in a contribution to the Mathematicians on the Philosophy of Mathematics Conference (1992) explicitly distinguished the Platonic world of mathematicians (Yes, that's what he called it) from, on the one hand, the real physical world, and, on the other, the world of conscious perception. Indeed, I suspect that nearly all mathematicians are, insofar as they give the matter conscious thought, unreconstructed Platonists.

All of which is fairly astonishing given the glaringly obvious philosophical problems such a position entails. Thus: how can mathematical reality be outside and objective and yet not be physical and, simultaneously, not be a creation of the human mind? The problem of the ontology of mathematical reality is accentuated if (like that militant atheist, Hardy) one eschews such theological explanations as mathematical reality being the creation of an omniscient God.

My point here is not that there is no such thing as Platonic mathematical reality but, rather, that this notion is highly problematic and, at the same time, highly popular amongst professional mathematicians (more so than such recent philosophical models as the quasi-empiricism of e.g., Tymoczko or the radical constructivism of e.g., von Glasersfeld).

One attraction of Platonism (the belief that there exists a mathematical reality independent of us) is that it gives an account of how (as, interestingly, Tymoczko has commented in POME Newsletter 6, p. 4) mathematical objects seem to exist. But more than this, Platonism seems to possess a positively affective power over the other philosophical models of mathematics. I will now anticipate my conclusion by saying that the affective power of Platonism results from an unknowing conflation of the philosophical with the psychoanalytical; or, to put it slightly differently, the philosophical notion of Platonism has deep psychoanalytic roots which are all the more powerful for being unarticulated.

Examining more closely the term mathematical reality we see that it suggests a psychic, or, if you prefer, a psychological, space which, on the one hand, is not purely private (like a dream) and, on the other, is felt to be a subject of exploration by one's own mind; and in this psychological space there are objects — mathematical objects — which, on the one hand, are manipulated with in accordance with the various, pre-existing results of mathematics, and, on the other, are imagined by you, the mathematician (all of which is evidenced by the common difficulty (mentioned by Zeeman in his contribution to the Mathematicians on the Philosophy of Mathematics Conference (1992) that we mathematicians have in deciding whether we discover or create mathematics). Mathematical reality and its objects seem both outside and inside at the same time. Significantly, in this evocation of mathematical reality and its objects no one specifically mathematical feature appears: indeed the very lack of mathematical features suggests that the concept I have been trying to evoke is really more general than, and subsumes, mathematical reality.

The psychoanalytically informed reader will by now have realized that the notion I have been trying to evoke is the so-called 'potential space' of D.W. Winnicott. Winnicott, who originally trained as a paediatrician, was a psychoanalyst who was one of the most innovative of the English 'object-relations' school. For the last twenty years of his life Winnicott became increasingly interested in the location of cultural experience (the title of one of his papers, see Winnicott, 1971, pp. 95–104). Of course, 'cultural' here embraces mathematics.

Winnicott's concept of potential space (and the related concept of transitional object) arose from his work on the very early mother–child relationship. For Winnicott, in the beginning 'the environment is holding the individual and at the same time the individual knows of no other environment and is at one with it'; that is, the infant is enveloped within, and conditional upon, the holding environment, i.e., the mother, of whose support he gradually becomes aware, a condition Winnicott said 'could be described at one and the same time as absolute dependence and absolute independence.'

Winnicott's concept of transitional object (and hence of potential space) arose from his attempting to explain how the infant made his first simple contact with external reality. The infant did this through 'moments of illusion' which the mother provided: these moments helped the infant to begin to create an external world and, simultaneously, to acquire the concept of an inside for himself. Such a moment of illusion might occur when the mother offered her breast at exactly the moment when the child wanted it (so that the infant's wish and the world's satisfying it coincided — which, of course, they never, in fact, do): the infant acquires the illusion that there is an external reality that corresponds to his capacity to create. Of course, there had to be a later converse process of 'disillusion' (not to be confused with the Kleinian Depressive Position) when the mother would gradually withdraw her identification with the infant and wean the infant away from her breast.

This led Winnicott to his concept of transitional object which belonged to an 'intermediate area of experiencing to which inner reality and external life both contribute'. Winnicott drew attention to 'the first possessions' — those rags, blankets, clothes, teddy bears and other objects to which young children become so powerfully attached: Winnicott called such objects transitional objects, so-called because their use belonged to an intermediate area between the subjective and that which is objectively perceived. The transitional object thus presents a paradox which cannot be resolved but must be accepted: the point of the transitional object is not its symbolic value but its actuality.

Winnicott came to posit what he called potential space. Potential space was the hypothetical area existing between the baby and the object (the mother) at the end of being merged with the object; that is, potential space arises at that moment when, after a state of being merged with the mother, the baby arrives at the point of separating out the mother from the self and the mother, simultaneously, lowers the degree of her adaptation to the infant's needs. At this moment, Winnicott says, the infant seeks to avoid separation 'by the filling in of the potential space with creative playing, with the use of symbols and with all that eventually adds up to a cultural life'.

For the inner and outer reality of the adult to be connected the continuance of an unchallenged intermediate area is necessary: the potential space, originally between baby and mother, is reproduced between individual and the outside world as it becomes the location of cultural experience. 'This intermediate area' Winnicott wrote 'is in direct continuity with the play area of the small child who is "lost in play" and was retained in the intensive experiencing that belongs to the arts, . . . and to creative scientific work.'

It follows from Winnicott's theory that for a mathematician mathematical reality is subsumed in potential space, that is, mathematical reality is an instantiation of potential space that occurs when one is doing mathematics; and that one's mathematical objects are transitional objects. This last claim that 'mathematical objects are transitional objects' may seem fairly preposterous in view of my earlier description of them as 'those first possessions . . . rags, teddy bears . . . to which young children become so powerfully attached'; yet this describes only the *first* transitional objects a child has: the notion of transitional object — like that of potential space — carries over into adult life (as is implicit in Winnicott's tantalizingly oblique paper on the location of cultural experience (Winnicott, 1971, pp. 95–104)). If we accept the view that one's mathematical reality is an instantiation of one's potential space that occurs when one is doing mathematics then the objects in this psychological space — the mathematical objects one plays with (tellingly, 'play' is a verb mathematicians often use to describe their activity) — are — or, more accurately, *function as* — transitional objects. From this perspective there is little psychological difference between, say, a teddy bear and a self-adjoint operator: for, whilst the former is a physical object and the latter isn't (although mathematics is recorded by physical marks on paper) both *function as* transitional objects at the appropriate stage of one's psychological development.

At this juncture a problem emerges. Why is it mathematicians are so conscious of what they call mathematical reality whereas, say, creative writers don't talk of 'literary reality' or engineers of 'engineering reality'. Part of the answer may be that mathematics is non-representational, that is, the meaning of a piece of mathematics is independent of the outside world – although in teaching a piece of mathematics one may (and often does) link it to the outside world so that it may more readily enter the pupil's potential space. Literature and engineering, on the other hand, are not non-representational. Given that mathematics is non-representational it is hardly surprising that mathematicians have a need to be able to posit, to be consciously aware of mathematical reality.

To be able to posit, to be consciously aware of — and to be able to visualize: these are all nearby concepts and, in turn, tie up with an oft-noted characteristic of mathematicians: namely, most visualize and regard mathematics (and not just those parts, like geometry or topology having obvious visual connotations) as an inherently visual subject. This last statement requires some discussion — and qualification, though. First, some mathematicians claim not to visualize: but the fact that surveys have been carried out investigating whether or not mathematicians think of themselves as visualizers would support the belief that mathematics is, at least, thought of as a visual subject. Second, the existence of mathematicians who went blind in later life — Euler and Pontdriagin are famous examples — does not contradict the view of mathematics as a visual subject: Euler and Pontdriagin may well have incessantly visualized. Third, there are examples of mathematicians blind from birth — I know of one, Paul Holliman who was a student at Middlesex University (Grattan-Guiness and Holliman, 1983). The existence of mathematicians blind from birth does not contradict the view of mathematics as being characteristically visual since that viewpoint pertains to the vast majority of born sighted mathematicians. In the case of Holliman it appears his 'imagery' (How the visual saturates language!) was aural — obviously; and tactile — the lecturer would 'draw' the shape of a graph, say, on his back (Grattan-Guiness and Holliman, 1983). Whilst it is impossible for us sighted from birth to enter the psychological world of the blind from birth it would seem that common to both is the concept of communication (although its mode — visual and aural in the one case; aural and tactile in the other — is necessarily different). And communication for the sighted infant, before language has asserted itself, takes place by looking: communication with the mother by looking into her eyes; communication with or more accurately, construction of — the self by looking into a mirror.

The reader may now realize that we have reached — with a bit of a jolt, perhaps — the notions of mirror role and mirror stage. There are two notions here, not one: Winnicott's concept of the mirror role of the mother (see Winnicott, 1971, pp. 111–19) and Lacan's concept of the mirror stage (Lacan, 1977, pp. 1–7, first published in 1939 although apparently originated in the mid-1930s, Benvenuto and Kennedy, 1986, pp. 47–63).

Though Winnicott's notion, as he acknowledged, was influenced by Lacan's somewhat different notion we deal with Winnicott's first. Winnicott's notion is based on the mutuality of the relationship between the mother and infant as manifested in their mutual gaze at each other. As has been confirmed by observational studies, the newborn infant is particularly attracted to, and fascinated by, the sight of the human face — especially the eyes and especially those of the mother. Significantly, the mother's eyes are at the appropriate distance for the baby to gaze into when being breast fed. As Winnicott pointed out when the infant gazes into the mother's eyes, the infant is reflected in them — and, mutually, the mother is reflected in the infant's eyes: it is as if the mother is acting the role of a mirror to the infant (Winnicott, 1971, pp. 111–19). Winnicott's notion of the mirror role of the mother thus provides an explanation for the mutual communication between mother and infant which, taking place as it does before the infant has acquired verbal language, is all the more powerful affectively.

Winnicott's concept of the mirror role of the mother thus gives an explanation of the ontogenesis of the link between gazing and communication in general. At the same time, it shows how characteristically visual a concept as that of gazing ties up with — or rather, includes — the idea of how mathematics seems so characteristically visual; for if we accept Winnicott's theory of transitional objects/ potential space (sketched earlier) — and its link with mathematics — it would seem that doing mathematics involves the gaze of the mind on transitional objects (here, mathematical objects) in potential space (here, mathematical reality).

But, it might be objected, Winnicott's notion of the mother's mirror role — linking, as it does, communication to gazing — isn't specific to mathematics; it could, that is to say, apply to any other cultural activity involving communication which had connotations of the visual. Now, one aspect of mathematical activity which — whilst not specific to it alone — is as characteristic of it as is the visual is the notion of making whole: e.g., of completing a pattern or e.g., of accounting for a special (or seemingly anomalous) case. And, as is elaborated below, it's Lacan's notion of the mirror stage which gives an explanation of the ontogenesis of this aspect of mathematical activity (Lacan, 1977, pp. 1–7).

Lacan's notion of the mirror stage describes a stage in the development of the infant (roughly 6–18 months) when the infant begins to recognize his or her image in a mirror (an event that has also been observed in studies of chimpanzees). At first, Lacan noted, the infant confuses the image with reality (and may try to grasp hold of the image behind the mirror); then comes the discovery of the image; and then, finally, the realization that the image is his or her own. To understand the developmental significance of the mirror stage one must realize that, at the same time, the infant is relatively uncoordinated, helpless and dependent on others. Now when the infant becomes aware of his or her image in the mirror the infant sees a totality which he or she can control through his or her own movement. The mirror image thus articulates a mastery of the body the infant has not yet objectively attained; and this imaginary (as Lacan dubs it) mastery anticipates the infant's later real biological mastery — and intellectual mastery, too, if e.g., a future mathematician. The mirror image, then, in representing to the infant a total and unified whole (caused and controlled by the infant) prefigures the mind's desire to make whole; and this latter desire whilst a characteristic of many activities, is, as commented earlier, a crucial aspect of mathematical activity.

The reader might, at this stage, be forgiven for thinking that the arguments presented in this chapter are a bit wild or fanciful (quite apart from any doubts the reader might have about the scientificity of psychoanalysis). This is entirely understandable: the intellectual discipline of reflecting on the activity of mathematics is at a nascent, primitive stage of development in which any intellectual moves must seem bold, unsupported as they are by any pre-existing *theory of mathematics* (in marked contrast to the intellectual sophistication of, say, literary theory which, interestingly, uses psychoanalytic concepts e.g., Belsey, 1980).

Part of the reason for the late development, as an intellectual discipline, of reflecting on the activity of mathematics may be the reluctance of professional mathematicians to write *about* mathematics as distinct from adding to mathematics by *doing* it (a view perfectly expressed by Hardy, with whom we began, who writes 'The function of a mathematician is to do something, to prove new theorems, to add to mathematics and not to talk about what he or other mathematicians have done' (Hardy, 1941, p. 1). Yet the situation becomes circular if professional mathematicians do not write about mathematics because it's not professionally respectable to do so: it won't become professionally respectable unless they do.

A further difficulty attends the use, as here, of psychoanalytic concepts in writing about mathematics: the use of psychoanalytic concepts to explicate the activity of mathematics has to be the work of psychoanalytically orientated mathematicians rather than psychoanalysts *per se*: for, apart from their lack of experience of mathematics, few psychoanalysts have written at all extensively on mathematics. (I say 'few' advisedly: Bion mentions mathematics in Bion, 1962; and Ferenczi has a late, unfinished and fragmentary essay in Ferenczi, 1955: both of these references are given in Chapman, 1972, pp. 206–16. One should also recall that Lacan often uses mathematics, and mathematical terminology, as a communicative metaphor e.g., Lacan, 1977).

Admitted, psychoanalysis, throws up various epistemological problems (some would claim to do with its scientificity). Yet despite these epistemological problems, psychoanalysis offers the most realistic insights into the workings of the human mind and hence into the experience, and activity, of mathematics.

References

BELSEY, C. (1980) *Critical Practice*, London, Routledge.

BENVENUTO, B. and KENNEDY, R. (1986) *The Works of Jaques Lacan: An Introduction*, London, Free Association Press.

BION, W. (1962) *Learning from Experience*, London, Heinemann.

CHAPMAN, L.R. (Ed) (1972) *The Process of Learning Mathematics*, Oxford, Pergamon.

DAVIS, M. and WALLBRIDGE, D. (1983) *Boundary and Space: An Introduction to the Work of D.W. Winnicott*, Harmondsworth, Penguin Books.

FERENCZI, S. (1955) *Final Contributions to the Problems and Methods of Psychoanalysis*, London, The Hogarth Press.

Grattan-Guiness, I. and Holliman, P. (1983) 'Undergraduate mathematics for the blind', *The Mathematical Gazette*, 67, pp. 77–89.

Hardy, G.H. (1941) *A Mathematician's Apology*, Cambridge, Cambridge University Press.

Lacan, J. (1977) (first published in 1949) *Ecrits: a selection*, translation A. Sheridan, London, Tavistock Publications.

Phillips, A. (1988) *Winnicott*, London, Fontana.

Winnicott, D.W. (1971) *Playing and Reality*, London, Tavistock Publications.

Chapter 12

Towards a Hermeneutical Understanding of Mathematics and Mathematical Learning

Tony Brown

Introduction

Pursuit of the notion of mathematical meaning has dominated much of the discussion to do with the philosophy of mathematics education. The flavour of current discussion generally seems opposed to notions of absolute meaning towards seeing meaning as a socially constructed phenomena. Such a move seems consistent with theoretical shifts in other academic fields but it still retains a tendency towards seeing meaning as in some ways independent of time and context, a notion associated with concepts rather than with conceiving, a fixity to which the learner converges. In particular, any attempt to assign a socially conventional meaning remains problematic since any account of what this might be is mediated by some symbolic system subject to the interpretation of any individual user.

Hermeneutics, the theory and practice of interpretation, attends to the process through which we develop an understanding of the world. Unlike post-structuralism which asserts 'a multiple play of meaning held in language' (Urmson and Ree, 1989), hermeneutic understanding is more governed by a belief that whilst the world may exist independently of humans, it cannot present itself directly to the human gaze. The hermeneutic task can then be seen as an uncovering of meaning, but a meaning dependent on the media and experiences through which it is observed.

The principal task of this chapter is to explore the consequences of asserting that mathematics is an essentially interpretive activity, comprising a system of symbols that is only activated within individual human acts. By seeing mathematical expressions as being used by humans in particular situations, rather than as things with inherent meaning, emphasis is placed on seeing mathematical activity as a subset of social activity, and as such, is subject to the methodologies of the social sciences. Firstly, I shall outline some issues arising through seeing mathematical expressions as being necessarily contained in action, resulting in meanings that transcend mathematical symbolism. This is followed by a discussion of how the emergence of mathematical phenomena in human understanding is a consequence of a linguistic process of classifying. After outlining certain ideas within hermeneutics, the method is discussed as an approach to describing mathematical learning and assessment of this.

Tony Brown

Speaking and Writing As Acting

In *Philosophical Investigations*, Wittgenstein (1958) suggested that the meaning of a word might be seen as its usage in language. Here, the notion of meaning as a phenomena subsides as a greater emphasis is placed on what an individual sees as being held in a given symbolic form in a particular time-dependent context. Within mathematics education this can facilitate a move away from seeing mathematics as something with an independent existence but rather as something that arises in the context of some particular social event; a style of activity rather than something to be learnt. A necessary characteristic of any social event however, is a style of symbolizing facilitating a system of exchange. In a situation concerned with mathematical learning this symbolizing will take many forms, using both inherited and original symbols in ways that endeavour to enable communication between participants. So viewed mathematical utterances, both spoken and written, can assume many attributes associated with any linguistic utterance in a social space. For mathematics, it is the symbols themselves that are the material reality since mathematical phenomena do not have a tangible existence outside of symbolization. However, the symbols are human creations that arise in speech or writing and are a subset of other human expressive acts and as such are subject to the methodology of social scientific enquiry. In particular, any issuing of a mathematical utterance into a social space needs to be seen as an action with a meaning that transcends any purely literal enquiry.

In his work on word usage Austin (1962) identifies three levels of effect arising from any sentence being uttered in a social context. Firstly, the locutionary effect is that of the literal meaning where the words are taken at 'face value'. Secondly, the illocutionary effect is that done *in* the utterance. For example, if I say 'I name this ship', the action is in the saying. Thirdly, the perlocutionary is that done *by* the sentence, for example, the effect on the atmosphere in a classroom as a consequence of the teacher saying something in a loud, disapproving tone. Ricoeur (1981) has suggested that these three effects are in a hierarchy according to the degree of interpretation required. The interlocutor can resolve the meaning of the locutionary effect by consulting a dictionary. However, to understand the perlocutionary effect one needs to have experience of being in the appropriate language using community where sentences are used alongside physical actions in a material environment.

In a social situation being directed towards mathematical learning, a variety of linguistic forms will be used within a broad communicative environment. Some of this language will be specifically associated with conventional mathematical ideas, but much will be less precise, supporting other aspects of the exchange. The exactitude often associated with mathematical symbolism does not characterize the communication taking place in many educational situations. Rather, in most learning situations we are concerned with activity taking place over periods of time where learning requires making sense of engagement in this activity and cannot be seen as the composite of getting to grips with a number of clearly defined 'mathematical' concepts. The placing of any written or verbal utterance into this influences communication in the way furniture might influence movement around a room. In this sense, Kaput (1991) who has explored the possibility of reconciling the use of inherited notations within a radical constructivist philosophy, has suggested that in this respect the placing of Cuisenaire rods in a particular arrangement might be seen as a form of writing.

Indeed there are many forms of mathematical discourses each flavoured by their particular social usage. For example, a university lecturer might speak as a platonist, where utterances are made as if they were extracts from a transcendental world of mathematics. Rather different would be the speech of a representative of the National Curriculum Council seeing mathematics in terms of how it might be partitioned for the purposes of testing. Different again would be someone who regards mathematics as a subject whose prestige is consequential to its ability to be applied in 'real life' situations. Richards (1991) offers other examples. Each of these people would enter into a dialogue placing varying stresses on their mathematically inclined utterances according to some ideological tainting. A string of symbols, no matter how neutral it may appear, cannot be seen independently of the context into which it is being issued. The selection of this string pre-supposes a selecting by someone with a particular purpose.

Mathematics As a Language

Can mathematics be seen as existing outside of the language that describes it? The question is related to the more general concern of how the world can be seen outside the language that describes it. In describing perception Peirce identifies three ascending levels; quality (the initial sense), fact (the identification of objects), law (the relations between objects). Perceiving the world requires categorization of it which involves differentiating and relating aspects of this world. It is towards this end we use language. Mathematical phenomena, those aspects of the world seen as pertaining to magnitude and number and the relations between them, emerge in this process, consequential to the way in which the world is subdivided into categories. However, this very categorization mediates and thus organizes our way of seeing.

The symbols and classifications of mathematics are historically determined. They are arbitrary in the sense that the symbols and classifications of any language are arbitrary. As such it can be viewed phenomenologically in that these symbols and classifications have particular meanings for any individual derived from that individual's experience of their usage. The field of mathematics only comes into being in its classification in language. The sense of this field can only be perceived retroactively and its existence presupposes a language. For someone learning mathematics there is a similarity with learning a language in that there is a need to grapple with an inherited mode of symbolization and classification, arbitrarily associated with some pre-existing world.

The style of describing mathematics is necessarily interpretive according to some mode of signification; a particular way of fitting words and symbols to experience. Barthes talks of certain styles of signification becoming naturalized in the sense that they become the culturally conventional way that seems to be entirely neutral, in the way that a realist painting might be regarded as the most straightforward representational mode. This is discussed by Coward and Ellis (1977). However, in learning mathematics, using alternatives might be seen as a way of initiating productive tensions for forcing awareness towards re-evaluating this supposed natural way. The linguistic overlay given to a situation can be seen as a way of introducing an interplay between the describing and that described. By asserting mathematical activity as essentially interpretive in nature, the production of meaning in this activity can be seen as deriving from a dialogue in a continuous process of introducing linguistic and symbolic form into the socially active space.

As an example, consider the flavours that can be given to a 4 × 3 rectangular lattice in the context of a particular activity, described in Brown (1990).

a)

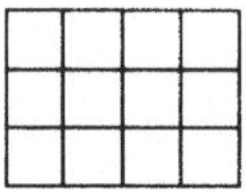

drawn on squared paper

b)

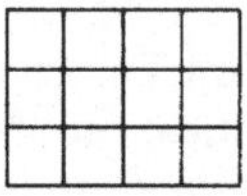

made out of plastic squares

c)

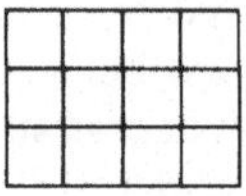

drawn on the chalkboard

d) Captured in writing or in spoken words:
e.g., ‘A rectangular garden lawn surrounded by a path comprising ten square paving stones.’
‘Two green squares side by side surrounded by red squares.’
‘Four squares in the top row, four in the bottom and one at each end of the lawn.’

e) Located on a table

garden	1	2	3	4	5	6	n
lawn	1	2	3	4	5	6	n
path	8	10	12	14	16	18	2n + 6

f) Pictured in the imagination

Each of these metaphorical representations open up a form of describing such lattices as they change dimensions. The play arising from making (metaphorical) leaps between such forms and (metonymic) moves within such forms results in successive acts of fitting and associating forms. By seeing equivalences between forms we can choose the form most suitable for our current purposes. For example, if we see a 153 × 3 rectangle as the 151st garden we do not need to build it with plastic squares and may prefer to deduce the information we need (e.g., how many red squares are needed for the path?) algebraically.

Saussure's influential work in linguistics, carried out at the turn of the century, was directed at the structure of the layer mediating experience. For him the signifier is the mental image or sound of a word or symbol. The signified is the mental concept with which the individual associates it. Together, the signifier and signified form the sign, a wholly mental phenomena constructed by the individual. Two forms of arbitrariness are implicit here. Firstly, if we take the signifier 'square', the word itself is quite arbitrary and is in fact different according to the language you are speaking (e.g., it is *kwadrat* in Polish). Secondly, a square is a special sort of rectangle, or a type of regular hexagon, or a type of rhombus. It is an arbitrary category and does not need a name of its own. We only introduce a name for convenience since in the way in which we operate in the world we use it a lot. Such a category may not be so crucial for an aborigine.

Saussure did little in investigating how such signs are associated with the real world, but rather saw meaning as being derived purely from the play of differences between signs. In this way, mathematics as a language, held in symbols, can be seen as independent of the real world, a mediating layer rather than a quality endemic in the physical world. So viewed categories of mathematics are cultural rather than transcendental, arbitrary rather than implicit. The play of meaning is consequential to sets of words being combined by humans in individual speech acts. In seeing mathematics as a language, as educators, we are not so much concerned with its qualities as a system (*langue*) but rather with its realization as discourse in the social environment (*parole*), i.e., with the way in which it is being used to signify, towards producing meaning.

The Phenomenology of Acting and Meaning

By seeing mathematics as something arising in social activity and the framing of mathematical statements as social actions, we are permitted the possibility of employing social scientific techniques towards establishing mathematical understanding. Much recent work in the human sciences has worked from the premise that the individual human subject perceives the world phenomenologically, that is, he or she sees the world comprising phenomena having particular meanings to him or her in particular contexts. Here individual objects are not seen as having meaning in themselves but only take on a meaning in the gaze of the individual who sees them from his or her own particular position and according to his or her current interest.

Underlying this view are specialized uses of the terms objective and subjective. Whilst there may be an independent material reality it only comes into a meaningful 'objective' reality when classified within the language of an individual human subject. I might talk about the situation I see myself in, as if it were independent of me, but I can only do this only after experiencing myself as part of it. In this way object and subject are in some sense part of each other. An object can only present itself to the gaze of the individual subject with his or her own particular phenomenology. The world of material objects only comes into being retroactively through being captured in language. An individual's consciousness is always a consciousness of . . . and is always intentional insofar as it seeks to make sense of that within its gaze according to some schema. A consciousness is always of an object and an object only presents itself to a consciousness. The nature of the objective is dependent on the way in which it is captured and accounted for in language by the subject. The mediating layer through which language is derived seems inescapable, brought into existence by consciousness itself and its need to organize that which it perceives. These ideas, generally accredited to Husserl, are discussed at length by Ricoeur (1966).

In making a mathematical statement I express certain intentions but am unable to guarantee that I communicate the meaning I myself attach to this statement. Such an action and its meaning to me are consequential to my categorization of the world in which I see myself as part. Whilst I may attempt to predict how my action might be read by others, the meaning of my action cannot be seen only in terms of my intention since it cannot be seen independently of the social environment into which it is issued. In addressing this Ricoeur (1966) differentiates between the 'voluntary' and 'involuntary' components of any action. The individual subject can assert himself or herself through the voluntary component of an action, i.e., that which he or she intends. The meaning of this action, however, only emerges as the resistances to this action take shape around it. These resistances, the involuntary component of the action, have no meaning in themselves, but rather are the contextual framing activated by the voluntary component. This implies a hermeneutic process where the subject voluntarily acts in the world he or she supposes it to be, but this in turn gives rise to (involuntary) resistances which are always at some distance from those anticipated. In order to act, however, there is a need for the subject to suspend doubt and act as if his or her reading is correct. This has been discussed in more depth by Brown (in press).

In later work, Ricoeur (1981) talks of the 'meaningful effect' of an action as being its 'objectification'; *the mark it leaves on time*. Thompson (1981), Ricoeur's translator, sees this as being related to the way in which the action might be described in retrospect, as if in some historical account. Ricoeur explores this in terms of an analogy with the objectification speech goes through in being committed in writing. By pursuing the paradigm of text interpretation he sees acting as analogous with writing and interpretation of this action as reading. It is through this sort of fixation that we can employ techniques of interpretation for both tasks of understanding (learning through signs) and of explanation (learning through facts), by seeing such tasks as necessarily intertwined. It is this relation between understanding and explanation that hermeneutic enquiry seeks to unfold.

Hermeneutic Understanding

Hermeneutics was originally developed and employed in the analysis of biblical texts but was extended, largely by Dilthey working at the turn of the century, to cover the whole of human existence. Leading modern exponents are Ricoeur and Gadamer who have developed it within phenomenology. Hermeneutics, whilst acknowledging that some interpretations are better than others recognizes that none is ever final. Hermeneutical understanding never arrives at its object directly; one's approach is always conditioned by the interpretations explored on the way. While one's understanding may become 'fixed' in an explanation for the time being such fixity is always contingent. In choosing to act *as if* my explanation is correct, the world may resist my actions in a slightly unexpected way, giving rise to a new understanding, resulting in a revised explanation, providing a new context for acting and so on. This circularity between explanation and understanding, termed the hermeneutic circle, is central to hermeneutic method.

Hermeneutics resists distinctions frequently made between the explanations of the natural sciences (knowing through facts) and the understanding of the human sciences (knowing through signs), preferring to see them both as subject to an interpretive framework. Within history, for example, whilst it may be possible to continue offering ever more interpretations of 'what happened?' if we are to act in the light of this knowledge we have to suspend doubt for the time being and assume a certain position towards getting things done. Conversely, taking mathematics, as an example from the other end of this scale, while we may have statements that 'on the surface' seem entirely incontrovertible, it is still necessary for an individual human to decide how such statements will be used in the social space or how they have been used. This is discussed in relation to mathematics teaching in Brown (1991).

In speaking of mathematics I cannot simply quote, in a neutral way, expressions as if from some platonic formulation. I am necessarily acting in time — whereas platonic mathematics is outside of time. Further, in doing this I refer, by implication (through the perspective I reveal), to myself, to the world I see, and to the person(s) to whom I am talking. Ricoeur (1981) emphasizes these discursive qualities of language usage in distinguishing *langue* and *parole*. In doing this he combines Saussure's linguistics with the speech acts described by Austin (1962) and Searle (1969). Mathematics is only shareable in discourse and the act of realizing mathematics in discourse brings to it much beyond the bare symbols of a platonic formulation of mathematics. The mathematics I intend to communicate is always mediated by the explanatory procedures of such a social event. My interlocutor is obliged to interpret my speech, reconciling parts with the whole, stressing and ignoring as he or she sees fit. The distinction between knowing through facts and knowing through signs becomes blurred in this process since the facts of mathematics are immersed in the usage of them. The expressions of mathematics are only arising within actions in social events.

Notions of hermeneutic understanding as applied to mathematics then require a shift in emphasis from the learner focusing on mathematics as an externally created body of knowledge to be learnt, to this learner engaging in mathematical activity taking place over time. Such a shift locates the learner within any account of learning that he or she offers, thus softening any notion of a human subject confronting an independent object. In this way positivistic descriptions that draw hard distinctions between process and content of learning mathematics are avoided since there is no end point as such but rather successive gatherings-together of the process so far, seen from the learner's perspective. In such an educative space, characterized by the communication of mathematical thinking, the introduction of different interpretations gives rise to the possibility of a productive tension between mathematical activity and accounts of it, enabling the very hermeneutic process of coming to know through juxtaposing varying perspectives as in the example above.

The exact expressions conventionally associated with mathematics only ever find expression in such activity, within the context of many other sorts of expression. Such statements are always, in a sense, offered as part of a distillation process;

a 'looking back' concerned with pinning down key points of the event. The reflective dimension inherent in this results in the active generation of mathematical expressions through time being part of the reality described. Similarly, the intention to learn is always associated with some presupposition about that to be learnt and learning is in a sense revisiting that already presupposed. This continual projecting forwards and backwards affirms an essential time dimension to mathematical understanding that can never be brought to a close by an arrival at a 'concept', since the very framing of that concept modifies the space being described.

Where then lie conventional notions of mathematical understanding? This issue seems problematic in that within hermeneutics understanding does not pertain to concepts with fixed meanings. Understanding is a process rather than a state. This clearly runs in conflict with sublime notions of understanding that suggest a state beyond that accountable in words. A more humble notion of mathematical understanding may be that it is simply the ability to tell a package of convincing stories generated by the learner himself or borrowed from the teacher. Further, this understanding is only proven if the learner can make use of certain aspects of the conventional, inherited system of exchange.

Describing and Assessing Mathematical Learning

As educators we are often not so much concerned with learning as with giving an account of learning. This might be the reproduction of a famous result, the application of a method in a particular real life situation or some verbal explanation of some work completed. Such an account is necessarily in some symbolic order associated with some over-arching system of exchange. For the student engaging in a mathematical activity there is a variety of ways of reporting back on the experience. The nature of any understanding demonstrated in this report, however, is always conditioned by the method of reporting chosen. But what can be captured in such a report? Is it the mathematics, the understanding, the activity? None of these can be described purely within the realm of mathematics, whatever that is. Some perspective of the describer is required and this depends on his position, his biographically defined background and his current motives.

By seeing the assessment of mathematics being directed towards the student making sense of his mathematical activity we overtly move in to the realm of interpretations. Assessment by the teacher can be directed towards participation in a dialogue involving the student in generating linguistic forms in respect of his view of the activity. A two-tiered interpretation is implied here; the student capturing his experience in symbolic form, and the teacher assessing this symbolic product as evidence of understanding. What is not implied here is any notion of a universal meaning to which both teacher and student converge, but rather '. . . objectivity is achieved through the coincidence of interpreting, that is, agreeing' (Brookes, 1977).

One might legitimately protest that there is a certain power relation here that creates a somewhat asymmetrical sort of agreeing, where the teacher, as representative of the conventional way of talking about things, sees his or her task as introducing this. Whilst the children may have the opportunity of offering some account of their understanding, within their own mode of signifying, the teacher, in entering any discussion, introduces a more conventional mode. The communication being sought in such an exchange brings into play some symbolic medium, comprising symbols, actions and words. But such is the power of the conventional mode of discourse that the quest for the learner may be to believe that he or she is joining the teacher in using the inherited language. This highlights a particular aspect of the teacher's power, consequential to the linguistic overlay he or she brings to the situation. The teacher's style of looking is accustomed to spotting concepts which are, after all, merely culturally conventional labellings. In this way the teacher's way of making sense of a student's work involves classifying this work as if looking to tick off categories on a National Curriculum checklist. The student's access to any notional transcendental mathematics is always mediated by a social pressure to capture this in the categories arbitrarily assigned by our ancestors. I would argue that much investigational work, such as that described in Brown (1990), permits the students to develop their style of signification more fully, prior to interception by the teacher introducing more conventional ways of describing the product, than might be possible in more traditional approaches.

Conclusion

By accepting a hermeneutic view of mathematical understanding we give primacy to the linguistic qualities of mathematical learning and so soften the distinction between mathematics and other disciplines. Mathematics becomes something held in the expressions of participants in mathematical activity, who are asserting their view of, and their relation to, some supposed mathematics. The reality of any transcendental mathematics relies on people acting as if it is there. The assertion that mathematics has no existence outside the material symbols that describe it echoes the 'lack' that Lacan describes as emerging after the layers of description are peeled away (Brown, Hardy and Wilson, 1993). Whilst this may be too extreme for many professional mathematicians, the transcendental mathematical truth that might be uncovered by hermeneutic enquiry cannot escape some flavouring from the process through which it is reached by the individual.

References

Austin, J.L. (1962) *How to do Things with Words*, Oxford, Oxford University Press.

Brookes, W. (1977) 'A hermeneutic approach to psychology and its applications to learning mathematics', Paper presented at the inaugural meeting of the *British Society for the Psychology of Mathematics Education*, 1977.

Brown, T. (1990) 'Active Learning Within Investigational Tasks', *Mathematics Teaching*, December.

Brown, T. (1991) 'Hermeneutics and mathematical activity', *Educational Studies in Mathematics*, 22.

Brown, T. (in press) 'Creating and knowing mathematics through language and experience', *Educational Studies in Mathematics*.

Brown, T., Hardy, T. and Wilson, D. (1993) 'Mathematics on Lacan's couch', *For the Learning of Mathematics*, Montréal, Québec.

Coward, R. and Ellis, J. (1977) *Language and Materialism*, London, Routledge.

Kaput, J. (1991) in von Glasersfeld, E. *Radical Constructivism in Mathematics Education*, Dordrecht, Kluwer Academic Publishers.

RICHARDS, J. (1991) in von Glasersfeld, E. *Radical Constructivism in Mathematics Education*, Dordrecht, Kluwer Academic Publishers.
RICOEUR, P. (1966) *Freedom and Nature*, Evanston, Northwestern University Press.
RICOEUR, P. (1981) *Hermeneutics and the Human Sciences*, Cambridge, Cambridge University Press.
SAUSSURE, F. (1966) *A Course in General Linguistics*, New York, McGraw-Hill.
SEARLE, J. (1969) *Speech Acts*, Cambridge, Cambridge University Press.
THOMPSON, J. (1981) *Critical Hermeneutics*, Cambridge, Cambridge University Press.
URMSON, J. and REE, J. (1989) *The Concise Encyclopaedia of Western Philosophy and Philosophers*, London, Unwin and Hyman.
WITTGENSTEIN, L. (1958) *Philosophical Investigations*, Oxford, Basil Blackwell.

Chapter 13

The Myth of Mathematics

Falk Seeger and Heinz Steinbring

The present chapter tries to explore some of the specific epistemological features of mathematical knowledge as it is constituted in communicative interactions between teacher and students.[1] A basic and often implicit underlying philosophy of mathematics in teaching is that of a 'sufficiently liberal platonism' which teachers have acquired during their university studies and then taken with them to school (Goodman, 1979). Introduced into mathematics teaching, this philosophy of mathematical knowledge undergoes certain changes because of specific conditions and requirements of interactive communication patterns and routines in the classroom discourse.

Obviously, one of the basic assumptions of the prevailing didactical contract is that in order to convey knowledge to the students it has to be based on intuitive or 'natural' ways of knowing and doing. Following Edwards and Mercer (1987), this forms the kernel of the didactical contract of 'progressive education'.[2] In the end this kind of 'naturalness' of mathematical knowledge together with the intentions of teaching to provide students with direct mathematical understanding leads students to a mythical view of mathematics.

Being theoretically interested when analysing communicative events in mathematics teaching, we try to better understand our role, that is the *role of the observer* in theory–practice encounters. Basically, the task to understand what is being observed is similar to that of an ethnographer who wants to understand an alien culture. The task is made only more complicated through the fact that usually researcher and teacher in mathematics education share the culture with the subject they attempt to understand. Thus, we meet two difficulties: the first one is to understand the strange being alien to the peculiar context, concepts and forms of consciousness one tries to understand, the second is to make the familiar strange being a member of the culture at large and being at the risk to take the familiar for granted.

A fundamental problem of understanding what is alien to oneself is illustrated in ethnological research: what could be the point of reference of an interpretation? What is understood as rational while other things appear as irrational? How is it possible to understand a different, and even alien, worldview without interpreting it within one's own worldview? The history of ethnology can be understood as the struggle to free itself from the idea that western society provides the criteria of rationality that could be used to assess non-western cultures, to refute those analyses that assess non-western cultural thinking as primitive, alogical and inferior to the western mode of thinking (Bourdieu, 1977; Habermas, 1981; Sahlins,

1976). The problem of the observer in ethnography and cultural anthropology will serve us as an analogy to rephrase the problem of understanding 'the other one' and his or her world view, be it the perspective of the practitioner or the perspective of the student. That is to say, we see some parallels in the relation between educational research and teaching practice and the relation between the teacher and the student. We started from an obvious parallel: that for both relations the view is anachronistic that knowledge is conveyed to practice/the student by theory/the teacher (Seeger and Steinbring, 1992).

The plan for the remainder of the chapter is the following one. We start from the empirical analysis of an episode in a mathematics classroom for learning disabled students. In this episode, the clash between the 'natural' and the 'formal' appears as the problem to pass in classroom discourse from intuitive ways of talking about and representing a mathematical task to the application of a mathematical rule. This problem is analysed in a first approximation as a problem of transition from an 'open' to a 'closed' context. In the second section this problem is rephrased in terms of the transition from everyday knowledge to scientific knowledge. The third section introduces in a more systematic way a central idea of the present chapter, namely that the closed context of mathematical knowledge could be characterized as mythical — at least from the perspective of the students. This section draws heavily on the ideas of Ernst Cassirer's theory of the evolution of relational versus substantial thinking. The fourth section applies some results first to mathematics education in general and second to the specific episode analysed in the first two sections in terms of a re-mythologization. The fifth, and last, section tries to give an outlook.

The examples we will refer to are drawn from the context of learning disability (LD). Our hypothesis is that with LD-students problems of knowledge understanding and development can be identified that are equivalent to those occurring in normal classes. In LD-classes these problems become even more obvious. One could argue that in LD-classes the didactical contract is reduced to the essentials of the broadcast metaphor. Thus, the study of the phenomena of teacher–student interaction in mathematics in LD-classes brings the basic problems of any kind of mathematics education to the fore.

An Exemplary Teaching Episode: Measuring the Height of a Tree

In the context of a research project on the use of open-curriculum material for students with learning deficits (LD), student–teacher interaction was observed and students were interviewed. The teaching episode we are going to look at was located within a unit on measurement in different applications. The main idea of the introduction of measurement referred to the concept of comparison. The material offered a number of worksheets with different open situations in the form of short stories, illustrated by pictures and diagrams. Students investigated the measurement of length, height, weight, time, volume, area, velocity, etc.

Generally, the aim of investigating a situation was to discuss how the measurement could be done, i.e., what kind of comparison had to be chosen in order to measure the given quantities. Some few situations referred also to the well-known means of measuring units, meterstick and ruler. Other situations required the use of indirect measurement. For instance, the ruler was used not only to measure length, but to support the measurement of reaction time: one student held the ruler above the open hand of another student who had to catch the ruler as soon as it was released and fell down. Here the measured length provided an indirect measure for measuring and comparing reaction time of students.

The main emphasis of the present material on measurement in different situations was on the conceptual relation between the quantity to be measured and the unit of measurement, which is not naturally given but has to be constructed, modified and changed. The exploration of this relationship was embedded in different concrete activities of measurement and comparison. There was no comprehensive formal evaluation of the measurement situation. Thus, these open situations did not aim at motivating the introduction of the calculus of decimal numbers, its addition and multiplication. The main object was to improve the concept of measurement — as an important example of the relational character of knowledge — not to perform too much formal calculations with the numbers generated in the measurement process. In this curriculum unit on measurement, estimating measures played a more important role than seemingly exact calculations.

In the following we shall describe in more detail an exemplary classroom episode, dealing with an interesting situation of measurement. This example will be used to explain how the open situation relates to the intended mathematical result. During the lesson preceding the one that is in focus now, the students had estimated and measured the heights of their fellow students and of the teacher. The lesson that we want to look at more closely began with a short repetition of what had been done before. Then a number of new measurement activities had to be performed: Estimating and later measuring the length, the width and the height of the classroom. The estimated and the measured results were written down on the blackboard.

During the discussion, a distinction was introduced between 'estimation', 'estimation by means of length of pace' and 'measurement by means of a meterstick'. Some students argued in this discussion, that they had not done 'pure' estimation of the length of the classroom, but they had used other objects of comparison, of which the measure had been determined before or which seemed to have a standard measure easily to be estimated. Some students used the tiles on the floor as units to determine the length and the width of the classroom. This is an example of constructing a new intermediate unit of measurement, adequate for the actual situation of measurement.

Subsequent to this situation of estimation and measurement, the teacher showed the following new situation on a transparency (see Figure 13.1):

> Irene says: 'The tree is 20 meters high.'
> 'No', says Karl, 'the tree is not 20 meters high.'
> Who is right?
> Look for a method to discover who is right!

The following excerpt from a lesson transcript gives an impression of how engaged the students discussed different possibilities of indirect measurement:

900 S.: Perhaps, if we know about the house, if we know how tall the house is,
901 then we can multiply this by 2.

Figure 13.1: Measuring the Height of a Tree

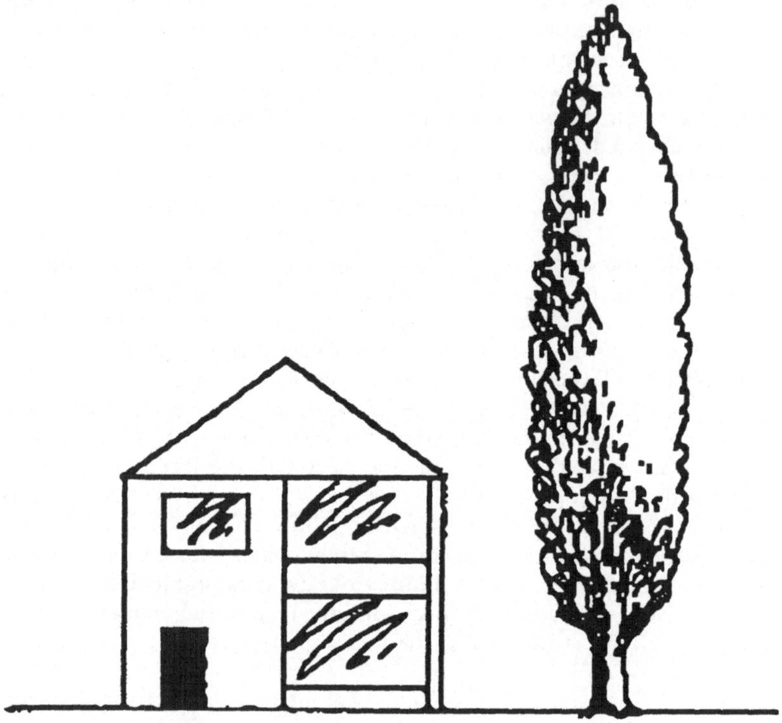

902 S.: Eh, by 2.
903 Several S.: Yes, the double.
904 S.: Well, while this is double the size of the tree . . .
905 T.: Well, there is much coming already, go on telling me, go on thinking
906 about it, I find this quite fascinating.
907 S.: Well, I would say the tree is . . .
908 T.: Murat, can you see well, I am obstructing your view, no?
909 S.: . . . 15 meters.
910 S.: 5 kilometers. (Laughter)
911 S.: No, I would say . . .
912 Milan (?): The tree is 4 meters.
913 S.: No-oh.
914 S. (outraged): 4 meters!
915 Murmuring.
916 T.: Milan has just made a guess.
917 S. (interrupting): 15 meters.

962 T.: Brahim, well, tell it aloud, I find your idea swell. Brahim, tell it
963 aloud to have everybody get it.

964 Brahim: That is, this eh . . . the yard–stick is 2 meters bi . . . eh, long. And if
965 you put that 2 times, then these are 4 meters. Thus I wouldn't say the
966 house is 4 meters, but taller.
967 S.: Yes.
968 several S. at once: I would 5, 6 meters, 5, 6 meters . . .
969 T.: Your guess is that house is not 4 meters, it is taller.
970 S.: Yes.
971 S.: 5, 6 meters.
972 T.: He has once again taken something known, the length of this
973 yard–stick, Marcel!
974 Marcel: Or here, a door is also 2 meters high, and then I would take it
975 more than 2 times, eh, the door fits in, the door drawn here.
976 T.: Which is the door?
977 Several S.: The black one.
978 T.: This one, no?
979 S. (interrupting): 2, 4, 6 meters.
980 T.: Exactly, the door is about, you say 2 meters high?
981 Marcel: Yes.
982 T.: We've seen just now, the rule hung here, you want to check, or shall we
983 just assume that what Marcel says is right?

In the first discussion quoted above the student offered answers, as for example:

- The tree is 4 metres high.
- The tree is twice as high as the house.
- We have to use the meterstick for solving this task.

During this discussion a student remembered that the door of the classroom was two metres high; he was convinced that the door in this picture also was two metres high. On the transparency the teacher marked the height of the door by a red line, indicating 2 metres. Then the ruler was used to estimate and to measure the height of the house. The question arose, how often the door fitted in the house. First the height of the door on the transparency is measured by the ruler and is determined as 1 centimetre. The height of the house on the transparency was determined as $4\frac{1}{2}$ centimetres, and consequently $4\frac{1}{2}$ times as high as the door. What is the height of the house, when the door is 2 meters high? In the discussion the mathematical problem arose to calculate $2 \cdot 4\frac{1}{2}$.

The students seemed to be surprised to be confronted with numbers, a fraction and a calculation within the frame of an open situation, and they started to guess. Some believed they had to calculate 4 times 4, some guessed that the house is 16 metres high, others estimated 20 metres. There seemed to be no connection between the open situation with the meaningful discussion on how to be able to estimate and to measure the height of the tree on the one hand and the formal calculation on the other.

The problem of a drastic rupture between a rather meaningful discussion of an open situation of measurement and an elementary formal calculation of some numbers and simple formulae (using multiplication and division) appeared repeatedly during the course of the whole teaching unit on measurement. The students' reaction indicated that they perceived a sharp rupture between a seemingly open and natural situation and the attempt to use some 'simple' formulae with numbers (fractions, decimal numbers), and the use of an elementary mathematical model for calculating some numerical results. They didn't seem to see any connection between these two domains, numbers with mathematical operations of a greater complexity than addition and simple subtraction seemed to constitute a totally different world, which had no relation to the open situation discussed before.

In the course of this actual teaching episode, the calculation of $2{\cdot}4\frac{1}{2}$ was done in the frame of a very strict 'funnel pattern' (Bauersfeld, 1978), with teacher questions narrowing the scope of possible answers increasingly. Step by step, the teacher questioned all elements which were necessary to do the calculation, when necessary she rejected false proposals of students to reach the intended goal directly. At the end of this funnel pattern, the teacher herself almost directly gave the answer to the problem of calculation, the students could only accept and confirm the suggested and restricted way of solution. Knowing now that the height of the house was 9 metres, and having estimated that the tree was twice as high as the house, the tree was 18 metres high. This calculation too, could only be performed with direct interventions of the teacher.

The described teaching episode on estimating and measuring the height of a tree, explicitly shows two very different phases of interaction between students and teacher. On the one hand, there was a meaningful and variable discussion on the problem of how to estimate and to measure the height of the tree; how to find a good unit of measurement in this situation; there were adequate estimations and comparisons, such as the tree is twice as high as the house. Students did not seem to have difficulties following the different scales of the picture with its objects on the transparency, on the projection or even in reality. On the other hand, a sharp rupture can be observed when $2{\cdot}4\frac{1}{2}$ came up in the discussion. The numbers presented with the mathematical operation seemed to provoke a total disconnection between the preceding phase of meaningful discussion and the beginning formal calculation. During the routine structure of the funnel pattern of interaction, the teacher concentrated on the numbers and on the mathematical operation, she implicitly reinforced the arithmetical frame of reasoning and of justification. There were no attempts to relate the numbers, the operation or the arithmetical relation inherent in this formula to the meaningful situation explored before. The arithmetical formula $2{\cdot}4\frac{1}{2}$ was 'solved' in an interactive process which concentrated more and more on the arithmetical elements alone without developing possible references to the actual situation of measurement.

The Transition from Everyday Knowledge to Scientific Knowledge

The process of knowledge development and of acquiring understanding in the above episode can be interpreted as an interplay between the open situation with its activities of measuring and the definiteness of the arithmetical formula. This relation between an open context of interpretation and the definiteness of the mathematical rule causes no problems as long as very simple mathematical operations and immediately conceivable mathematical concepts are involved.[3] Thus for instance, addition with concrete numbers can be integrated into the open situation very naturally and produces nearly no problem of understanding. As soon as the complexity of the mathematical operations grows and the theoretical nature of mathematical concepts increases, the problem of integrating the mathematical formal knowledge (some mathematical formula) into the open situation becomes more severe.

The separation of the interpretative context of an open situation from the formal rule begins with the operation of multiplication (especially when this operation cannot easily be reduced to a kind of repeated addition) and becomes even more difficult when fractions are involved. Now, for many students in this class a technical complexity of mathematical operations arises, which is too great to be balanced in a natural and direct way. This problem arises in principle in 'normal' mathematical classrooms too; here the rupture between open situation and definiteness of the mathematical rule comes into being with the fraction calculus and finally with elementary algebra and the concept of variable.

Teachers traditionally try to cope with this rupture through technical means; they tend to provide students with clear procedures and unambiguous rules for applying and using mathematical operations and concepts. The clear procedure is a methodical help to provide students with a secure ground on which they are supposed to construct their knowledge and try to integrate the formal knowledge with the open situation. For example, the teachers in the project emphasized that especially the weak students had to master elementary mathematical procedures as routines. In the course of a lesson, students had to calculate $6.5 \cdot 10$, a task they could not solve directly. Consequently the teacher asked them to make a written calculation, which at the end produced the same problems of conceptual understanding as before. The formal procedure itself provided no conceptual help.

There are traditional forms of overcoming the rupture between open situations and the definiteness of the mathematical rule in classroom interaction. An important pattern of interaction is the 'funnel pattern'. In the course of understanding new knowledge by integrating this new knowledge into the domain of existing natural knowledge, the question–answer interplay between teacher and students narrows down step by step until there is no other possibility left and the students virtually have to understand and solve the problem. In the course of the funnel pattern the new knowledge is totally reduced to the old knowledge, what is unknown is integrated into what is already known. This reduction process of new knowledge conceives mathematical knowledge first of all as a procedural or as an algorithmic matter, which can be dissolved into small steps and could be given to the student in this way (Chevallard, 1985).

The funnel pattern interaction in the mathematics classroom produces not only an algorithmization of mathematical knowledge; at the same time the mathematical knowledge is deprived of its theoretical character in the course of the interactive reduction process (Steinbring, 1988, 1989).

From Open to Closed Context: Towards the Myth of Mathematical Knowledge

In this section we will try to describe in very general terms from an 'anthropological' or 'ethnographic' perspective what happens with the transition from an open to a closed context.

We start from the fact that there is a series of noteworthy parallels between the situation of a cultural anthropologist or ethnographer where she or he tries to understand an alien culture and the situation of a researcher in mathematics education who tries to understand communicative processes in the classroom. It may seem strange to adopt such a perspective *vis-à-vis* the reality of mathematics education. To adopt such a decentred perspective in our view, however, seems to be a necessary step in a process of understanding that can best be described as 'making the (all too) familiar strange'.[4] Obviously, the attempt to understand cannot make a halt at a decentred perspective but needs the complementarity of the strange and the familiar. Here, however, we put an emphasis on the first step.

The analysis of the teaching episode above has shown that with the clash of the 'natural', everyday and the 'formal', mathematical context two world views meet whose relation is problematic insofar the 'formal' world view is not simply a prolongation of the 'natural' one. While students move within the open context of everyday problems safely and with considerable creativity, their behaviour within the 'closed' context of mathematics can be seen as 'superstitious' or 'mythical'. World views can be said to be more 'open' or 'closed', which suggested to look at the problem of 'myth and knowledge' more closely.

The concept of a mythical world view has been used in cultural anthropology in order to explain the obvious difference between the perspective of the (western) observer and a culture that he or she tried to make sense of. An early attempt was made by Lévy-Bruhl, who tried to explain the difference between the 'primitive' and the 'advanced' culture with the assumption that 'primitive' thinking rested on a pre-logical level (Lévy-Bruhl, 1925). In quite a similar way the concepts of children have been interpreted in developmental and educational psychology. Even in the work of Piaget and Vygotsky there are tendencies to view children's thinking as having deficits because it seems not to have certain elements of adult thinking at its disposal.

In the history of cultural anthropology it increasingly turned out to be inadequate to interpret alien cultures and world views within the frame of reference of the occidental world view. Evans-Pritchard (1977), for instance, found that the view of the prelogical nature had to be refuted, because 'primitive' thinking turned out to be in no way inferior to western thinking in regard to the use of logic. With that the issue of an adequate description of the mythical world view was put on the agenda: any description had to come to an understanding of this world view in its own terms and had to refrain from attributing a principled inferiority.

In his studies on changes in the historical development of thinking, the transition from the concept of substance to the concept of function, Ernst Cassirer had drawn attention to the fact that the relation between speech/language and thinking was of peculiar relevance to the description of the mythical world view. In his view, mythical thinking rests on the identification of word and object.

Mythical thinking, thus, withdraws in a surprising way from those explanatory models that implicitly or explicitly rest on the adult occidental word view. The identification of word and object, as it were, does not utilize a basic structural moment of occidental thinking: the division of subject and object, of the cartesian worlds of *res cogitans* and *res extensa*. With that it is a wrong question to ask: 'What does mythical knowledge represent?', because it presupposes the division between the world of objects and the world of symbols that is not made in the mythical world view. This kind of 'amalgamation' of the world of symbols with the world of facts has impressively been described by Sahlins (1976, 1981) not only for so-called 'primitive' societies but also for the present-day society of the United States.

What kind of 'representational' quality does a world view incorporate in general? If a world view expresses a certain quality of 'being-in-the-world' (Heidegger, 1927) and with that articulates a totality of understanding, they are akin to representation in the sense of a portrait that represents the person as a whole plus a certain perspective on that person:

> In a similar way worldviews lay down the basic conceptual frame within which we interpret anything happening in the world as something in a peculiar way. Worldviews like portraits cannot be true or false. (Habermas, 1981, p. 92; our translation, F.S./H.S.)

If the rationality of world views cannot be determined in regard to their competing truth, how then could they be compared? Habermas' (1981) view is that openness versus closedness is a dimension that allows for a comparison.

He starts from an application of the principle of cognitive development as it has been elaborated by Piaget with the concept of *decentration*.[5] Cognitive development here generally means the 'decentration of an egocentric worldview' (Habermas, 1981, p. 106). Correspondingly, cognitive development proceeds from a closed world view as it is given in egocentrism to an open decentred world view. Now, it was especially Lev Vygotsky who challenged this basic assumption of Piaget's theory and proposed an alternative model on the basis of his own studies on the role of language in cognitive development, followed by others, e.g., J.S. Bruner. In Vygotksy's view the development of the child does not start with an egocentric individualism but with a socially distributed perspective, as can be shown in the transition from outer speech to inner speech (Vygotsky, 1987).[6] Children's thinking is 'centred' indeed, but not on their own, yet undeveloped Ego, but on the 'other one', as he or she is met in the persons of parents, siblings, peers and teachers. As in the theory of G.H. Mead (1934) it is exactly *the other one* who provides the possibility for the *Ego* to develop.

With that, some reasons are presented to view the process of socio-cognitive development as one that starts from 'openness' and develops increasingly into 'closedness'.

Actually, it seems plausible to assume that the closedness of the 'mathematical world view' as students come to meet it in the shape of formal mathematical knowledge has some features of a myth. In what follows, we would like to give an account of mythical thinking especially as it is given in the version of Ernst Cassirer.

Mythical thinking is characterized by Cassirer essentially on three levels:

- Knowledge and concepts as a specific kind of object;
- myth and causality; and
- 'right' or 'wrong'

Knowledge and Object

An important feature of mythical thinking is that knowledge becomes a specific kind of object.

> . . . the mythical *form of thought*, which attaches all qualities and activities, all states and relations to a solid foundation, leads to the opposite extreme: a kind of materialization of spiritual contents. (Cassirer, 1955, p. 55)

There is no sharp distinction between 'image' and 'object' in mythical thinking.

> Only observers who no longer live in it but reflect on it (the opposition of 'image' and 'object') read such distinctions into myth. Where we see mere 'representation', myth, insofar as it has not yet deviated from its fundamental and original form, sees real identity. The 'image' does not represent the 'thing'; it *is* the thing; it does not merely stand for the object, but has the same actuality, so that it replaces the thing's immediate presence. Consequently, mythical thinking lacks the category of the ideal, and in order to apprehend pure signification it must transpose it into a material substance or being. (*ibid.*, p. 38)

Mythical thinking transforms ideas and concepts into something 'real', into an object. This type of materialization is not simply a form of concrete empiricism, it does not transform mental ideas simply into particular concrete things. These 'real' objects actually embody higher-order generalization. In the history of the number concept, for example, there has up to modern times always been a mixture of mythical and of relational or conceptual thinking.

> But in all these cases such a historical mixture of forms would not have been possible if the forms did not agree, in both content and systematic significance, in at least one characteristic motif, one fundamental tendency. Mythical number stands then at a spiritual turning point — it too strives to escape from the narrowness and confinement of the immediate sensuous-material world view to a freer, more universal view. However, the mind cannot apprehend and penetrate this new universality as its own creation but sees it as a foreign, demonic power. (*ibid.*, p. 144)

With regard to the object it is important to notice, that mythical thinking performs a materialization of ideal contents and conceptual relations; this process of materialization contains special aspects of generalizations and universalizations.

Myth and Causality

Further, the 'logical' connection of knowledge, the 'laws' and the forms of justification obtain a specific manner within mythical thinking.

> Mythical thinking is, in general, distinguished from a purely theoretical world view as much by its *concept of causality* as by its *concept of the object.* For the two concepts condition each other: the form of causal thinking determines the form of objective thinking, and vice versa. Mythical thinking is by no means lacking in the universal category of cause and effect, which is in a sense one of its very fundamentals. (*ibid.*, p. 32)

> This isolating abstraction, which singles out a specific factor in a total complex as a 'condition', is alien to mythical thinking. Here every simultaneity, every spatial coexistence and contact, provide a real causal 'sequence'. It has even been called a principle of mythical causality and of the 'physics' based on it that one take every contact in time and space as an immediate relation of cause and effect. The principles of *post hoc, ergo propter hoc* and *juxta hoc, ergo propter hoc* are charactristic of mythical thinking. (*ibid.*, p. 45)

Comparing theoretical thinking with mythical thinking, Cassirer explains the difference of both conceptions of causality.

> It is as though the conceptual consciousness and the mythical consciousness applied the lever of explanation at entirely different points. Science is content if it succeeds in apprehending the individual event in space and time as a special instance of a general law but asks no further 'why' regarding the individualization as such, regarding the here and now. The mythical consciousness, on the other hand, applies its 'why' precisely to the particular and unique. It 'explains' the individual event by postulating individual acts of the will. (*ibid.*, p. 48)

The mythical form of causality, which gives explanations and justifications according to spatial or temporal proximity, represents a kind of argumentation often to be found in classroom interaction. Students do not ask for theoretical connections, they don't pose the question of 'why' in cases when there is seemingly a simultaneousness and a nearness on the level of real objects. A formula is justified because of its formal functioning, there seems to be no need for theoretical justification.

'Right or Wrong'

The third fundamental aspect of mythical thinking refers to the antithetical judgment of 'right and wrong'. In the frame of myth this is called the *mana-tabu rule*, which expresses the contrast of the holy and the profane, of light and dark etc. (Cassirer, 1955, Chapter 1) At first this aspect seems to be only partly applicable to mathematical knowledge. But it is well-known how dominant in everyday teaching the contrast of right and wrong is for an understanding of school mathematics. This rule seems to be close to the familiar appraisal of the value of mathematical knowledge as it is expressed in the dialectics of symmetry, beauty, ideal and distortion, fuzziness etc. For mathematics teaching the original contrast of right and wrong plays an important role.

The emergence of theoretical thinking and reasoning is dependent on mythical thinking and at the same time has to overcome the simple forms of mythical thinking.

> But perhaps we can best appreciate the meaning and origin of this way of thinking if we consider that even in scientific knowledge the sharp distinction between thing on the one hand and attribute, state, and relation on the other results only gradually from unremitting intellectual struggles. Here too the boundaries between the 'substantial' and the 'functional' are ever and again blurred, so that a semimythical hypostasis of purely functional and relational concepts arises. (Cassirer, 1955, pp. 58–59)

Methodical Forms of a Re-mythologization of Mathematical Knowledge in the Mathematics Classroom

Traditional teaching processes in the mathematics classroom tend to introduce (implicitly) from the beginning a universal epistemological basis for the status of mathematical knowledge. All mathematical knowledge, all concepts and operations are situated on the same level; there seems to be no requirement for changing the status of knowledge during its development; to acquire new knowledge simply means to accumulate new subject matter of the same type. This conception of the character of knowledge transforms mathematics for the students to an externally given structure with materialized concepts and rules. There are objects, arranged according to strange rules, objects to operate with according to mysterious regulations. This external structure is organized according to obscure principles of cause and effect.

This *a priori* fixed linear and unique structure of knowledge materializes mathematical concepts and operations for students, because this predetermined structure denies the theoretical or self-referential nature of mathematical knowledge and it prevents self-organized processes of learning and understanding. In this way learning is not understanding, learning reduces to collecting new things.

Teaching and learning is closely linked to 'demonstrating' and to 'seeing', for which things and pictures are essential to point at and to look at. These important aspects of teaching and learning support methodical forms of a re-mythologization of mathematical knowledge. In the primary grades, methodical forms of a re-mythologization of mathematical knowledge are very obvious; the natural numbers (reduced set of numbers) and the natural arithmetical operations (reducible to direct conceptions and to immediate concrete operations) are subjected to such forms of mythologization. For instance, the introduction of the elementary natural numbers in the first grade often uses pictures of everyday situations to illustrate the corresponding number, for example 5 times 'anything' (as for example 5 birds). In the classroom interaction of these and similar pictures of everyday situations are used to exercise in a routine way how to decode the right number and how to read off the only accepted arithmetical operation presented in this picture. The picture of an everyday situation has to be formally interpreted by the corresponding number sentence which codifies the meaning of the materialized numbers and the naturally given operation (for example addition as a kind of 'putting things together', for more details see Voigt, 1990).

The elementary natural numbers and the natural operations are embedded in forms of teaching and social interactions, which emphasize the direct meaning, immediateness, naturalness etc. of mathematical knowledge. There seems to be no problem for young pupils to understand directly what these elementary elements of mathematical knowledge mean.

The growing complexity of technical and operational aspects of mathematical concepts and operations (for instance in fractional calculus, or in elementary algebra) aggravates the methodical possibilities of reducing this knowledge to natural and immediate conceptions as they are dominant in mythical thinking. Teaching processes, however, tend to maintain the forms of direct instruction, of providing immediate meaning by transforming mathematical concepts and operations into natural things. Finally, the concept of variable seems to be not transformable to a natural thing; but classroom observation shows that students cope with the variable concept in forms of universal concretizations: Variables are general things, they are general but can only be understood when a concrete number is substituted.

Then, a technical complexity like this can no longer be treated by methodical reduction to natural things and direct understandable operations. When faced with serious problems of understanding and with misinterpretations, teachers consequentially try to simplify even more and reduce knowledge to smaller pieces. Simultaneously this methodical transformation of knowledge changes the social role of the classroom interaction: The social context of teaching–learning processes with its frames and patterns for constituting new knowledge replaces more and more the constraints of theoretical mathematical knowledge (Bauersfeld, 1988).

Teaching processes tend to produce a two-fold methodical re-mythologization of mathematical knowledge:

- a re-mythologization related to knowledge:
  reducing knowledge to natural things and suggesting a direct justification of theoretical relations in knowledge (variable as a place holder for numbers, the reduction of new knowledge to old knowledge, the reduction of mathematical operations and relations to concrete actions); and
- a social re-mythologization:
  in the course of social classroom interactions and in the social construction of mathematical meaning the knowing teacher alone has the justification and the appraisal of knowledge at his or her disposal; the epistemological constraints of mathematical knowledge are more and more suppressed.

The specific nature of school mathematics and the social frame conditions of teaching–learning processes with their interactive patterns of communication constitute a basis for processes of re-mythologization of theoretical mathematical knowledge. Such processes of re-mythologization are partly necessary in teaching processes, they provide a link between the new and the old knowledge, they partly transform 'strange' knowledge to 'natural' knowledge. At the same time, re-mythologization tries to integrate the new knowledge into the familiar social communication of the classroom discourse, in order to make the new knowledge understandable in terms of the implicitly constituted and socially accepted form of 'language' of communication.

On the one hand there are legitimate needs for naturalness of knowledge and of communication about this knowledge, on the other hand the theoretical nature of mathematical knowledge gets into conflict with methodical re-mythologization as soon as the transformation process of theoretical knowledge to natural knowledge is considered as a universal and absolute means to prepare knowledge for teaching.

The emerging conflict between theoretical and natural aspects of mathematical knowledge can be observed in the short teaching episode on 'Measuring the height of a tree'. During classroom observation it was surprising, how effectively the students could cope with the picture on the transparency, with the relations between 'house', 'tree' and 'door'. From an external observation, this situation and the classroom interaction offered a multitude of conceptual relationships. The height to be estimated and to be measured requires the establishment of direct and indirect relations and comparisons: tree with house or with an other mentally represented tree; tree with some remarkable objects on the house; house with door; etc. This frame of possible relations seemed to be even more complex with regard to the means of visualizing the situation on the worksheets, on the transparency, and even with the overhead-projection on the screen. In principle, it seemed to be necessary to balance a great variety of relations between different objects which themselves were differently represented by the change of scale. How was it possible that the students could use this situation without any severe problem? In which perspective did the students organize this complex of possible relations?

The frame of the methodical re-mythologization of knowledge offers possible explanations for this situation. One could imagine that this complex of possible relations in the problem situation shows the theoretical character of knowledge, this means especially for measurement problems the requirement of indirect measurement by means of constructing adequate units and establishing suitable comparisons. The students didn't take such a theoretical perspective on the problem. They looked at the problem from a 'mythical' perspective; the objects presented in this situation were taken as natural things; the students operated with the thing more or less directly. There was no question, whether the door on the picture was 2 centimetres high and at the same time was about 15 centimetres high on the enlarged projection. For the students, there was only the door, the natural object, and the other pictorial representations showed the same door, they *were* the door. The students felt no need to distinguish between the door and its different representations. They did not have to rethink this natural situation nor to introduce theoretical relations into this picture.

The problematic rupture in the interaction arises, when the arithmetical calculation $2 \cdot 4\frac{1}{2}$ had to be done. It seemed to be impossible to integrate this arithmetical formula into the natural context of the 'house and tree' picture. Essentially, the arithmetical formula with the operations of multiplication and the fraction is a symbolic representation of a theoretical relation. In certain situations it seems to be possible to re-construct the 'natural meaning' of this formula, to integrate this formula into the natural context of communication and of the specific manner to interpret knowledge. A major methodical goal of mathematics teaching is the integration of new, theoretical knowledge into the natural context of communication. Sometimes students are successful in integrating this knowledge, but often there are fundamental obstacles in the theoretical knowledge itself which prevent this integration.

The conflict between the naturalness of everyday knowledge and the systemic and self-referential character of theoretical mathematical knowledge constitutes a basic problem for teaching–learning processes. On the one hand, the social context of teaching requires a certain amount of naturalness to make communication and understanding in the classroom possible. On the other hand, the theoretical nature of new knowledge is in conflict with immediate naturalness of theoretical relations. In which way does teaching cope with this conflict? Traditionally, classroom observations of everyday mathematics teaching show that intentions of integration prevail. The methodical aim is to re-mythologize theoretical knowledge as far as possible. When there is no more integration possible, a fundamental rupture arises: there seems to be a total disconnection between the natural situation and the formal mathematical relation, as it can be observed in many teaching episodes with slow and handicapped learners.

To really establish a fruitful balance between the needs of social communication and the theoretical nature of knowledge requires to organize the social teaching–learning process itself as a self-referential process (Steinbring, 1991). Essentially this means to question in the course of knowledge development the 'natural situation of communication and knowledge' itself, to initiate modifications and changes and not to take these situations as unchangeable or as a fixed basis for the process of acquisition of new knowledge.

Concluding remarks

We have started our paper with the relation between the nature of mathematical knowledge and ways of communicating this knowledge in school. In learning situations with handicapped students this relation lies at the heart of the didactical contract in action. This describes a vicious circle, a paradox. We have argued that all learning processes essentially are processes of sharing: Sharing a general structure — be it characterized as mythical knowledge or as theoretical knowledge. Language (or more generally a symbol system) plays a decisive role for the problem of sharing. For the concluding exposition of this problem, the relation between language and thinking provides important insights.

In his very interesting book 'Seeing voices' (Sacks, 1989), Oliver Sacks has collected many informative observations on the relation between language/symbol systems and thinking of deaf persons. We cannot pursue all the productive analogies between speechlessness (muteness) of deaf persons and speechlessness of students in mathematics teaching which are possible. We shall only provide a few general suggestions.

A number of autobiographical reports of deaf persons are especially interesting because they give indirect hints about the time in which these persons could neither write, speak (by gesture) nor could understand speech. With great agreement these persons characterized themselves as 'animals', as 'dogs' etc. And with the same agreement, they experienced the acquirement of speaking by gesture and thereby in particular the (mostly sudden) comprehension of a more general and abstract concept as a kind of a 'second human birth'.[7]

In the reports as well as in the observations made by Sacks it is surprising that for deaf persons posing questions belonged to the most difficult discursive practices to be learnt. To be able to pose a question meant in a certain sense, to give a voice to one's own person, to oneself. And the ability to ask seemed to be a progress in coping with deafness as essential as the insight that general and abstract concepts exist and that they can be named. It is very difficult to explain with a few words how deeply these moments affected the life of a deaf person. This was really the moment when a deaf person was able to share the human community and to comprehend himself or herself as a human being. Sack's book offers a multitude of exciting descriptions of such situations.

What could be the analogies for the situation of LD-students? And what could be the applicability to 'normal' teaching? The key problem seems to be that the didactical contract permits particular discursive practices and legitimates only them positively. 'Natural' discursive practices as 'dialogue', 'conversation' and 'discussion' are suspicious from the perspective of the didactical contract. Especially the student-question seems to be seen as problematic. If this analogy is accepted, the students in the mathematics classroom must be able to 'speak' with a symbol system which allows the participation in the meaning of the mathematical discourse. If this participation does not succeed, it is experienced as a personally caused subjective failure. Treating such unsuccesses in the school, in the first place consists in using remedial strategies which reinforce the conception of failed participation by presupposing it, this means not dare to expose students to complex requirements, but to reduce them to aspects thought to be essential (for instance the automatization of the basic arithmetical operations).

A fundamental mechanism increasing the effectiveness of these processes is the broadcast metaphor according to which the teacher can hand over the new knowledge to the student like an object. In the zone of proximal development marking the transition from old and known knowledge to new knowledge yet to be learnt, this metaphor only aims at the transport of knowledge from the 'already knowing' teacher to the 'ignorant' student; only the student has to acquire the knowledge provided by the teacher. Didactical principles striving for the participation of students in the mathematical discourse in contrast have to take care for a reciprocal acquisition of knowledge: teachers have to understand the existing — especially the mythical — students' conceptions of knowledge (not only to treat them as misconceptions), while students acquire teachers' conceptions of knowledge. The basic figure of this developmental process to a large extent corresponds to the 'hermeneutic circle'. The determining and the determined are situated in a circular relationship to each other and it is impossible to decide here with certainty the cause and the effect. The figure of the hermeneutic circle is an adequate means to examine the importance of the theoretical nature of knowledge in mathematics teaching.

In understanding mythical thinking, to understand the role of speech is a key factor. If mathematical knowledge undergoes a re-mythologization the result is a 'transformation of history into nature' (Barthes, 1970): what in fact is a product of human discourse is given all the qualities of a 'natural' object that resides outside the spheres of social construction. In the rationalist tradition of Cassirer the identification of word and object, of symbol and referent looks like a deficit. If, however, the stance of the hermeneutic tradition is taken, as it is articulated by

Gadamer in that all being that can be understood is given as language, a different picture emerges. We could then argue that mathematics can only be understood to the degree that it is spoken about. The myth of mathematics is the product of a certain history of speaking (or being silent) about mathematics. In that it shares the characteristics of other myths, as Lévi-Strauss (1969) gave them: myths are stories with a history, where the myth is a kind of tinkering with language elements ('*bricolage*').

It is important to note that the contribution of the cultural-historical school of psychology, as it is given in the work of Lev Vygotsky, Alexander Luria and Alexej N. Leont'ev is highly relevant to this issue. The characteristic features of mythical thinking as identification of object and word or '*bricolage*' with language elements can also be found in the reports of Luria and Vygotsky about the thinking of adults in pre-industrial societies. In their work, the characteristic features of mythical thinking were interpreted from another vantage point. The cultural-historical approach is based on the idea that psychological functions have to be interpreted as developmental processes on three different levels: on the level of phylogenesis, social-cultural history and ontogenesis. Accordingly, the transition of thinking from elementary to its more 'advanced' forms is a major topic.

In the 1930s A.R. Luria and L.S. Vygotksy conducted an expedition to Central Asia with the aim to do research on the reasoning processes of 'primitive' people. They found characteristic differences between the reasoning processes of people living in pre-industrial settings as compared to the standards of more advanced western societies. One of the striking findings, for instance, showed up with typical classification tasks: four objects, a hammer, a saw, a spade and a log of wood were shown to the subjects and they were asked which three were similar and which one did not belong to the group because it was different. The subjects regularly responded that the hammer, the saw and the log belonged together or that all four belonged together. Their responses showed that they primarily focused on practical situations or on properties of the objects, and they never used abstract concepts like 'tools' to group objects. Generally, their reasoning processes focused on the objects and not on word meanings.

In a similar way, these persons from a pre-industrial society could not make sense of a syllogistic reasoning task like the following one: 'Cotton can only grow in a hot and dry climate. In England it is cold and humid. Can cotton grow there?' The subjects usually responded that they had never been in England and thus could not answer the question, or that a person who had been travelling a lot possibly could give an answer.

In summary, the reasoning processes were highly situated which is also true for the use of word meanings. No reasoning with decontextualized word meaning could be found which is so typical for occidencal thinking. In a certain way, this finding fits nicely to our above characteristic of mythical thinking, especially the identification of object and word. In the light of these findings, the reasoning processes that seem so typical for mythical thinking appear to be closely connected to literacy. That is to say, the peculiar quality of mythical thinking is largely a result of the fact that these persons have not been introduced into the use of decontextualized symbol meaning as a result of learning to read and write.

The world of mathematics remains closed, a myth, to an understanding if discourse in mathematical classrooms only identifies the world view of children as different and not compatible to the mathematical world view. To really understand the world view of children as the voice of 'the other one' in a dialogue is as necessary for the 'decolonization' of the child as it is for opening up mathematics.

Notes

1. We cordially thank Ute Waschescio for critical comments and editorial improvements when preparing the final version of this chapter.
2. It seems, however, erroneous to assume that Piaget could be held responsible for this development as he was seen to emphasize that 'operational thinking' needs concrete operations and activities in order to develop. In fact, the view that any thinking has to travel through the empire of the senses has a much longer philosophical tradition.
3. Of course, what is 'simple' is very different in different settings. In the definition of what is 'simple' the didactical contracts of an LD-school and a normal school radically differ.
4. Margaret Mead is said to have noted: 'If a fish were to become an anthropologist, the last things he would discover would be water' (quoted by Spindler and Spindler, 1982, p. 24).
5. Piaget referred to egocentrism for instance in the following way: 'Thought, springing from action, is indeed egocentric at first for exactly the same reasons as sensorimotor intelligence is at first centred on the particular perceptions or movements from which it arises. The construction of transitive, associative, and reversible operations will thus involve a conversion of their initial egocentricity into a system of relations and classes that are decentralized with respect to the self, and intellectual decentralization (not to mention the social aspect) will in fact occupy the whole of early childhood' (Piaget 1950, pp. 122–3).
6. Carl Ratner (1991) has shown that this idea is already present in the developmental psychology of James Mark Baldwin.
7. This notion originates from Leont'ev (1981), characterizing the phase of the second year of life, in which children start learning to speak.

References

Barthes, R. (1970) *Mythologies*, Paris, Editions du seuil.

Bauersfeld, H. (1978) 'Kommunikationsmuster im Mathematikunterricht — Eine Analyse am Beispiel der Handlungsverengung durch Antworterwartung', in Bauersfeld, H. (Ed) *Fall-studien und Analysen zum Mathematikunterricht*, Hannover, Schroedel, (pp. 158–70).

Bauersfeld, H. (1988) 'Interaction, construction and knowledge: Alternative perspectives for mathematics education', in Grouws, D.A., Cooney, T.J. and Jones, D. (Eds) *Effective Mathematics Teaching* Hillsdale, Lawrence Erlbaum, pp. 27–46.

Bourdieu, P. (1977) *Outline of a Theory of Practice*, Cambridge, Cambridge University Press.

Cassirer, E. (1955) *The Philosophy of Symbolic Forms. Volume Two: Mythical Thought*, New Haven, Connecticut, Yale University Press.

Chevallard, Y. (1985) *La transposition didactique — Du savoir savant au savoir enseigné*, Grenoble, La Pensée Sauvage.

Edwards, D. and Mercer, N. (1987) *Common Knowledge. The Development of Understanding in the Classroom*, London, Methuen.

Evans-Pritchard, E.E. (1977) *Witchcraft, Oracles and Magic Among the Azande*, Oxford, Clarendon Press.

Goodman, N.D. (1979) 'Mathematics as an objective science', *The American Mathematical Monthly*, 86, pp. 540–51.

Habermas, J. (1981) *Theorie des kommunikativen Handelns. Band 1: Handlungsrationalität und gesellschaftliche Rationalisierung*, Frankfurt/Main, Suhrkamp.

Heidegger, M. (1927) *Sein und Zeit*, 16th ed. 1986, Tübingen, Niemeyer.

Leont'ev, A.N. (1981) *Problems of the Development of Mind*, Moscow, Progress Publishers.

Levi-Strauss, C. (1969) *La penseé sauvage*, Paris, Plon.

Lévy-Bruhl. L. (1925) *La mentalité primitive*, 4th ed. Paris, Alcan.

Lurija, A.R. (1974) *Ob istoriceskom razvitii poznavatel'nych processov* (The historical development of thought processes), Moscow, Nauka.

Mead, G.H. (1934) *Mind, Self, and Society. From the Standpoint of a Social Behaviorist*, Chicago, University of Chicago Press.

Piaget, J. (1950) *The Psychology of Intelligence*, London, Routledge and Kegan Paul.

Ratner, C. (1991) *Vygotsky's Sociohistorical Psychology and its Contemporary Applications*, New York, Plenum Press.

Sacks, O. (1989) *Seeing Voices. A Journey into the World of the Deaf*, Berkeley, California, University of California Press.

Sahlins, M. (1976) *Culture and Practical Reason*, Chicago, University of Chicago Press.

Sahlins, M. (1981) *Historical Metaphors and Mythical Realities*, Ann Arbor, Michigan, University of Michigan Press.

Seeger F. and Steinbring H. (1992) 'The practical phase in teacher training: Preparing for professional practice under changing conditions', *Zentralblatt für Didaktik der Mathematik*, 24, 7, pp. 280–6.

Spindler, G. and Spindler, L. (1982) 'Roger Harker and Schönhausen: From familiar to strange and back again', in Spindler, G. (Ed) *Doing the Ethnography of Schooling. Educational Anthropology in Action*, New York, Holt, Rinehart and Winston, pp. 20–46.

Steinbring, H. (1988) 'Nature du savoir mathématique dans la pratique de l'enseignant', in Laborde, C. (Ed) *Actes du Premier Colloque Franco-Allemand de Didactique des Mathématiques et de l'Informatique*, Grenoble, La Pensée Sauvage, pp. 307–16.

Steinbring, H. (1989) 'Routine and meaning in the mathematics classroom', *For the Learning of Mathematics*, 9, 1, pp. 24–33.

Steinbring H. (1991) 'The concept of chance in everyday teaching: Aspects of a social epistemology of mathematical knowledge', *Educational Studies in Mathematics*, 22, pp. 503–22.

Voigt, J. (1990) 'Der interaktiv hergestellte Bezug von Rechensätzen zu Sachverhalten — eine thematische Prozedur im Unterrichtsalltag', *Die Zukunft des Mathematikunterrichts*, Landesinstitut für Schule und Weiterbildung. Soest, Soester Verlagskontor, pp. 120–23.

Vygotsky, L.S. and Lurija, A.R. (1930) *Etyudy po istorii povedeniya — Obez'yana, primitiv, rebenok.* (Essays on the development of behavior, Ape, Primitive, child) Moscow, Gosizdat.

Vygotsky, L.S. (1987) 'Thinking and speech', in Rieber, R.W. and Carton, A.S. (Eds) *The Collected Works of L.S. Vygotsky. Vol. 1. Problems of General Psychology*, New York, Plenum Press, pp. 37–285.

Part 3

Enquiry in Mathematics Education

> Thinking is more interesting than knowing, but less interesting than looking. (Goethe, in Auden and Kronenberger, 1970, p. 350)

Enquiry, investigation and problem-solving have been at the heart of the human search for knowledge and understanding since the Presocratic Philosophers or earlier. The Ancient Greek speculations, wild as they were, gave rise to the foundations of science, logic and most academic knowledge including that jewel in the crown of knowledge, Euclidean geometry. Since those days, the history of mathematics and science has also been a history of problem posing and solving. Pappus, Bacon, Descartes, Whewell, Polya, Lakatos and many others have reflected deeply on the methodology of knowledge creation in these fields. For example, Bacon (1960) in 1620 proposed a method of induction for arriving at hypotheses, which were then subjected to testing. In order to facilitate the genesis of inductive hypotheses, he proposes the construction of systematic tables of results or facts, organized to show similarities and differences. Descartes (1931) published in 1628 a work embodying twenty-one 'Rules for the direction of the mind' that included many explicit heuristics, such as 'sequential enumeration of examples to facilitate inductive generalisation' and 'representation of relationships by algebraic equations', which have a profoundly modern ring to them. Such methodological concerns have reappeared in research in mathematics education as part of a new awareness of the import of problem-solving in mathematics (Schoenfeld, 1992).

In the modern era, numbers of philosophers have identified enquiry and problem-solving as lying at the heart of the philosophic and scientific enterprises. Dewey (1920) and the Pragmatists viewed enquiry and the search for knowledge to be the central function of epistemology. Collingwood (1939), like the Hermeneutic philosophers, proposes a logic of enquiry as the centrepiece of a philosophy of history and interpretation. In the philosophy of science, a glance at some of Popper's book titles, such as *The Logic of Scientific Discovery* (1959), *Conjectures and Refutations* (1963), and *Unended Quest* (1976) reveals the central place of enquiry in his epistemology. Of course Popper distinguishes the 'context of discovery' carefully from the 'context of justification', but both pertain to scientific enquiry, encompassing the construction and validation of knowledge. Laudan (1977) explicitly proposes a 'Problem Solving Model' of scientific progress. He argues that provided it occurs in contexts permitting critical discussion, problem-solving is the essential characteristic of scientific rationality and methodology. In mathematics too the whole maverick tradition in the philosophy of mathematics including Davis, Hersh, Kitcher, Lakatos, Putnam, Tymoczko, Wang, Wittgenstein and others is concerned with rehabilitating and accommodating the practices of mathematical enquiry within the philosophy of mathematics.

This quick trot through 2500 years of mathematics, science and philosophy reveals that enquiry is the basis all knowing, for *enquiry is knowledge in the active voice.* In addition, an ever-present and growing body of opinion also locates it at the heart or epistemology, the theory of knowledge. Many other schools of thought and movements could be cited in support of this claim, as a glance through Volume 3 and 4 quickly reveals. But it is time to look more specifically at enquiry in educational research, and in mathematics education in particular.

Enquiry is the process of finding out, that is knowledge making or getting, and this process enters into mathematics education and mathematics at a variety of levels. There is first, the object of enquiry. This is the problem or task that engages the student in enquiry as a means of learning. Much has been written on the import of this area (e.g., Bruner, 1960). The problem is often the starting point for enquiry in research mathematics, too, as the above discussion indicates. Second, there is the process of enquiry. Research itself is systematic and critical enquiry carried out with the aim of producing knowledge. This describes the process of enquiry both in mathematics and in mathematics education (not to mention elsewhere). See e.g., Ernest (1993). In each of these fields, enquiry is part of a collective human endeavour, with shared goals and criteria for success. Typically, the outcome of enquiry is expected to:

- link with and build on existing knowledge;
- engage in theory building;
- use systematic methods of enquiry;
- justify knowledge claims; and
- result in a systematically organized text.

An enquiry needs to result in a public text so that others can access its results and evaluate them. Indeed it can be argued that such outcomes do not become knowledge until publicly scrutinized and accepted. Certainly this is the philosophers' view.

Thus enquiry, whether in mathematics or mathematics education, raises the issue of text. This is not a coincidental matter. Mathematics is a discipline that is perhaps uniquely textual. It relies on a formal symbolic apparatus that does not describe or refer to the experienceable world (except at the level where it is part of everyday language and practices). Consequently various problems arise for learners in attempting to master this uniquely imaginary symbolic field. Many of these problems are concerned with signification, since meaning is defined internally. Further problems arise from the strong objectivity of mathematical language that excludes all reference to persons, places, events and time. Instead, the reader of many mathematical texts meets a sequence of imperatives, and expects to do little more than obey them, or not (Rotman, 1988).

Enquiry starts with a problem or question, perhaps implicit, and ends with text. This description doubtless fits the process of writing all of the chapters in the two volumes. But in this section, enquiry, problems and language are the explicit subject matter explored in the contributions, each bringing its own emphasis, style and perspective. In the modern trend towards reflexivity in research, it is inevitable that more attention will be devoted to enquiry and methodology. So it does not take a crystal ball gazer to predict that this will be a growth area in research. But what the chapters here also indicate is a growing awareness of language, and its role in the genesis, framing and constitution of knowledge. Many perspectives beginning to be noted in mathematics education share this emphasis, including Vygotsky and Activity Theory, socio-linguistics and psycholinguistics, semiotics, social constructivism and constructionism, post-structuralism, post-modernism and hermeneutics, as well as logic, philosophy and epistemology. I have no doubt that this too will be an increasingly important area of research in mathematics education.

References

AUDEN, W.H. and KRONENBERGER, L. (Eds) (1970) *The Faber Book of Aphorisms*, London, Faber.

BACON, F. (1960) *The New Organon*, Indianapolis, Bobbs-Merrill.

BRUNER, J. (1960) *The Process of Education*, Cambridge, Massachusetts, Harvard University Press.

COLLINGWOOD, R.G. (1939) *An Autobiography*, Reprinted 1944, London, Penguin Books.

DESCARTES, R. (1931) *Philosophical Works*, 1, Cambridge, Cambridge University Press (reprinted by Dover Press, New York, 1955).

DEWEY, J. (1920) *Reconstruction in Philosophy*, Reprinted in 1950, New York, Mentor.

ERNEST, P. (1993) 'Epistemology and the relationship between Subjective and Objective knowledge of Mathematics', Paper presented at Conference on the Cultural Context of the Mathematics Classroom, Osnabrück, Germany, 11–15 October.

LAUDAN, L. (1977) *Progress and Its Problems*, Berkeley, California, University of California Press.

POPPER, K. (1959) *The Logic of Scientific Discovery*, Hutchinson, London.

POPPER, K. (1963) *Conjectures and Refutations*, London, Routledge and Kegan Paul.

POPPER, K. (1976) *Unended Quest*, Glasgow, Fontana-Collins.

ROTMAN, B. (1988) 'Towards a semiotics of mathematics', *Semiotica*, 72, 1/2, pp. 1–35.

SCHOENFELD, A. (1992) 'Learning to think mathematically', in GROUWS, D.A. (Ed). (1992) *Handbook of Research on Mathematics Teaching and Learning*, New York, Macmillan, pp. 334–70.

Chapter 14

The Problem of the Problem and Curriculum Fallacies

Stephen I. Brown

Problem-solving and Curriculum

With the same gusto that ushered in the 'new mathematics' on a cushion of 'understanding', 'structure' and 'learning by discovery' four decades ago, 'problem-solving' has emerged as the dominant theme in mathematics education for the close of the twentieth century. Beginning in the 1980s the theme is closely aligned with 'real world' applications. The twin concerns are expressed in the National Council of Teachers of Mathematics (NCTM) brief document entitled *An Agenda for Action: Recommendations for School Mathematics for the 1980s* (1980). There we find:

> Problem solving must be the focus of school mathematics for the 1980s . . . Performance in problem solving will measure the effectiveness of our personal and national possession of mathematical competence . . . Problem solving involves applying mathematics to the real world, serving theory and practice of current emerging sciences. (NCTM, 1980, p. 2)

The theme of problem-solving is reiterated in *The Curriculum Evaluation Standards For School Mathematics* (1989) and other related documents of the 1990s. We are told that:

> Problem solving should be the central focus of the mathematics curriculum. As such it is a primary goal of all mathematics instruction and an integral part of all mathematical activity . . . Ideally, students should share their thinking and approaches with other students . . . In addition, they should learn to value the process of solving problems as much as they value the solution. (NCTM, 1980, p. 23)

Problem-solving thus begins to acquire a social context — frequently referred to as 'communication' in addition to a continued focus upon 'real world' applications.

The gusto is clear, not only in the above document, but in the recent cascade of articles, year books, national and international conferences, special issues of journals and the like. Granted that we need to exercise caution in advocating total elimination of such activities as drill and forms of practice for the purpose of solidifying ideas, and granted that there is no well-spring to eliminate all forms of explanation by teachers, what fault could we find in placing priority on an activity that reflects the integrity of the subject itself? After all, mathematics as a discipline has emerged through efforts at solving problems. Furthermore, the conception of problem-solving being proposed in most quarters is broad enough to include not only routine tasks and typical 'world problems' but problems that are non-routine as well.

Consistent with advocating non-routine problems, there is an increased tendency to be concerned with problem-solving in a way that connects that activity not only with the product but the process of thinking as well. Rooted in a conception of thinking compatible with Dewey's (1933) 'reflective thought', mathematicians and educators borrow from, and expand upon, a model of problem-solving that is associated with the work of Polya (1954, 1957, 1962). Essential elements of the model are:

1. Gain an awareness or understanding of the problem.
2. Consider possible strategies for solving it.
3. Choose a strategy.
4. Carry out the strategy.
5. Verify the solution.

What is missing from such a perspective? In coming up with a partial answer to that question, we focus on two fallacies: the fallacy of 'the problem as given' and the fallacy of 'scientific context'.[1] We end this essay with two grandparent fallacies that may in fact be responsible for the previous two.

The Fallacy of 'the Problem As Given'

Though instructional programmes, curriculum and research in the domain of problem-solving differ considerably in terms of methodology, intent and focus, and though there is a considerable interest in use of the language of exploration, they tend to share an assumption regarding the sequence of problem-solving events. That is, they assume that a problem is to be given to a student and the associated task is for the student to attempt a solution, granted that heuristic strategies may be encouraged and honoured in such a scheme. It is depicted in Figure 14.1. At first glance it may seen peculiar to question the sequence in Figure 14.1 for if one is interested in improving problem-solving competence what would seem more on-target than practising the activity?

There are difficulties embedded in the assumption that a goal is best achieved by a kind of practice that exemplifies all the elements of that goal. Kamikaze pilots (as an extreme case) would have very little opportunity for practice under such an assumption. In the case of problem-solving, there are conceivably numerous associated competencies, feelings and skills that require attention that do not bear directly on the act of problem-solving *per se*. Activities akin to those which inculcate patriotism for kamikaze pilots (such as training in perseverance) for example, might be called for.

Though interesting for the pedagogical possibilities that come to mind in actual 'training' of problem solvers an even greater number of options are suggested

Figure 14.1: The Standard Problem-solving Paradigm

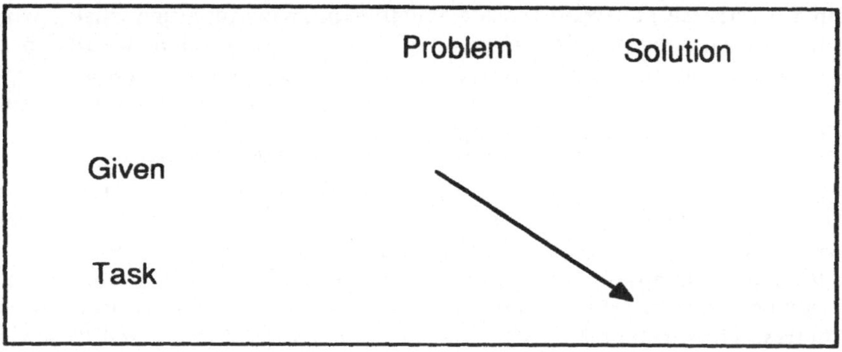

if we focus on the intellectual activity of problem-solving that is overlooked in the simple scheme above. As a start let us accept the concept of problem, as a departure point but challenge the view that it must be 'given'. We shall in this section eventually reverse the order of acceptance and challenge.

Where does this view of the problem as 'given' come from? In a narrow sense all mathematical proofs begin with 'the given'. Mathematical arguments *per se* are reducible to the form 'p implies q', and efforts to demonstrate that form then begin with the *assumption* that p is the case, i.e., p is assumed as true or p is given.[2]

That the activity of proving (or disproving) conjectures in mathematics has a particular form says a great deal about what is published in professional journals, but it says very little about how a problem evolves in a discipline. Though a focus upon the heuristics of problem-solving tells us something about the process of trying to solve a problem once it is acknowledged as a problem, it says nothing about the process involved in deciding what problems one ought to try to solve in the first place.

It is perhaps a useful insight to observe that in any discipline there are potentially an infinite number of problems from which one might make selections. Of that number some are selected as worth pursuing and others are deemed as inappropriate for any number of reasons (e.g., perceived as too hard, too simple, outside of the confines of the discipline, unenlightening). The centrality of the problems or the questions one asks in a particular discipline is well captured by Toulmin (1977) who argues that it is *not* content *per se* but rather the questions asked that separate fields. He comments:

> If we mark off sciences from one another . . . by their respective domains, even these domains have to be identified not by the types of objects with which they deal but rather by the questions which arise about them . . . Any particular type of object will fall in the domain of [say] biochemistry only insofar as it is a topic for correspondingly biochemical questions. (Toulmin, 1977, p. 149)

What Toulmin says about disciplines can be said as well about subspecialties within a discipline and even about a chunk of knowledge within a subspecialty.[3]

Instead then, of only giving people problems to solve it would seem to be appropriate (especially if one takes what Toulmin suggests about the nature of disciplines seriously) to spend considerable time in trying to decide what is worth solving and what the benefits might be of doing so. How might we re-conceptualize the relationship between 'the given' and 'the task' so as to suggest strategies for including even the neophyte in something akin to a kind of aesthetic exploration that is *behind* problem-solving?

Brown and Keren (1972) and Silver and Smith (1980) suggest ways of comparing problems in terms of their deep and surface similarities in order to gain some clarity on the essential similarity of problems that may appear to be different as well as those that are different although they may appear to be the same.

It would seem appropriate then for us to consider the possibility that students not only be given the problem but that they *choose* to explore some problems and not others. There is a great deal in the way of curriculum that could be devised to enable the student not only to choose but to establish criteria for selection among competing problems. Such criteria as the following might be considered in a student's decision to pursue a small number of problems from a larger potential population:

- anticipated difficulty of solution;
- relationship to problems already understood or solved;
- potential for the problem to open new territory;
- embeddedness within a particular branch of mathematics;
- potential for the problem to clarify on what is not well understood; and
- similarity of problem(s) to those already defined in the field (*à la* Toulmin).

Of course students would differ among themselves both in their selection of problems and in their selection of relevant categories. The important educational task would be not only to select categories but to discuss why particular categories are invoked by particular students and how their choices related to those that have been selected in the history of the discipline.

It is important to appreciate that the several 'choice' categories suggested above do not have an implied directionality. For example some people in considering anticipated difficulty of solution might be prone to opt for problems that are perceived to be harder rather than easier or similar to already solved problems. Some people might select problems that muddify rather than clarify what is already understood. It is this implied freedom of choice which most dramatically illuminates a kind of self-understanding that is denied by Polya-type problem-solving strategies — heuristics which would guide us to select what is relatively simple, already known and in general, more comfortable.

Though it is perhaps possible to conceive of students choosing from an array of already created problems and participating in a dialogue about their choice, it is also possible for even relatively 'naive' students to participate in the reversal of what is depicted in Figure 14.1 — to in fact *create* problems. That is, the solution to a problem (regardless of whether it is the student or the teacher who provides such a solution) can become the inspiration for posing new problems. This phenomenon, depicted in Figure 14.2, has more curriculum potential than is generally acknowledged.

Figure 14.2: A Reversal of the Standard Problem-solving Paradigm

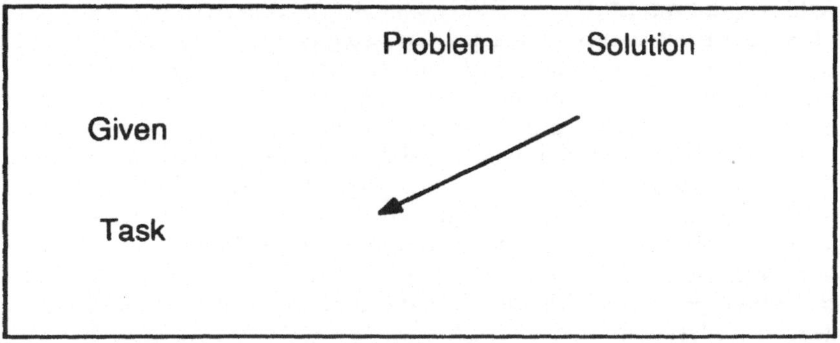

Consider the following problem described in detail in Brown and Walter (1990):

> Given two equilateral triangles of length a and b respectively. Find the length c, of a third equilateral triangle whose area equals the sum of the other two. (Brown and Walter, 1990, p. 105)

It turns out that a solution to this problem is very peculiar. We find out that c must be such that $c^2 = a^2 + b^2$. But armed with the realization that a form of the Pythagorean theorem has emerged in an unexpected place, that solution (perhaps even given to students) beckons the creation of new problems, An example of a new problem might be:

> I know that for a right triangle, the square on the hypotenuse equals in area the sum of the squares on the other two sides, Why does the problem with equilateral triangles reduce itself to the same kind of result? What's going on here?

This impulse to pose a new problem based upon a solution of a previous one is not an impulse that is acquired automatically. It is one that must be nurtured. What is particularly interesting, however, is that the sites from which such inspiration might be developed are much less precious than one might imagine. They abound as soon as one who is knowledgeable (e.g., a teacher) becomes aware of the fact that everything we teach must have been surprising to those who uncovered the ideas earlier in their evolution. To re-create the sense of surprise, those who arrange for the unfolding of curriculum need to appreciate the near truism that what is generalizable unites what was previously perceived to be disparate. Teaching which is slow to point out commonalties among disparate experiences that have a unification allows the students to be surprised by these commonalties and to search for the unification through the posing of problems.

In the above example, a curriculum which explicitly attempted to teach that the Pythagorean relation was not really a statement about squares on right triangles, but one about the relationship among any three similar figures on the sides of such a triangle, would rob the student of the impulse to pose new problems (many of which might be 'off base') that might lead to the generalization as something quite special.

It should be clear that what I am discussing is not 'discovery learning', nor trying to justify a curriculum which moves temporally to generalizations only after establishing many examples. It may very well be the case that something with a prematurely high degree of abstraction has the potential to generate surprise because of its misfit with what is expected. I am essentially addressing questions about the nature of mind rather than narrowly construed pedagogical ones. In asking about the role of surprise, I am wondering what kind of view of inquiry we wish to inculcate rather than how we might efficiently teach any sequence of subject matter. See Brown (1971, 1973, 1991), Cooney *et al.* (1994) and Movshovits-Hadar (1988) for further elaboration of the concept of surprise and its place in mathematics education.

How else might we suppress 'problem' as a first step in inquiry even when problem-solving is viewed as a worthwhile activity and goal? A significant site can be found in 'situations'. Given a triangle or a billiard ball table, or a bicycle wheel, or the layout of streets in some pattern, there is, strictly speaking, no problem stated at all. That situations can yield problems and that students may be taught to generate problems from situations is part of the story that relates problems to situations.

A particularly powerful 'situation' site is a definition. That is, a definition is obviously not itself a problem (though in the history of ideas definitions are frequently singled out only after an array of problems indicate the saliency of a particular theme). Take for example the definition that a number is prime if it has exactly two different factors.

That definition is a challenge (as are all definitions) to figure out why it is so special. One reason is that, based upon years of research, one knows that there is a lot to say about what flows from the concept of prime. Standard curriculum defines the concept and then shows off all that is known and unknown (unsolved) about prime numbers. As significant, however, for one who is beginning to understand the power of the concept is how prime numbers relate not to what flows from singling them out, but to 'near relatives'. So, a definition of 'prime' inspires a number of challenges such as:

- What if a number has only one factor? three? an odd number? an even number of factors?
- What if we focus not on 'factor' but 'addends' of a number? Are there some numbers that have exactly two 'addends'?
- What if the number system is not the set of natural numbers, but the set of integers or rational numbers? Are there prime numbers in that system? (See Brown, 1978, for elaboration of these questions in the context of 'almost' even numbers.)

A careful development of inquiry based upon such 'challenges' has been deemed the 'What If Not' Scheme. Brown and Walter (1969, 1970) developed this scheme as a significant and essentially neglected aspect of problem-posing a quarter of a century ago. There they suggest how a robust application of the scheme incorporates five or six levels of inquiry that are captured roughly by 'challenging the given' in a way described above. There of course are many other problem-posing strategies that are more 'accepting' than 'challenging' what is given in a situation, some of which are summarized and illustrated with many examples in Brown and Walter (1990, 1993).[4]

The relationship between 'situation' and 'problem' moves in the other direction as well. That is, we might ask how a problem might be used not a directive to 'solve' as in Figure 14.1, but might be 'neutralized' to create a situation. This task is suggested by an interesting phenomenon in the history of mathematics. Euclid's parallel postulate states, 'Through a given outside point, there is exactly one line parallel to a given line.'

For hundreds of years, mathematicians believed that this postulate was so much more complicated than the other incidence postulates about lines and points in the plane that it ought to be capable of proof. The 'problem' that persisted for hundreds of years was: How can one prove the parallel postulate from the others?

It took the genius of people such as Reimann, Lobachevsky and Bolyai to reveal that the difficulty in solving the above problem was that it was not solvable at all as stated. A major difficulty in solving it was that the 'how' at the beginning of the sentence represented excess baggage. It was the creation of non-Euclidean geometry which established that the parallel postulate was in fact independent of the others — thus a *bona fide* postulate.

One way of perceiving the historically interesting dialogue generated by efforts to prove the postulate from others is that mathematicians saw a problem so intimately embedded in a situation that they were incapable of 'neutralizing' it. What we see as a problem in a situation, and what it takes for us to neutralize a problem is something that has the potential to illuminate a great deal for us as individuals as well as for us as a culture. Forty years ago, the fact that women and non-white males were paid less than white males for comparable work was viewed more as a situation than as a problem — at least by white males. Our ability to 'see' problems in a situation is very much dependent upon the kinds of paradigms we hold with regard to a situation (writ 'theory' or 'political/social philosophy'). It would seem to be a valuable educational goal for us to explore in all disciplines how it is that problem and situation reverberate to let us know something about our commonalties and differences as individuals and about our assumptions as a culture as well.

We schematically summarize in Figure 14.3 the several alternatives (and surely there are many more) to the dominant scheme described in Figure 14.1. In doing so, it is worth keeping in mind that what we have been exploring is more a conceptualization of the educational breadth of problem and problem-solving than its strictly pedagogical dimensions.

The Fallacy of Scientific Context

The models of problem-solving in mathematics education are ones that derive primarily from the discipline of mathematics. The heuristics that we discussed earlier that view problem-solving as essentially impersonal (despite a renewed interest in social context) and scientific would have us understand the problem, create a plan for solving it, carry out the plan and so forth. Now while this kind of activity reflects a view of rationality, there are reasonable alternative views of

Figure 14.3: A Summary of Some Non-standard 'Problem' Paradigms

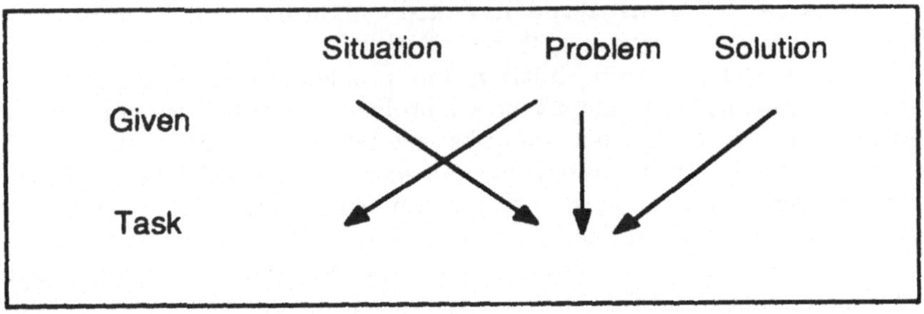

rationality which are left out. In particular such a model neglects to take into consideration both problematic aspects of the nature of mathematical thought and a realization that matters of education *per se* have a humanistic dimension that are not captured by a concept of problem-solving as scientific. Thus in viewing problem-solving as a component of mathematics education, the field is unfortunately driven by a limited view of the former (mathematics) rather than a robust view of the latter (education).

The very conception of the relationship of a problem to its solution establishes what will count for acceptable exploration. Sarason (1978) describes the relationship between a problem and its solution when viewed scientifically as follows:

> A problem has been solved (a) when it does not have to be solved again because the operation that led to the solution can be demonstrated to be independent of who performs them (b) when the solution is an answer to a question or set of related questions and (c) when there is no longer any doubt that the answer is the correct one. (Sarason, 1978, p. 374)

Indeed it is this conception of problem-solving that cuts through all the differing programmes for problem-solving in mathematics education. What is left out by a programme of this sort? When one either produces or views a mathematical solution to a problem there are all kinds of questions one might ask that would be considered irrelevant to the context of an essentially scientific paradigm. Consider questions like the following:

1. Of the many different solutions that have been offered for this problem which do you find the most attractive?
2. How does your view of what is attractive in problem-solving today compare with what it was five years ago?
3. What do you find out about yourself as you listen to your response about the concept of attractiveness?

The view of problem-solving as scientific not only controls the kinds of pedagogical questions and tasks one is encouraged to engage in, but also directs the kinds of mathematical problems to be explored. Though problems may be difficult and non-routine there is the expectation that with appropriate background and motivation they will eventually (an hour, a week, a month or two) be solved.

This point of view is in sharp contrast with much of the humanities. It would seem ludicrous for example to have students examine the nature of loneliness or the meaning of life in the context of death with the intention of *solving* the problem tomorrow or perhaps next week. The concept of solvability in the sciences differs considerably from that in the humanities as Sarason's comment implies.

Ironically enough we can locate the roots of a more humanistic conception of solution even within the domain of mathematics. Beginning with the philosophical work of Gödel a half century ago, our view of the discipline as firm and embodying our best hopes for the glory of the 'work ethic' was shattered. Using deduction Gödel demonstrated some essential limitations of deductive thinking in mathematics. (Popularized descriptions of Gödel's work can be found in Nagel and Newman, 1958 and in Hofstadter, 1979.) Up until 1940 there was embedded in the psyche of practising mathematicians the belief that any well-stated problem could be solved eventually if one were only intelligent and diligent enough. In 1940 Gödel sounded a kind of death knell for such a view of mathematics. He demonstrated that in any 'interesting' mathematical system (meaning at least as complicated as arithmetic) there had to exist truths that could *not* be demonstrated as such.

So far so threatening, but things get worse. It might be frustrating to know that the bounds of our inquiry are circumscribed but it might not be quite so devastating if those bounds were knowable. How can we identify mathematical statements that exhibit the property of being true but unprovable? The answer is that we cannot. If one works a lifetime on a problem and reaches no solution, there is no way of determining whether the difficulty is a consequence of our human 'frailty' or a result of the problem's essential unprovability (in the Gödel sense).

If the implications of Gödel's findings for the mathematics community have yet to be fully appreciated, there are surely problems in drawing pedagogical conclusions regarding problem-solving. Nevertheless there are directions one might pursue that allow mathematics to be more closely aligned with the humanities than has heretofore been the case. One educational fallout of Gödel's work might be that we push the bounds of taking process seriously by honoring incomplete or partial solutions, approximations, use of methods that combine mathematical and scientific thinking, probabilistic solutions and the like, some of which have been suggested by educators for reasons unrelated to Gödel. There are other possible directions, however, that have yet to be worked through. One might not only wish to implicitly apply possible consequences of Gödel's meta-mathematics, but one could also *refer* directly to his work in teaching mathematics, indicating where it came from, some essential elements (teaching such powerful concepts as that of self-referentiality) of his argument and exploring with students its possible implications both personally and from the perspective of the discipline.

A middle ground between applying pedagogical consequences of Gödel's meta-mathematics and actually teaching some form of his work, might be to raise in a hypothetical mode the essential philosophical issues that generated and surrounded his findings. Questions like the following are suggested:

- Do you think all questions are answerable?
- What do you think it means for a question to be answerable?
- What does it mean for a question to be well (poorly) formulated?
- What does it mean to say that something is *proved*?
- What does it mean to say that something is *provable*?
- Do you think all well-formulated statements should be provable?
- How would you feel if you found out that not all well-formulated statements were provable?
- If not all well-formulated statements in mathematics were provable, how would the concept of proof in mathematics affect your belief in the provability of proof in other areas of your life?

Such a programme would considerably expand not only what counts for problem-solving but would also transform what is involved in 'applying mathematical thinking to the real world' so that such application would point a mirror not only towards the discipline but towards the student in an effort to understand self as an integral part of the real world. In taking such philosophical issues seriously, students would be led to re-examine their most fundamental assumptions regarding the values they attach to thinking and the sense in which certainty, clarity and personal detachment function in thinking about any aspect of human experience. Mathematical thinking and problem-solving then become not an isolated self-contained vehicle (with occasional safe forays in applications to the sciences) for conveying a distorted conception of knowledge, but rather beckon us to entertain the senses in which all knowledge and exploration are both personal and problematic.

Two Grandparents of Problem-solving Fallacies and Curriculum

There are two grandparent fallacies that seem to contribute to much of what we have discussed above. One (let us call it a grandpa fallacy) relies upon a confusion of conception of problem with educational implications; the other (a grandma) depends upon a confusion of thinking and problem-solving. First to the grandpa fallacy:

A Grandpa Fallacy

Though the literature in mathematics education tends not to confront directly what it means for something to be a *problem*, there is a mischievous linkage between problem and problem-solving that is consistent with much of what is written about the concept of problem in philosophy of logic and philosophy of science. Furthermore, though the most recent international documents suggest a 'softer' view of pedagogy in the teaching of problem-solving, they all fall prey to this fallacy. Consider the following conceptions of problem and question (a near cousin of problem in the philosophical literature) by Cohen (1929), Schlick (1935), Nickles (1981) respectively:

> As a logical entity, the question is the clear embodiment of the characters by which the [associated] propositional function has been defined. (Cohen, 1929, p. 354)

> There are many questions which are empirically impossible to answer, but not a single real question for which it would be logically impossible to find a solution. (Schlick, 1935, p. 25)

> What [is a . . . problem]? My short answer is that a problem consists of *all* the conditions or constraints on the problem plus the demand that the solution (an object satisfying the constraints) be found. (Nickles, 1981, p. 109)

Now these quotes are not all of the same era. They span a half a century. But it is important to appreciate that not all of these people were hard-nosed logical positivists. It turns out, for example that Nickles, known as a 'friend of discovery' (meaning that he is interested in preserving problem creation as an entity for philosophical inquiry rather than as something to relegate to psychological inquiry alone) was really trying to do something a bit extraordinary with his definition of problem. That is, he is (among other things) challenging the view of Popper that components of a problem such as its history are not part of the problem itself. Popper wanted to consider the history of a problem to be part of the background of the problem rather than the problem itself.

So, Nickles is expanding a conception of problem in some potentially interesting ways for educators. But . . . and here is the rub: By so closely connecting a problem with its constraints, *Nickles is essentially defining a problem in terms of its solution.* That is, if the concept of solution did not exist, then the concept of problem would have no meaning. In its most extreme form, what this requires is that for something to be problem, it has to be posed in such a way that its solution is built-in (exactly what that means differs from among philosophers).

Now what precisely is the grandpa educational fallacy? It is to observe that even if we buy into this hard-nosed conception of problem in relation to solution, we are not obligated to treat problems only (or even primarily) in terms of their solution. That is, the conceptual connection between problem and solution does not dictate how a problem may be used for educational purposes. That is, to use Nickles' conception of problem, even if it follows that a problem has a demand built in that its solution be sought in order for it logically to be a problem, it does not follow that one is obligated to seek a solution when one is presented with or creates a problem.

Once we break this connection, we are in a position to think much more imaginatively about the place of problems in educational contexts. Once we have a problem, it takes on a life of its own. We can ask questions like:

- Who created that problem and why?
- What is assumed implicitly about the background of the field and of the person when they pose the problem that way?
- In what form if any might this problem have been posed 100 years ago? 1000 years ago?

One can come up with better questions than these, but my point is that it is by thinking seriously about the disassociation of problem from solution as an educational phenomenon that we may be able to come up with alternatives for the uses of problems that we have not yet imagined. Furthermore, there is some interesting irony here. That is, if we read some of the writings of philosophers who are most intent upon coming up with a definition of problem in relation to solution, we may very well be able to adapt categories they introduce for our educational purposes that do not draw those connections so tightly. The example of Nickles and the concept of the history of a problem is a case in point. Regardless of how tightly we want to build the history of a problem into the very meaning of a problem from a philosophical point perspective, there is much we can do from an educational point of view by making efforts to connect the two.

There are of course other places we can look in order to acquire a more personal and robust notion of problem regardless of its connection with the concept of solution, but it is consoling that the most hard-nosed of analysts can provide us with machinery that has the potential to liberate us from their tight grip as well.

Earlier we quoted the work of Sarason, a psychologist, in trying to understand what is wrong with problem-solving models as they are used in social action. A point of view that is consistent with his, but that is considerably more moving and that raises interesting options is the following by someone in the humanities, Richard Wertime (1979):

> A problem is . . . a project for the future we commit ourselves to by an act of will. This means by implication that a problem entails risks, since all future projects — to use Hannah Arendt's term — involve uncertainty. There is . . . an essentially . . . promissory dimension to the act of facing a problem. A problem is not an entity which has existence independent of a person . . . They involve a significant act of self-surrender which can seriously jeopardize the individual's sense of self. (Wertime, 1979, pp. 192–3)

A Grandma Fallacy

Ironically enough, another fallacy that may in fact contribute to the first two world views we criticized earlier (the fallacy of the 'problem as given' and the fallacy of 'scientific context'), may in fact be attributable to Dewey and some of the other pragmatists such as Peirce and William James. Dewey has modelled most of his view of inquiry upon the scientific model and it is not surprising that there is an essentially isomorphic relationship between Polya's and Dewey's models of thinking. Driven by a desire for inquiry to be genuine and not *a priori* or predetermined as was the case with much a great deal of nineteenth-century philosophy, Dewey was determined to debunk traditional dualities that had persisted for generations before him. He saw no sharp divide between thinking and action, mind and body, knowing and doing, moral and scientific ways of thinking.

Given such an orientation, it was a small step for Dewey to assimilate all thinking into a problem-solving model. In *How We Think*, Dewey (1910) comments,

> The two units of every unit of thinking are a perplexed, or confused situation at the beginning and a cleared-up, unified, resolved situation at the close. The first of these situations may be called pre-reflective. It sets the problem to be solved . . . In the final situation the doubt has been dispelled; the situation is post reflective. (Dewey, 1910, p. 106)

But even if one uses scientific thinking as the model of thinking in all domains, this does not reduce all thought to a problem-solving mode of operating, as if one's primary concern is that of meeting and changing some 'real world' problem. Even science is not in a constant state of problem-solving in that sense. Scheffler (1973) makes the point well when he says,

> [Dewey's] conception of the nature of empirical control is . . . unduly narrow, and fails to do justice to abstract, theoretical considerations in the scientific assessment of evidence. Ideas in science are not all of one kind and only certain simple types can be analyzed as instruments for transforming the world. (Scheffler, 1973, p. 154)

His criticism of the problem-solving mode as the only mode within which thinking takes place is that much more poignant when applied outside the realm of science. In the same work, he asks:

> But what reason do we have for assimilating all reflective thinking to the problem-solving model in general? In ordinary speech, for example, the poet is thinking in the process of composition, the artist in creation, the translator in attempting a translation, and yet none is seeking the answer to a question. Though subsidiary questions need to be answered in the course of each activity, no *answer* or *set of answers* as such brings each activity to a unified and resolved close.; only a satisfactory poem, painting, or translation will do. (*ibid.*, p. 157)

As educators, we need to ask ourselves what some aspects of genuine thinking may be that we wish to encourage among our students that are not in and of themselves problem-solving in either a narrow or a broad sense. To a large extent, it is that question that this chapter has attempted to address. The list is obviously incomplete, and I look forward to further exploration by my colleagues.

Conclusion

The arena of problem-solving illustrates how much educational mischief can result when one acts as if the logical analysis of a concept dictates its educational application. Not only in *solving* problems, but in *choosing*, *creating* and neutralizing problems one has the potential to find out a great deal about one's own world view and even about the explicit and implicit assumptions of a smaller community (of mathematicians or of students of mathematics) and of a society in general. In what sense does one search for simplification? complexification? In what sense does one reach for better understanding by attempts at clarifying, 'muddifying'? In what ways should we re-frame these same kinds of questions if our interest is in understanding society at large? What kinds of situations am I incapable of seeing as problems? What does that tell me about myself, about my contemporaries?

There is nothing about the standard body of knowledge of mathematics nor about the concept of 'problem' nor about the activity of problem-solving within mathematics that particularly encourages the asking of these kinds of questions. On the other hand, none of these questions are incompatible with such terrain. They are, however, the kinds of questions which one is tempted to ask in conjunction with inquiry about a discipline if one is inclined to take seriously issues of mind and of personhood that outstrip the domain within which the discipline is normally defined.

Notes

1. Some of the discussion of the fallacy of the problem as given and the fallacy of scientific context in this piece elaborates upon an earlier essay of mine which focused on the question of how the concept of a liberal education applies to mathematics. See 'Liberal education and problem solving: some curriculum fallacies' in Brown, 1986.
2. That all *statements* are reducible to the form 'p implies q', does not of course suggest that all *proofs per se* must begin with p in the argument of the proof.
3. Such an observation of course needs to be tempered by the realization that there are occasions upon which early proofs of elusive mathematical conjectures are found in fields that appear not to have anything to do with the conjecture under question. The production of two different proofs of the famous prime number theorem (how many primes are there less than, or equal to, any natural number?) in 1896 by Hadamard and Valée-Poussin that required the domain of complex numbers for their analyses is a case in point. For further discussion of this particular problem, see *Some 'Prime' Comparisons* in Brown, 1978. For a discussion of the more general phenomenon, see Gian-Carlo Rota's essay, 1993.
4. Questions related to the logic of problem-solving in relation to problem-solving in particular can be found in Brown (1984) and Brown and Walter (1990). Issues related to the social/political significance as well as philosophical implications are discussed in Brown and Walter (1988), Ernest (1991) and Freire (1970).

References

Brown, S.I. (1971) 'Rationality, irrationality and surprise', *Mathematics Teaching*, 55, pp. 13–19.

Brown, S.I. (1973) 'Mathematics and humanistic themes: Sum considerations', *Educational Theory*, 23, 3, pp. 191–214.

Brown, S.I. (1978) *Some 'prime' comparisons*, Reston, Virginia, National Council of Teachers of Mathematics.

Brown, S.I. (1984) 'The logic of problem generation: From morality and solving to deposing and rebellion', *For the Learning of Mathematics*, 4, 1, pp. 9–20.

Brown, S.I. (1986) 'Liberal education and problem solving: some curriculum fallacies', *Proceedings of the Philosophy of Education Society*, pp. 299–311.

Brown, S.I. (1987) *Student Generations*, Arlington, Massachusetts, Consortium On Mathematics and Its Applications (COMAP).

Brown, S.I. (1991) 'Yet one more revolution in school mathematics', *Newsletter of the Graduate School of Education*, University of Buffalo, Spring, 1, 3, 4.

BROWN, S.I. and KEREN, G. (1972) 'On problems in *gestalt* psychology and traditional logic: A new role for analysis in the doing of mathematics', *Journal of Association of Teachers of Mathematics of New England*, Fall, pp. 5–13.
BROWN, S.I. and WALTER, M.I (1969) 'What if not?', *Mathematics Teaching*, 46, pp. 38–45.
BROWN, S.I. and WALTER, M.I (1970) 'What if not?: An elaboration and second illustration', *Mathematics Teaching*, 51, pp. 9–17.
BROWN, S.I. and WALTER, M.I. (1988) 'Problem posing in mathematics education', *Questioning Exchange*, 2, 2, pp. 123–31.
BROWN, S.I. and WALTER, M.I. (1990) *The art of problem posing* (second edition), Hillsdale, Lawrence Erlbaum.
BROWN, S.I. and WALTER, M.I. (1993) (Eds) *Problem Posing: Elaborations and Applications*, Hillsboro, New Jersey, Lawrence Erlbaum.
COHEN, F. (1929) 'What is a question?', *The Monist*, 39, pp. 350–64.
COONEY, T.J., BROWN, S.I., DOSSEY, J., SCHRAGE, G. and WITTMANN, E. (1994) *Integrating Mathematics and Pedagogy*, Portsmouth, New Hampshire, Heinemann.
DEWEY, J. (1910) *How We Think*, New York, Heath.
ERNEST, P. (1991) *The Philosophy of Mathematics Education*, The Falmer Press, London.
FREIRE, P. (1970) *Pedagogy of the Oppressed*, New York, The Seabury Press.
HOFSTADTER, D. (1979) *Gödel, Esher, Bach*, Boston, Birkhauser.
MOVSHOVITS-HADAR, N. (1988) 'School mathematics theorems: An endless source of surprise', *For the Learning of Mathematics*, 8, 3, pp. 34–39
NAGEL, E. and NEWMAN, J. (1958) *Gödel's Proof*, New York, New York University Press.
NATIONAL COUNCIL OF TEACHERS OF MATHEMATICS (1980) *An Agenda for Action: Recommendations for School Mathematics for the 1980s*, Reston, Virginia Author.
NATIONAL COUNCIL OF TEACHERS OF MATHEMATICS (1989) *Curriculum Evaluation Standards for School Mathematics*, Reston, Virginia, Author.
NICKLES, T. (1981) 'What is a problem that we may solve it?', *Synthèse*, 47, pp. 85–115.
POLYA, G. (1954) *Mathematics and Plausible Reasoning: Vols. 1 & 2*, Princeton, New Jersey, Princeton University Press.
POLYA, G. (1957) *How to Solve It*, New York, Anchor Books.
POLYA, G. (1962) *Mathematical Discovery: On Understanding, Learning and Teaching Problem Solving*, 1, 2., New York, John Wiley and Sons.
ROTA, GIAN-CARLO (1993) 'The concept of mathematical truth', *Essays in Humanistic Mathematics*, Alvin White (Ed), Washington, DC, Mathematical Association of America.
SARASON, S. (1978) 'The nature of problem solving in social actions', *American Psychologist*, 33, 4.
SCHEFFLER, I. (1973) *Reason and Teaching*, London, Bobbs-Merrill Company.
SCHEFFLER, I. (1974) *Four Pragmatists*, New York, Humanities Press.
SCHLICK, M. (1935) 'Unanswerable questions?', *The Philosopher*, 13, reprinted in MICHAEL MEYER (Ed) (1988) *Questions and questioning*, Berlin, Walter de Gruyter.
SILVER, E.A. and SMITH, J.P. (1980) 'Think of a related problem', in KRULIK, S. and REYS, R.E. (Eds) *Problem Solving in School Mathematics: 1980 Yearbook of the National Council of Teachers of Mathematics*, Reston, Virginia, National Council of Teachers of Mathematics, pp. 146–55.
TOULMIN, S. (1977) *Human Understanding*, Princeton, Princeton University Press.
WERTIME, R. (1979) 'Students, problems and courage', in LOCHHEAD, J. and CLEMENTS, J., *Cognitive Process Instruction*, Philadelphia, Pennsylvania, Franklin Institute Press.

Chapter 15

Enquiry in Mathematics and in Mathematics Education

John Mason

Introduction

Enquiry in mathematics appears at first to be unproblematic: the mathematician explores ideas, develops notation, defines terms, and proves theorems. Of course, when examined more closely, it becomes rather more difficult to describe what mathematicians do in a way which they acknowledge and recognize (Polya, 1962; Tall, 1980; Davis and Hersh, 1981; Kitcher, 1983; Mason *et al.*, 1984). Mathematicians build on each other's work, using results and techniques, and following up variations and generalizations. Tracing references through mathematical reviews, citation indices, and conferences, provides a skeletal picture of the development of ideas and notation.

Enquiry in mathematics education also appears at first to be unproblematic: the educator studies difficulties that pupils and teachers experience, proposes alternative approaches, and validates those approaches. Of course, when examined more closely, it also becomes rather more difficult to describe what researchers do in a way which they acknowledge and recognize. Educational researchers tend to augment and vary the frameworks, instruments, and approaches of colleagues, rather than building on their results. They are more likely to refer to others in passing, to legitimate their assertions through reference to others with similar thoughts, than they are to develop from, and extend, each other's work. Tracing references through citation indices and conferences reveals more about what the author has been reading recently than anything else.

The audience for mathematicians is other mathematicians, and scientists in cognate disciplines. The audience for educational researchers is teachers, and other researchers. Whereas mathematicians are unified by common technical terms, educators are divided by the disparate use of loosely defined words which are multiply interpretable. Mathematical conclusions include statements of the conditions under which they are valid, whereas educational assertions are highly context-dependent and the domain of applicability hard to state: changes in teacher awareness and sensitivity changes the phenomenon itself. For example, as attitudes and approaches vary, the nature of what is perceived as problematic alters. Whereas mathematicians assert what *is* the case, and others can then employ these to justify yet further assertions, educators assert what *was* the case or theorize about what might be the case, leaving others to integrate the analysis, rather than the content, into their own practice.

Educational research is deeply self-referent, unlike mathematics. Remarks made about educational writing apply also to that writing, whereas mathematics is rarely (except in set-theoretical paradoxes and Gödel's theorem), self-referent. Mathematics education includes not only reflection upon and analysis of learning and teaching, but also upon the conduct of researching itself, while mathematics is concerned with mathematics, not with the conduct of mathematics.

So much is obvious. I shall argue that there is more in common, at least potentially, than there is different, between enquiry in mathematics and enquiry in mathematics education, and that it is to the advantage of mathematics education to maintain and develop that link.

Finding a Place to Stand

In mathematics, it is possible to state or refer to the axioms one is accepting and to build an edifice of deduction. It is this aspect of apparent certainty and lack of freedom which has attracted philosophers to use mathematics as a paradigmatic example for epistemology, and which continues to attract some adolescents. 'In mathematics, you know when the answer is correct.' The same quality may also repel other adolescents, because logical deduction is seen to conflict with the development of free-flowing intuition.

In mathematics education, it is much harder to find any firm ground on which to stand. As soon as you think about why certain students find a particular topic difficult, you are led through a web of influences: how the topic is taught, the attitudes of the text-authors and the teachers to learning and doing mathematics, relevance of the topic to the interests and concerns of students, political and economic conditions in the community, and so on. Each of these is problematic, and interconnected with the others.

Much has been written about philosophies of mathematics, and considerable effort has been put into finding connections between the overtly expressed philosophies of (and attitudes to) mathematics among trainee and practising teachers and the success, confidence and attitudes of their students. Contrasts have been noted between what teachers say they do, and what they are observed to do (Lerman, 1986; Ernest, 1989; Thompson, 1991); or rather, contrasts emerge between what researchers hear teachers say, and what researchers say they see teachers do. In each case, the researcher's perspective, values, and interests, play an important role in influencing 'what is observed'. The phenomena of education, the 'objects of study', seem much less well-defined than in mathematics. This may be one reason why those who feel secure in mathematics often look down on those who inhabit more amorphous and ambiguous worlds such as the world of education. Yet even in mathematics there are crises in what constitutes proof, and in how firm the foundations really are.

Individuals search for a perspective that grounds their educational discussion firmly, and enables progress, but rarely is there much commonality between different schools of research, much less between individual researchers. What is for one, definite and 'well-defined', is for another problematic. One effect is that researchers constantly look to other fields for a fresh perspective on what seem to be intractable problems. For example, constructivist perspectives in education draw on psychology to emphasize the role of the individual in making sense and constructing the world in which they operate; the social constructivist movement draws on sociology to emphasize the role of enculturation and peer influence in how knowledge is constructed. How are these to be reconciled? Must one dominate the other? (von Glasersfeld, 1984; Mellin-Olsen, 1987; Vygotsky, 1978; Lerman, 1993; Ernest, 1991, 1993).

The psychological-sociological constructivist debate can be seen as a manifestation of nature-nurture debates, of the desire to find somewhere stable to stand and view the world. Varela *et al.* (1991) stress that separation of organism and environment and attempts to apportion influence between them only makes sense in a Cartesian frame, founded on a separation of knower and known. When the gerund *knowing* is stressed rather different perspectives become available.

It is as if commentators are driven to find a single all-encompassing framework, a single viewing platform from which to observe and comment on the flora and fauna of the mathematics education pond. Fortunately, there is at the same time a growing realization that single perspectives are neither possible nor desirable.

In any discussion, it is necessary to find some place to stand. Certain words have to be taken as having shared meaning; certain relationships have to be taken as appreciated. And yet it is the fluidity of meaning, the shifting sands of interpretation as perspectives are modified, which makes the whole enterprise of educational enquiry at once so frustrating and so exciting. To progress, mathematics education needs a recognized means for negotiating shared meanings for technical terms.

Multiple Perspectives

Multiple perspectives is no new idea. Galileo, drawing upon Copernicus, contributed substantially to a shift from single to multiple perspectives: what you see depends on where you stand. Newtonian mechanics exploited this shift of basepoint, only to run into difficulty in explaining light. Physicists finally had to accept that light sometimes acts as a particle, and sometimes as a wave, and the task becomes trying to decide which perspective is most helpful in which contexts. The discovery of non-Euclidean geometries provided a mathematical stepping stone, showing that again there is more than one place to stand. Biologists have debated nature-nurture at length, with no firm conclusion, and even the once stable notion of evolution through survival of the fittest is in question as sole explanatory mechanism. Current discussions stimulated by a psychoanalytic perspective reveal a debate between generative and *post hoc* sources for metonymic connections. The inadvertent remark, the Freudian slip, may signal a deep-seated connection or may trigger new directions of thought. Are individuals responsible for their own constructions of their own reality, or are they (merely) the product of the culture? Does language form and limit, even constitute, thought, or is language but one form of externalizing thought? If all I can observe is behaviour, and all I can assess is behaviour, is teaching simply a matter of training behaviour?

In all these cases, a reasonable alternative to polarized debates is to grasp both poles, to argue that where you stand determines to some extent what you can see; that there can never be a universal platform, a single all-embracing, all-explaining perspective. Rather than deciding on one or another, it is usually most fruitful to grasp them both, to see both poles of a tension as releasing energy for deepening appreciation of the situation (Mason, 1986) rather than directing it into discussions about which pole is the more critical. An ancient Armenian proverb sums it succinctly: every stick has two ends.

Poles of Enquiry

The desire to stand apart, to study from the outside, may perhaps be a product of an Archimedean-Cartesian programme, which has reached its apotheosis in a western scientific-engineering tradition. As long as enquiries are not pursued too deeply (for example to the quantum level, or into personal equilibrium and harmony), western material science, based on tracing cause-and-effect chains employing logical deduction, is the source of much that is positive about present society. Campbell and Dawson (in press) use the image of 'an Archimedean point' as the lodestone for philosophers seeking firm, stable ground. They suggest that the entire Cartesian programme depends on a separation of the knower and the known, and that all other dualities which characterize western philosophies follow from it. Keller (1983, p. 190) has argued that this separation is intimately bound up with a Baconian metaphor of sexual relations in which males (mind) dominate servile females (nature).

By contrast, eastern tradition has focused on merging, on studying from the inside. It has provided access for many to personal fulfilment, harmony, and centredness. Varela *et al.* (1991) offer a western view of an eastern perspective based on mutual specification and codetermination (p. 198), in which the very existence of a unique self, so essential to cognitive science, is challenged. Even in drawing attention to a potential distinction, I betray western dichotomy-driven roots, for in distinguishing, I blur a long tradition of attempts to transcend duality (Husserl, Hegel, Peirce, Bateson, Gilligan, Varela among many others), and the deep roots of mathematical science to be found in Arab, Indian, Chinese, and Japanese cultures. To express is to overstress. To make a distinction is to stress difference and appear to ignore similarity. Temptation to decide between, to choose one perspective over another is falling into the single-perspective, simplicity, 'place-to-stand' position. It is not a matter of deciding between, but rather of drawing upon the strengths of both depending on the context.

To find out how pupils respond to certain stimuli, to chart progress and development, it is sensible, even necessary, to make distinctions, to characterize, to craft questionnaires and test items, to validate them on large populations, and then to employ them in longitudinal studies. To become more sensitive to pupils' needs, and to be able to respond in the moment, it is sensible, even necessary, to examine one's own experience rigorously, and to use that as a stepping stone to appreciating the experience of others. The former I call *outer* research, the latter, *inner* research.

Legacies of the outer pervade educational discourse, and can be found in the unchallenged 'primitive' terms that arise in most discussions of children learning mathematics. For example, the word 'understanding' comes from the outer approach, separating knower from known literally by the one 'standing under' the other. The fact that it is not easy to find a suitable replacement for 'understanding' is what characterizes it as an educational primitive. An alternative, such as 'centre-merging' could be used to mean something similar but from an alternative perspective, carrying (literally, in the metaphor) a sense of integration and oneness with what is known.

Mechanisms of Reasoning

Underlying any form of enquiry there are assumptions about how one reasons. In mathematics, the norm is deductive justification, following up intuitive insight and direct perception of patterns and forms. I associate logical mathematical deduction with cause-and-effect chains, since these form the heart of scientific reasoning about the material world. They might be called 'macro-physical and chemical reasoning', since at non-quantum levels these domains typically involve chains of logical deduction to explain phenomena.

But I have already indicated that education may not be best served by continuing to employ a solely cause-and-effect perspective. Making a little change (or a big one, as governments are now doing) may influence an educational situation, but rarely in predictable ways. It is very hard to justify an agricultural-treatment perspective in which a single change of treatment causes an observed effect, nor is it plausible that if all other factors could be held constant, that any observed effect would remain robust, since the many impinging factors are constantly adjusting to each other. In scientific enquiry, all factors are held as constant as is possible; in education, no factor remains stable when another is perturbed.

I suspect that the robustness of cause-and-effect thinking in education is due to the absence of a strong biological metaphor which supports and illustrates an alternative perspective of, in the words of Varela *et al.* (1991), mutual specification and codetermination. Our scientific metaphors all stem from small-scale phenomena obtained by Cartesian simplification through breaking things down and assuming that they are easily and sensibly glued back together, rather than from large-scale phenomena which permit us to retain complexity without feeling lost. Mutual interaction and adjustment is a large-scale phenomenon, so examples are likely to be complex ones, like the Earth's atmosphere, or a social grouping of animals or people, and although it can sometimes help to look at components, simplification can also miss the essence of complex phenomena. Campbell and Dawson (in press) offer the metaphor of a tree growing in which new layers are added to the old; the old determine the new, but the new also alter the old.

If deductive reasoning accompanies simplification, what accompanies complexity? In-dwelling in experience, allowing situations to speak, and working on refining one's own sensitivities, produce a programme which is rather different from dissecting phenomena which are assumed to be stable and repeated, but which may turn out to be constituted by perspective and not independent of observer. In-dwelling acknowledges mutual specification and codetermination.

Effective enquiry enables people to shift from 'talking-about' (using vocabulary in discourse but not in-dwelling what they are saying), to 'seeing through' (using the vocabulary as a means of describing what they experience). In mathematics, symbols become windows (Griffin and Mason, 1990) to a mathematical world of experience (Davis and Hersh, 1981), effective notation being that which enables the researcher to see through to essential patterns. In mathematics education, terms are more often walls than windows, blocking us from appreciating what others are saying.

Similarities

Having stressed differences between mathematical enquiry and mathematics-educational enquiry, I now want to stress potential and putative similarities, and to indicate ways in which mathematics education would be improved if it maintained close links with its parent discipline.

Processes of Enquiry

Mathematics makes great uses of the twin processes of specializing and generalizing (Polya, 1962; Mason *et al.*, 1984). Special and extreme cases are used to test conjectures, and more importantly, to enable the researcher to become thoroughly familiar with the phenomenon before being able to describe or express it generally, much less make any conjectures. Generalizations are sought, along with different representations, to try to find the 'right language', the 'right context' which will simplify the problem in hand. Mathematics abounds with examples of this process: linear algebra and change of basis, and representation of a geometrical configuration in different geometries are two of many examples. Lakatos (1976) suggested that definitions of terms usually emerge as a means for simplifying the statement of results and techniques, as opposed to being natural or God-given. Conjectures are justified by chains of reasoning which enable the reader to see, literally (*theorem* comes from the Greek meaning *seeing*), what the author saw, but only in retrospect, when the pieces fit together.

Mathematics education also makes use of specialization, but often in a rather more haphazard fashion. To study the general one must specialize, but re-generalizing is much more problematic. Research programmes often begin with very general aims, such as to study the effect of teacher attitudes on children's learning, but by the time the questions have been refined and made practical, the possibility of re-generalizing is severely reduced. Teachers and researchers love quoting particular incidents to colleagues in order to exemplify a point, either from their formal research or from their own children, and this was made into a discipline by Piaget and Freudenthal among others. By building up a collection of brief-but-vivid descriptions of incidents that others can recognize, and using these to sensitize themselves to further similar incidents, practitioners and researchers alike can improve their awareness, can extend what they are attuned to notice (Gattegno, 1970; Williams, 1989). By locating themselves in a constant, ongoing process of specializing, illustrating the use of technical terms by brief-but-vivid descriptions of incidents which speak to colleagues, teachers and researchers alike could improve the quality of communication, could ground discussion and provide stability in a manner much closer to mathematics than is presently the case.

Generalization in mathematics education is a major issue. How can you generalize from observations of a handful of pupils, even a handful of classes in a handful of schools? A generalization is useful only if it speaks to people's experience, providing a label around which numerous specific and abstracted experiences crystallize. The power of a generalization lies in the extent to which it helps people make sense of the past, helps them prepare for the future, or helps them act in freshly in the moment. If generalities are separated from the examples which they label than they lose their ability to trigger metonymic associations and so inform sensitivities in the moment by moment unfolding of teaching or researching. Thus generalization which does not resonate with special instances remains abstract; generalization arising from generic, immediately recognizable instances can be as powerful as mathematical generalizations based on similarly generic examples.

The distinction between naturalistic enquiry which values 'thick' descriptions of particular cases, and scientific enquiry which values the articulation of general laws and replicability (Ernest, 1991) appears to separate two independent research paradigms. And yet, every stick has two ends. Each has elements of the other embedded within its practices. The clear distinction between phenomenon studied and theory, which dominates scientific enquiry, is mirrored in educational enquiry when distinguishing between *accounts of* incidents, and explanations or justifications which *account for* them (Davis and Mason, 1989). Attending to one's own experience of participation in process, and looking out for potential analogies with student experience, which constitutes much educational research, are mirrored in scientific enquiry when researchers attend to the conduct of their research, and the changes that take place in their own thinking.

The scientific enquirer spends time reading other people's results, interpreting, internalizing, assimilating and accommodating, making sense from their own perspective. Although their practice is not an object of scientific research, their experience as a researcher corresponds to the objects studied naturalistically in education. Similarly, the naturalistic enquirer locates specific relevant distinctions, and then uses these to categorize and explain observations in ways which sometimes conform metaphorically with conduct of scientific enquiry.

Using the root meaning of *theorem* as *seeing*, mathematics and mathematics education can both be looked at as being about seeing:

- Mathematicians try to see pattern and structure in the material world and the mental world of forms and relationships, and try to convince others that they too can (and in some sense, must) see the same thing.
- Mathematics education researchers try to see pattern and structure in the socio-psychological behaviour of students and teachers, test whether such seeing informs future actions, and try to provide frameworks for enabling others to see them too.

The differences lie in the necessity of seeing, the ways one seeks validity for oneself, and the methods for convincing others.

Beyond Duality

In this section I propose ways of transcending Cartesian dualities and grasping the stick of enquiry. It is a commonplace (St. Maurice, 1991) for mathematics education to be conducted by reference to the triad of

- mathematical content;
- student; and
- teacher/author/tutor.

While acknowledging the socio-political-economic context of school, community, and government. But in such acknowledgment, attention still focuses on one or other aspect of the potential interactions. Content is analysed, and students and teachers are observed, recorded, and analysed, as objects. Tensions are considered between student and content (for example study of epistemological obstacles), between student and teacher (for example by analysing power relations, discipline, and control), and between teacher and content (What does the teacher know about mathematics, and how does that influence their actions?). Rather less attention is devoted to the actions which take place as a result of the triad itself.

Triadic Action

The triad as a critical element in the structure of the psyche is described in the *Baghavad Gita* (Mascaró, 1962) in terms of three *gunas*, variously interpreted as materiality, energy, and spirit, or passivity, activity, and presence. The *gunas* provide one root for the modern triad of enactive, cognitive, and affective aspects of the psyche. But there is more to structure than mere components. The *gunas* form three interwoven strands, and the interweaving is critical. Plato expresses the essence of the triad in the Timaeus:

> But two things cannot be rightly put together without a third; there must be some bond between them. And the fairest bond is that which makes the most complete fusion of itself and the things it combines, and proportion is best adapted to effect such a union. (Hamilton and Cairns, 1963, 31d, p. 1163)

Here mathematics is used as a paradigmatic experience for appreciating the more general and philosophical role of reconciliation or mediation. Any action is seen as requiring the presence of three interwoven impulses:

- the acted upon;
- the acting; and
- the reconciling or mediating.

It is the third impulse which is so often ignored in research, and which thereby impoverishes the exposition of findings. For example, exposition is commonly thought of as a teacher standing up and telling students things, in some sort of lecture format. Constructivist rhetoric superficially adopted enables researchers and teachers to justify their dissatisfaction with normal practice of exposition, despite the fact that a constructivist perspective has no judgments about exposition as a mode of interaction (Ernest, 1993). Many educators espousing constructivism have been known to attend lectures on constructivism, and even to have enjoyed them! The most that constructivism can say about lecturing is that it is unwise to assume that students know what the lecturer told them.

A triadic view of exposition observes that when preparing a lecture, it is the fact of the imminent audience which enables the lecturer to contact the content in fresh ways, in a state conducive to creativity and connection-finding. When the event itself happens, it is the presence of the audience which enables the lecturer to contact the ideas, as if drawing the audience into the world of experience of the lecturer. The lecturer acts upon the content, mediated by the student. Not all lectures have this quality of course. The action can be aborted by many different factors, from lack of preparation, to loss of contact with the audience, to failure to appreciate the nature of exposition.

Explanation is another form of student–teacher interaction in which the content is often ignored. Whereas in exposition the lecturer draws the audience into the lecturer's experience of the world of mathematics, in the action of explanation, the teacher enters the world of the student. The initiative lies with the student, and the content mediates, brings the teacher and student into contact. As soon as the teacher thinks 'Ah, now I see what the difficulty is,' explanation switches into exposition. The initiative moves from the student to the teacher. Further analysis of the six different modes of interaction which are possible can be found in Mason (1979), and the triad can be extended to larger structures (Mason, 1987; Bennett, 1956).

Action seen as the mutual operation of three forces is quite different from deductive reasoning, and prompts different methods of research, different types of findings, and different forms of presentation of conclusions. It supports an influencing rather than forcing perspective; it promotes supporting development rather than causing change; it looks to mutual specification and codetermination rather than cause-and-effect.

Learning from Mathematics

Mathematics is validated by acceptance by colleagues that the reasoning is correct, though it is said that most papers have one or two repairable errors, and many have at least one serious error which escapes both the author and the referees as well as the editor. But mathematical results are not used by other researchers until the arguments have been worked through, filling in details and re-integrating the elements into a sense of the whole. In this respect, validation of mathematical results are very similar to educational ones. It is not enough in either case simply to announce results. One must seek the agreement of a wider community (Ernest, 1990) One must offer justification, in a form which permits others to re-experience, to re-construct the reasoning for themselves. Only then do the proposals become 'results'. In either case, an assertion only becomes a result for an individual when they have integrated the thinking behind the assertion into their own perspective, made it 'their own' as the cliché goes.

Mathematics education is validated by a community response in which awareness, perspectives, and techniques, are integrated into practice and discourse. This only happens where the researcher or surrogate is able to devise activities which enable colleagues and/or practitioners to experience directly or vicariously the sensitivity which has enabled the researcher to notice. The researcher is not trying to communicate what has *been* noticed, but drawing attention to the release of energy contained in the potential for noticing in the future.

The mathematics-education research community does not often take the time to present reports in a form which enables others to re-construct the awarenesses and sensitivities developed during the research. Without these awarenesses there is little to be gained, apart from an interesting wrinkle in methodology, or a nice quote from a student to be used in some other context. Since the content of mathematics education is less precise, less definitive than the apparent content of mathematics, it means we have to work harder in order to enable research and teacher colleagues to appreciate our findings and integrate our approaches into their own work.

In order to instil some order, some structure into mathematics education, it might be worthwhile cleaving more closely to mathematical enquiry, by taking the effort to negotiate meaning of terms being used. Gattegno, who tried to develop a science of education (Gattegno, 1990), emphasized the necessity for a disciplined approach to educational interchange (Gattegno, 1970), and the discipline of noticing (Davis and Mason, 1989) proceeds along similar lines. By taking specializing and generalizing more seriously, mathematics education could benefit from reflecting the essence of its parent discipline.

Mathematical generalization, based on sensitivity to perceived patterns arising from selective stressing and ignoring (Gattegno, 1970) can be mirrored in mathematics education by using special cases to resonate personal experiences to develop an intricate web of interconnections and triggerable actions (meaning). By distinguishing between brief-but-vivid accounts of incidents and explanatory, justificatory accounts for actions, educational researchers could build an organic network of sensitivities and sensitizing experiences, ever forming and reforming in mutual specification and codetermination. By presenting research in forms which promote personal construal, in which readers find themselves seeing their past experiences in fresh light, and sensitizing them to potential incidents to notice in the future, researchers could invoke mathematical and educational experience to communicate with teachers and researchers in ways which would foster a more disciplined and productive approach. By maintaining contact with the mathematical roots of the educational topics being explored, and seeking analogous structures in their own experience at their own level, teachers and researchers could build a discipline combining the best of mathematical and educational methodology.

References

Bennett, J. (1956) *The Dramatic Universe*, 1, London, Hodder and Stoughton.

Campbell, S. and Dawson, A. (in press) 'Learning as embodied action', in Sutherland, R. and Mason, J. (Eds) *Nato Advanced Workshop on Exploiting Mental Imagery with Computers in Mathematics Education* (in press), London, Springer Verlag.

Davis, P. and Hersh, R. (1981) *The Mathematical Experience*, Brighton, Harvester.

Davis, J. and Mason, J. (1989) 'Notes on a radical constructivist epistemethodology applied to didactic situations', *Journal of Structural Learning*, 10, pp. 157–76.

Ernest, P. (1989) 'The impact of beliefs on the teaching of mathematics', in Ernest, P. (Ed) (1989) *Mathematics Teaching: The State of The Art*, pp. 249–54, London, The Falmer Press.

Ernest, P. (1991) *The Philosophy of Mathematics Education*, London, The Falmer Press.

Ernest, P. (1992) 'The relation between the objective and subjective knowledge of mathematics', in Seeger, F. and Steinbring, H. (Eds) *The Dialogue Between Theory and Practice in Mathematics Education: Overcoming the Broadcast Metaphor*, Materialien und Studien Band 38, Institut für Didaktik der Mathematik der Universität Bielefeld.

ERNEST, P. (1993) 'Constructivism and the psychology of learning and the nature of mathematics: Some critical issues', *Science and Education*, 2, 2, pp. 87–93.

GATTEGNO, C. (1970) *What We Owe Children: The Subordination of Teaching to Learning*, London, Routledge and Kegan Paul.

GATTEGNO, C. (1990) *The Science of Education*, New York, Educational Solutions.

GRIFFIN, P. and MASON, J. (1990) 'Walls and windows: A study of seeing using Routh's theorem and related results', *Mathematics Gazette*, 74, 469, pp. 260–9.

HAMILTON, E. and CAIRNS, H. (1963) (Eds) 'The Timaeus', in *Plato: the collected dialogues*, Bollingen Series LXXI, Princeton, Princeton University Press.

KELLER, E. (1983) 'Gender and science', in HARDING, S. and HINTIKKA, M. (Eds) *Discovering Reality: Feminist Perspectives on Epistemology, Metaphysics, Methodology, and Philosophy of Science*, Dordrecht, Reidel, pp. 187–205.

KITCHER, P. (1983) *The Nature of Mathematical Knowledge*, New York, Oxford University Press.

LAKATOS, I. (1976) *Proofs and Refutations*, in WORRAL, J. and ZAHAR, E. (Eds) Cambridge, Cambridge University Press.

LERMAN, S. (1986) Alternative Views of the Nature of Mathematics and Their Possible Influence on the Teaching of Mathematics, Unpublished PhD dissertation, King's College, London.

LERMAN, S. (1993) 'The "Problem" of Intersubjectivity in Mathematics Learning: extension or rejection of the constructivist paradigm?', preprint submitted.

MASCARÓ, J. (1962) *The Baghavad Gita*, Harmondsworth, Penguin (translation).

MASON, J. (1979) 'Which medium, which message', *Visual Education*, pp. 29–33.

MASON, J. (1986) 'Challenge For Change: To Stay Alive, Every Teacher Must Become A Researcher', Paper in Mathematics Teaching: Challenges For Change, AAMT, 211th biennial Conference Brisbane, pp. 9–30.

MASON, J. (1987) 'Pythagorean perspective', *Mathematics Teaching*, 117, pp. 8–9.

MASON, J., BURTON, L. and STACEY, K. (1984) *Thinking Mathematically*, London Addison Wesley.

MELLIN-OLSEN, S. (1987) *The Politics of Mathematics Education*, Dordrecht, Reidel.

POLYA, G. (1962) *Mathematical Discovery: on understanding, learning, and teaching problem solving* (combined edition), New York, Wiley.

ST. MAURICE, H. (1991) 'A Guide to commonplaces: On the use of loci in educator's discourse', *Journal of Curriculum Studies*, 2–3, 1, pp. 41–54.

TALL, D. (1980) 'The anatomy of a discovery in mathematics research', *For the learning of mathematics*, 1, 2, pp. 25–34.

THOMPSON, A. (1991) 'The development of teachers' conceptions of mathematics teaching,' *Proceedings of PME NA*, Underhill, R. (Ed), 2, pp. 8–14.

VARELA, F., THOMPSON, E. and ROSCH, E. (1991) *The Embodied Mind: Cognitive Science and Human Experience*, Cambridge, Massachusetts, MIT Press.

VON GLASERSFELD, E. (1984) 'An introduction to radical constructivism', in WATZLAWICK, P. (Ed) *The Invented reality: how do we know what we believe we know — contributions to constructivism*, London, Norton.

VYGOTSKY, L. (1978) *Mind in Society: The Development of Higher Psychological Processes*, Cambridge, Massachusetts, Harvard University Press.

WILLIAMS, H. (1989) *Tuning-in to Young Children: An Exploration of Contexts for Learning Mathematics*, M.Phil thesis, Milton Keynes, Open University.

Chapter 16

Demystifying Mathematics Education Through Inquiry

Marjorie Siegel and Raffaella Borasi

Introduction

The mathematics instruction that most American students experience in today's classrooms embodies pervasive cultural myths that misrepresent the nature of mathematics as well as what it means to learn mathematics. Mathematics textbooks, pedagogical practices, and patterns of classroom discourse, especially, work in concert to perpetuate the idea that mathematics is the 'discipline of certainty'. Together with a behaviourist view of learning, this myth has led students and teachers alike to reduce mathematical learning to the acquisition of ready-made algorithms and proofs through listening, memorizing, and practising. However comfortable these myths may be, they have debilitating consequences for students as they invite students to develop beliefs like 'There is only one correct way to solve any mathematical problem[;] . . . mathematics is a solitary activity done by individuals in isolation[;] . . . formal proof is irrelevant to processes of discovery or invention' (Schoenfeld, 1992, p. 359) as well as learning behaviours that are not conducive to success in mathematics (Borasi, 1990). Many students, in short, have come to perceive mathematics as a 'stainless steel wall' (Buerk, 1981) — cold, hard, and unapproachable, a mysterious activity quite distinct from their everyday lives and reserved for people with special talents.

In recent years, the mathematics education community has proposed recommendations for reconceptualizing school mathematics (i.e., HMSO, 1982; NCTM, 1989, 1991; NRC 1989) that challenge these myths. Fueled by psychological research on the processes students actually use to approach mathematical tasks, philosophical discussions on the nature and development of mathematical knowledge, and debates about the economic needs of today's technological society, many mathematics educators have called for the creation of learning environments that position students as mathematicians who produce rather than consume mathematical knowledge. In this chapter, we wish to contribute to this conversation about transforming mathematics education by looking more closely at what it means to produce mathematical knowledge. To do so, we will focus on the concept of *inquiry* and the implications it has for rethinking both the nature of mathematical knowledge and the goals and practices of mathematics education. Our analysis will focus, in particular, on the contributions that the American pragmatist Charles Sanders Peirce and scholars working in an emerging field of sociology, known as the 'sociology of scientific knowledge' (SSK) have made to the concept of inquiry. While these perspectives have not yet found a place in the mathematics education literature, together they offer a rich and complex portrait of inquiry that further dispels the myth of mathematics as the 'discipline of certainty'. At the same time, they offer new insights into the nature and practice of inquiry that may support mathematics educators as they begin to imagine what future mathematics classrooms could look like.[1]

Perspectives on Inquiry

The first perspective on inquiry we will examine was developed by Charles Sanders Peirce (1839–1914) over 100 years ago, yet continues to shed light on contemporary problems in the philosophy of science. Though Peirce is probably best known as one of the modern founders of semiotics, the study of signs as social forces (Eco, 1976), he spent thirty years as a research scientist with the US Coast and Geodetic Survey, the chief scientific agency of the US government at the time. His commitment to developing a theory of inquiry grew out of his belief that the scientific method offered the best approach to the producing truth. We will return to this point later to clarify Peirce's approach to truth, but for now it is important to note that he regarded the scientific method as superior to other methods of producing knowledge because of its potential for self-correction.

Peirce's theory of inquiry begins with his rejection of the Cartesian approach to the problem of knowledge, which proposed that knowledge depended on a firm foundation of first premises which are direct or unmediated. Peirce argued that absolute knowledge is unattainable on the grounds that all knowledge is indirect since we know the world through signs. And, since signs must be interpreted by other signs, there is no guarantee that the knowledge we have is absolute. Thus, Peirce rejected the idea that knowledge is stable and certain and proposed, instead, that knowledge is processual, ever open to doubt. This does not mean that Peirce was a skeptic; rather, he maintained that we could not live with doubt on a daily basis and we should not, therefore, continuously doubt that which we hold to be stable. At the same time, we should recognize the contingency of that stability and be ready to revise or reject knowledge found to be false (Skagestad, 1981). Peirce posited a view of knowledge that was radically different from the foundational metaphor that had held sway in his time:

> To Peirce, knowledge is no longer regarded statically as a body of propositions resembling a more or less finished building, but dynamically as a process of inquiry. (Skagestad, 1981, p. 18)

The metaphor he chose to encapsulate his sense of knowledge was that knowing is walking on a bog.

> [W]e never have firm rock beneath our feet; we are walking on a bog, and we can be certain only that the bog is sufficiently firm to carry us *for the time being*. Not only is this all the certainty that we can achieve, it is also all the certainty that we can rationally wish for, since it is precisely the tenuousness of the ground that constantly prods us forward, ever

closer to our goal. Only doubt and uncertainty can provide a motive for seeking new knowledge. (*ibid.*)

This metaphor captures well the central role doubt plays in Peirce's theory of inquiry. In his well-known article, 'The Fixation of Belief', Peirce writes: 'The irritation of doubt causes a struggle to attain a state of belief. I shall term this struggle *inquiry*, though it must be admitted that this is not a very apt designation' (1877/1982, p. 67). In this quote we can see that it is doubt which sets the process in motion. Most often this feeling of doubt arises when an anomaly — something that doesn't make sense in light of existing beliefs — is encountered. The struggle to replace doubt with belief is carried out through a cycle of reasoning that involves three forms of inference: abduction, deduction, and induction. Peirce argued that hypotheses are generated through abduction; the possible consequences of these hypotheses are developed through deduction; and, finally, these consequences are tested against experience through induction. This three-fold division may appear unnecessary at first as we are accustomed to the more familiar idea that induction involves a move from data to hypotheses or tentative conclusions and deduction a move from conclusions to data. But Peirce argued that induction actually involves two different processes: abduction, in which an hypothesis is put forward as a tentative framework, and induction, in which this framework is tested by selecting particular instances suggested by the framework (Deely, 1981). In abduction, then, the hypothesis that is generated provides a tentative explanation of what is, at first, unexpected and puzzling. On the basis of incomplete information, one offers an explanation that, if true, would show that the surprising event is a matter of course (Smith, 1988). Anomalies thus play a crucial role in inquiry inasmuch as they motivate the search for new connections among prior knowledge in order to generate an hypothesis that gives meaning to an unexpected event. This is not an insignificant move; indeed, Peirce claimed that abduction was the only form of reasoning capable of starting new ideas (Skagestad, 1981).

Up to this point it may appear as if Peirce were describing the activities of individual scientists, but the scientific community figured prominently in his theory of inquiry. Peirce regarded inquiry as a 'public process carried out by a community of investigators' (Skagestad, 1981, p. 24) and depended on this community to distinguish purely private thoughts from those which would hold up under public scrutiny; knowledge that could pass the test of the community was taken as real. From this perspective, truth is that upon which the community will converge in the long run. Yet, both the notion of 'long-run' and 'community' are ideals. The 'long run' is an infinitely long run and the 'community' is a hypothetical community; both constructs allowed Peirce to formulate a concept of truth as absolute and objective but only hypothetically so. Over any *finite* long run, absolute truth would never be attained. This notion of truth as attainable in the long run (an ideal-limit theory of truth) seems to contradict Peirce's view of knowledge as mediated unless his pragmatic realism is considered. Though he has often been miscast as a logical positivist, Skagestad (1981) suggests that Peirce's realism is in fact 'pragmatic realism'. In other words, realism functions as an empirical hypothesis that explains why scientific inquiry is successful in dealing with the world (p. 5). Thus, far from having a positivistic approach to truth, Peirce uses the ideal-limit theory to remind the scientific community of the fallibility of knowledge and thus keep the road of inquiry open.

Anthropologists and sociologists working from the perspective of the sociology of scientific knowledge have taken the notion of community a step further and conducted ethnographies of laboratory life (Latour and Woolgar, 1979; Knorr-Cetina, 1981, 1983) in order to move beyond *post hoc* accounts of inquiry to a better understanding of what scientists actually do when working on such problems as plant proteins and the brain's endocrine system. From this perspective, doing science is a thoroughly social practice in the sense that 'reality' is constructed through the discourse and rhetorical practices of the research community. Hence, for SSK scholars the community is primarily a discourse community in which 'truth' is established through a rhetorical contest. This is a somewhat different position than the one Peirce took, although when we later discuss the notion of the symmetry break or inversion, we will see that both perspectives take up the spirit of pragmatic realism. The difference may derive from the fact that Peirce, working in the early days of modern science, sought to establish science as a superior method for finding truth in the long run, whereas SSK scholars, working at the apex of science, are interested in deconstructing and demystifiying science, that is, they wish to shake the myth that science is a privileged and especially credible method for producing knowledge (Woolgar, 1988).

Unlike earlier accounts of the sociology of science which suggested that social factors affected factors external to the intellectual activities of the scientist, researchers working within SSK argue that the *technical* findings of scientists working at places like the Salk Institute are constituted by the social context. In other words, the findings are understood as the product of a series of choices and decisions occasioned by the circumstances in which they are situated, rather than as something that exists independently of the inquirer waiting to be 'discovered'. Scientific instruments, which are characterized in this literature as inscription devices in the sense that the output is a written mark, can give us a good look at how content and context become fused. Scientists acknowledge that such apparatus represent the end result of debates in the field and do not provide direct readings of the phenomenon being studied. And yet, once the inscription is read, the mediating status of the instrument and its dependence on representation is forgotten. As Latour and Woolgar (1979) note, there is a transformation of the inscription into the terms of the 'mythology' of the culture of science so that a particular curve on a readout from an instrument might constitute a breakthrough on a particular problem. This move, which SSK researchers call a symmetry break (Knorr-Cetina, 1981) or an inversion (Latour and Woolgar, 1979), is central to the practice of science. In the context of the research lab, scientists understand the constructed nature of their work, but when they write papers they proceed as if their work involved mere description. The rhetorical practices scientists use are what turn inscriptions into 'truth'; scientists work hard to convince the reader that their findings are not representations at all but unmediated nature and, moreover, that the reader has not been convinced by anything other than the elegant logic of the inquiry. From this perspective, scientific knowledge is constituted by the context, an artifact (Neilsen, 1989), yet it achieves its privileged status as fact from a process in which dialogue and debate among colleagues in the lab is transformed into a statement with quotation marks around it and, finally, into a flat factual statement in a textbook (Latour, 1987).

As noted earlier, the notion of a symmetry break or inversion is related to Peirce's pragmatic realism. The idea that scientists construct ways to mediate 'nature' and then treat their constructions as real is a perfect description of what Peirce meant by pragmatic realism. Recall that what makes Peirce's realism *pragmatic* is that he regarded it as an hypothesis that would explain why scientists had found so much success using the scientific method. As a scientist himself, he understood the role signs played in mediating the world but also understood that scientists had to believe their findings were real in order to proceed with their work. In other words, they could not function as skeptics and doubt that which they took to be true, which in this case was the existence of reality (Skagestad, 1981). Indeed, this point is illustrated in Jonas Salk's forward to Latour and Woolgar's ethnography of the Salk Institute, *Laboratory Life* (1979). Though supportive of their enterprise, he is not quite convinced that they have captured doing science as *he* experiences it, a point which suggests that scientists will not be the ones to deconstruct their discipline!

Demystifying the Nature of Mathematical Knowledge

Mathematics has long been held up as the 'queen' of the sciences, the paradigm case of absolute knowledge inasmuch as it was thought to be derived deductively from axioms, that is, first premises which served as an unshakable foundation. However, surprising developments such as the realization that geometries based on different sets of axioms than those chosen by Euclid could be legitimate, or Gödel's proof that any formal system of a certain complexity contains some undecidable propositions, have eroded the foundational status of mathematics as a discipline. Indeed, historical accounts of the development of key mathematical topics, such as those offered by Kline (1980) and Lakatos (1976), should dispel any doubt that mathematics might not be affected by the epistemological considerations developed in the previous section.

Once we consider it legitimate to apply the view of 'knowledge as inquiry' developed by Peirce and SSK scholars to the field of mathematics, a number of important implications about the nature of mathematical knowledge and activities, many of which resonate with contemporary movements in mathematics education, can be derived. Specifically, an inquiry epistemology challenges popular myths about the truth of mathematical results and the way in which they are achieved, and suggests, instead, that: mathematical knowledge is fallible; mathematical knowledge is created through a non-linear process in which the generation of hypotheses plays a key role; the production of mathematical knowledge is a social process that occurs within a community of practice; and the truth value of mathematical knowledge is constructed through rhetorical practices. In what follows, we will briefly expand on each of these points.

Mathematical Knowledge is Fallible

A first consequence of looking at mathematics in a spirit of inquiry requires that we accept the idea that achieving absolute truth is an illusion. Rather, mathematical results can only be sanctioned by the mathematical community of the time on the basis of existing knowledge and evidence as well as agreed upon criteria (including adherence to specific rules of logic and coherence with the existing system). The history of mathematics has provided some notable examples of how these criteria, as well as the knowledge and evidence upon which the decision is based, may change with time and thus may cause mathematicians to revise some of their assumptions, definitions, and/or results. Consider, for instance, Lakatos' reconstruction of how Euler's original theorem on the 'characteristic' of polyhedra was challenged by some counter-examples and revised more than once through an iterative process of 'proofs and refutations' involving several mathematicians over a considerable period of time. This historical event cannot be easily reconciled with the view that once 'proved' a mathematical result can be considered absolutely true and uncontestable. Another interesting example is provided by the creation of the first non-Euclidean geometry, an event that made mathematicians realize that their belief in Euclidean geometry as the only, and true, way to describe spacial relationships was mistaken (a belief they had held for about 2000 years!). Finally, we would like to mention that the notion of what constitutes an acceptable mathematical proof is itself far from being stable and non-controversial, as shown by Kline's historical analysis (Kline, 1980, Chapter 14).

However surprising these conclusions might seem to students and teachers who idealize mathematics as the 'discipline of certainty', the notion that mathematical knowledge is fallible is an accepted position among mathematicians and mathematics educators alike. Radical constructivists in mathematics education, for example, take a similar position on knowledge, one that parallels the epistemological assumptions Peirce articulated:

> [Radical] constructivism can be described as essentially a theory about the limits of human knowledge, a belief that all knowledge is necessarily a product of our own cognitive acts. We can have no direct or unmediated knowledge of any external or objective reality. We construct our understanding through our experiences, and the character of our experience is influenced profoundly by our cognitive lenses . . . Mathematicians act as if a mathematical idea possesses an external, independent existence; however, the constructivist interprets this to mean that the mathematician and his/her community have chosen, for the time being, not to call the construct into question, but to use it as if it were real, while assessing its worthiness over time. (Confrey, 1990, pp. 108–9)

As Confrey points out, the rejection of absolute knowledge does not mean that 'truth' is abandoned; it is understood, instead, as a product of social negotiation within a community of inquirers. In fact, the role of the community in the construction of knowledge is an important theme in radical constructivism that points to the nature of a discipline as a collective body with a set of shared norms, values, beliefs, and practices. At the same time, some mathematicians and mathematics educators have objected to the 'skepticism' that could be associated with a radical constructivist position (see, for example, Wheeler, 1987). Peirce's 'pragmatic realism' offers some resolution to this problem in the sense that mathematicians, like other inquirers, use the concept of reality to explain why they are able to indeed use mathematics to make sense of the world. From this perspective, reality has the status of an hypothesis, one which makes it possible to engage in mathematical thinking. If radical constructivists were complete skeptics, it is unlikely that they would see any value in rethinking the purpose and practices of mathematics education. Since this is not the case, it is possible to conclude that radical constructivism, like the field of SSK, is not so much a skeptical position as it is a rhetorical strategy for shaking up the school curriculum.

How Mathematical Knowledge Is Created

Mathematical knowledge is produced through a non-linear process in which the generation of hypotheses plays a key role. Peirce's characterization of inquiry as a struggle to settle doubt and fix belief through a cycle of abduction, deduction, and induction challenges the common-sense notion that inquiry is a linear process. This view has much in common with Lakatos' thesis that significant mathematical knowledge is created through an iterative process of 'proofs and refutations', as noted earlier. The 'zig-zag' path that Lakatos describes is a result of the fact that mathematicians must put forward a tentative explanatory framework — a conjecture — *before* they have sufficient evidence that the framework or hypothesis they are proposing will indeed be acceptable. Contrary to the belief that the production of mathematical knowledge is a 'sure thing', the process begins with what Lakatos called 'conscious guessing'. As Peirce noted, it is this guess, or abductive inference, that sets the process of knowledge production in motion; without it, mathematicians are left with only the knowledge that has been agreed on by the community. In a sense, then, the non-linear nature of inquiry, with all the risk-taking, doubt, and uncertainty it engenders, works against the conservative force of the community's norms and as such serves a far more valuable function to mathematicians than the more familiar belief that the evaluation of knowledge claims is the key move in mathematics.

What the Production of Mathematical Knowledge Is

The production of mathematical knowledge is a social process that occurs within a community practice. While many students believe that mathematics is an activity undertaken chiefly in isolation, as noted in the introduction, the perspectives on inquiry explored thus far suggest that the creation of mathematical knowledge is situated within a community of practice. This position is once again supported by historical evidence. It is worth noting that the successive refinement and elaborations of Euler's initial theorem on the 'characteristic' of polyhedra involved the concerted effort of several mathematicians, who carefully examined each other's conclusions, looked for potential counter-examples, and suggested ways to deal constructively with them. Kline's account of the debates generated by the creation of the first non-Euclidean geometries in the mathematical community of the time is even more revealing of the role played by the community of practice in establishing the truth of a mathematical result, as it shows how judgments regarding the 'new claim' may be affected by the reputation of those supporting contrasting positions as well as the strength of existing beliefs. Kline writes:

> That Euclidean geometry is the geometry of physical space, that it is the truth about space, was so ingrained in people's minds that for many years

> any contrary thought such as Gauss's were rejected . . . For thirty years after the publication of Lobatchevsky's and Bolyai's works all but a few mathematicians ignored the non-Euclidean geometries. They were regarded as a curiosity. Some mathematicians did not deny their logical coherence. Others believed that they must contain contradictions and so were worthless. Almost all mathematicians maintained that the geometry of physical space, *the* geometry, must be Euclidean . . . However, the material in Gauss's notes became available after his death in 1855 when his reputation was unexcelled and the publication in 1868 of Riemann's 1854 paper convinced many mathematicians that a non-Euclidean geometry could be the geometry of physical space and that we could no longer be sure what geometry was true. (Kline, 1980, p. 88)

Several mathematics educators (e.g., Lave, 1988; Resnick, 1988; Schoenfeld, 1988, 1992) have also recognized that mathematics is a social practice. Elaborating further on the significance of this point for mathematics education, they have argued that 'becoming a mathematician' does not mean simply acquiring a certain body of established knowledge, skills, and abilities, but, more importantly, taking on a set of beliefs, norms, world views, and practices characteristic of the mathematics community.

The Truth Value of Mathematical Knowledge

The truth value of mathematical knowledge is constructed through rhetorical practices. If the production of scientific and mathematical knowledge involves the negotiation of meanings within a community of practice (Lave, 1988), then communication becomes integral, if not central, to the process of inquiry. SSK scholars take the relation between language and community even further and argue that the social negotiation of meaning and 'truth' is accomplished through rhetorical practices. In other words, agreement by the community that a particular statement is 'true' is the result of writing in such a way as to convince others that the statement is not itself a construction but exists independently of the researcher. Lakatos' description of how proofs are constructed is a good example of the rhetorical nature of 'true' mathematical knowledge. When the mathematician is engaged in the actual process of constructing a proof, he or she is caught up in making conjectures, testing them against counter-examples, and then going back to revise the initial conjectures. But when he or she publishes the proof, the tentative, non-linear, and constructed nature of the proof is edited out so that what remains is an elegant series of seemingly deductive moves that denies its constructed status through its rhetorical structure. If other members of the community accept this argument, the mathematician can be said to have been successful in winning the rhetorical contest and the proof will be taken as true. SSK scholars thus argue that the key move in inquiry is transforming a *post hoc* account into a flat factual statement.

Richards (1991) highlights this point in his examination of mathematical discussions. He suggests that there are four kinds of mathematical discourse: research mathematics, inquiry mathematics, journal mathematics, and school mathematics. The first two are characterized by the logic of discovery whereas the second two feature reconstructed logic. Notice that although there are distinctions (chiefly in the professional mathematician's heavy reliance on proofs) between the discourse of professional mathematicians (who speak research mathematics) and that of mathematically literate adults (who speak inquiry mathematics), they have much in common in the sense that both use discourse patterns associated with the logic of discovery, that is, conjectures and arguments. Notice, too, the similarity between journal mathematics (the language of mathematical publications) and school mathematics (traditional classroom discourse and the language of textbooks); both feature reconstructed logic with the main difference being that students who encounter the discourse of school mathematics usually have no idea that the formalized proofs in their textbooks were once as full of zigs and zags as their everyday cognition (Rogoff and Lave, 1984). In short, the discourse of school mathematics truncates the actual practice of mathematical communities and in so doing denies students a meaningful entrée to mathematical learning.

Reconceiving Mathematics Instruction in the Spirit of Inquiry

The discussion of the nature of mathematical knowledge, developed in the previous section, has shown that the idea of mathematics as the 'discipline of certainty' misrepresents the field. If mathematics instruction were conceptualized in such a way as to highlight this fact, students might begin to recognize the limitations and 'human' elements present in the discipline and start to see mathematics as more attractive and accessible, as shown by the results of a few instructional experiences involving math avoidant women (see, for example, Borasi, 1992; and Buerk, 1991, 1985). More specifically, reconceiving mathematics instruction in the spirit of inquiry suggests fundamental changes that may help demystify mathematics for students. This means that students, like professional mathematicians, would construct their own mathematical knowledge by actively engaging in mathematical inquiries, an assumption supported by empirical studies on problem-solving conducted from a constructivist perspective (Baroody and Ginsburg, 1990; Ginsburg, 1983, 1989; Steffe, von Glasersfeld, Richards and Cobb, 1983). Moreover, when inquiry is understood as a social practice, 'socializing' students into a mathematical community becomes a major goal of school mathematics. Resnick (1988) notes the radical nature of this change in the following statement:

> If we want students to treat mathematics as an ill-structured discipline — making sense of it, arguing about it, and creating it, rather than merely doing it according to prescribed rules — we will have to socialize them as much as to instruct them. This means that we cannot expect any brief or encapsulated program on problem solving to do the job. Instead, we must seek the kind of long-term engagement in mathematical thinking that the concept of socialization implies. (Resnick, 1988, p. 58)

Taken together, these considerations suggest different activities, goals, values, and social norms than those characteristic of traditional classrooms, or even classrooms informed by a 'problem-solving' or 'discovery-learning' approach. In what follows, we will point out some of these differences by discussing what mathematics classrooms might look like if the focus of mathematics instruction were reinterpreted as stimulating and supporting students' engagement in mathematical inquiries within a community of practice (see Borasi, 1992; Borasi and Siegel, in press, for some concrete images of instructional experiences informed by the principles identified here).

First of all, inquiry classrooms emphasize the full complexity of knowledge production and expect students to see the doubt arising from ambiguity, anomalies, and contradiction as a motivating force leading to the formation of questions, hunches, and further exploration. Teachers in inquiry classrooms, therefore, are less inclined to take the role of the expert and clear things up *for* the students and more interested in helping students use this confusion as a starting point for problem posing and data analysis. Furthermore, students, acting as inquirers, come to see the way in which their knowledge is contingent upon the values and context that frame the problem as well as the choices and decisions they make in the course of their investigations. No longer are students outsiders who regard knowledge as something handed down from high and preserved as flat factual statements in textbooks. As members of a community of practice they understand that what ends up in textbooks is often the endpoint of a long debate, which may have been decided on the basis of the argument made and not the correspondence between the phenomenon and the statement describing it. In this way, the authority of textbooks is deconstructed and they simply become artifacts of a particular conversation. Second, unlike traditional classrooms where students learn in the presence of but not *with* others, learning in inquiry classrooms is collaborative. Value is placed on the creation of a community of practice within which meanings are negotiated. This does not mean that students need not construct personal understandings, but rather that collaboration and social interaction provide support for that process. At the same time, the moral imperative of inquiry classrooms respects diversity and individual voice so that the production of collective meanings does not become a tool for silencing difference.

Third, student generativity is a distinct feature of inquiry classrooms. The understandings that students construct as a result of their mathematical inquiry cannot be reduced to the duplication and display of ready-made knowledge so common in traditional classrooms, but rather involves the generation of new meanings and connections. This means that instead of always addressing mathematical problems and questions set by the teachers, students become involved in setting directions for their own mathematical inquiries, posing new problems, and reformulating and/or expanding upon existing ones. Several mathematics educators have proposed instructional strategies that can support students in these 'unusual' activities in the context of mathematics instruction (see, for example, Brown and Walter, 1990, 1993; Borasi, in press).

Fourth, teachers in inquiry classrooms no longer transmit information through the channels of talking or reading but take up the much more challenging job of supporting student inquiry. This means establishing a radically different set of social norms and values in the classroom as well as finding ways to invite students into the inquiry process, and support them as they engage in the process. This support comes in the form of demonstrations of how to approach various aspects of the inquiry process, invitations to students to engage in the process, and explicit and implicit messages that help students value all that is involved in being inquirers. Drawing on the work of whole-language educators, inquiry teachers understand the importance of demonstrations, engagement, and values in creating a community of learners that supports student inquiry (Harste, Short and Burke, 1988). Perhaps one of the most important things inquiry teachers do is listen to students (Confrey, 1991; Harste, Woodward and Burke, 1984; Paley, 1986). They listen so as to understand their beliefs about learning and the background experiences they bring to specific inquiries, to invite their involvement in the framing and organization of the inquiry, to gain insight into the meanings and connections students construct, and to hear their concerns and questions about the inquiry process itself and the changes in classroom norms and values it engenders.

Finally, students take centre-stage in inquiry classrooms and become active members of a community of practice who share the responsibility of planning, conducting, and reflecting on their inquiries with other community members, namely, their peers and the teacher. This role requires a much greater degree of risk-taking on the part of the student as well as a willingness to listen and negotiate with others (as noted, for example, by Lampert, 1990). Risk is woven into all aspects of the inquiry process, from the initial formation of the problem through to the final presentation and evaluation of end results. Given their prior school experiences, many students may be hesitant to immediately trust that teachers are really interested in hearing their ideas and not just waiting to catch them making a mistake. Hence, the first days, weeks, and months of the class take on special significance as students become socialized into practices that depend on their contributions and learn that a range of responses, not just one, is valued.

Conclusion

Throughout this chapter, we have argued that school mathematics misrepresents the nature of mathematical knowledge and, in so doing, teaches students that mathematics is all about certainty, not inquiry. The result is that in many classrooms, mathematics is like an iceberg of which students can see only the tip; the bulk of it, dynamic and complex, is hidden from view. The elegant solutions and 'logical' proofs they are taught seem to spring ready-made from their textbooks without a hint of the struggle that characterizes the actual process of knowledge production. Schwartz (cited in Richards, 1991) comments on this limitation of school mathematics in the following statement:

> There is something odd about the way we teach mathematics in our schools. We make little or no provision for students to play an active and generative role in learning mathematics and we teach mathematics as if we expected that students will never have occasion to invent new mathematics. (Schwartz, in Richards, 1991, p. 27)

We suggest that if we want students to have an opportunity to invent new mathematics, we must demystify mathematics by valuing what lies beneath the tip of the iceberg. Conceptualizing mathematics instruction as inquiry offers one route to take in order to accomplish this goal.

Note

1. This study was partially supported by a grant from the National Science Foundation (award #MDR-8850548). The opinions and conclusions reported in this chapter, however, are solely the authors'.

References

BAROODY, A.J. and GINSBURG, H.P. (1990) 'Children's learning: A cognitive view', in DAVIS, R.B., MAHER, C.A. and NODDINGS, N. (Eds) *Constructivist Views on the Teaching and Learning of Mathematics,* Reston, Virginia, National Council of Teachers of Mathematics, pp. 51–64.

BORASI, R. (1990) 'The invisible hand operating in mathematics instruction: Students' conceptions and expectations', in COONEY, T.J. and HIRSCH, C.R. (Eds) *Teaching and learning mathematics in the 1990s*, 1990 Yearbook of the National Council of Teachers of Mathematics, Reston, Virginia, NCTM, pp. 174–82.

BORASI, R. (1992) *Learning Mathematics Through Inquiry*, Portsmouth, New Hampshire, Heinemann.

BORASI, R. (in press) 'Capitalizing on errors as "springboards for inquiry": A teaching experiment', *Journal for Research in Mathematics Education.*

BORASI, R. and SIEGEL, M. (in press) 'Reading, writing and mathematics: Re-thinking the basics and their relationship', Subplenary lecture delivered at the International Congress in Mathematics Education, Québec, Québec, August 1992 (to be included in the Proceedings in ROBITAILLE, D.F., WHEELER, D. and KIERAN, C. (Eds)).

BROWN, S.I. and WALTER, M.I. (1990) *The Art of Problem Posing*, 2nd ed. Hillsdale, New Jersey, Erlbaum.

BROWN, S.I. and WALTER, M.I. (Eds) (1993) *Problem Posing: Reflections and Applications*, Hillsdale, New Jersey, Erlbaum.

BUERK, D. (1981) 'Changing the conception of mathematical knowledge in intellectually able, math-avoidant women', Unpublished doctoral dissertation, State University of New York at Buffalo.

BUERK, D. (1985) 'The voices of women making meaning in mathematics', *Journal of Education*, 167, 3, 59–70.

CONFREY, J. (1990) 'What constructivism implies for teaching', in DAVIS, R.B., MAHER, C.A. and NODDINGS, N. (Eds) *Constructivist Views on the Teaching and Learning of Mathematics*, Reston, Virginia, National Council of Teachers of Mathematics, pp. 107–22.

CONFREY, J. (1991) 'Learning to listen: A student's understanding of powers of ten', in VON GLASERSFELD, E. (Ed) *Radical Constructivism in Mathematics Education*, Dordrecht, Kluwer, pp. 111–38.

DEELY, J. (1981) 'The relation of logic to semiotics', *Semiotica*, 35, 3–4, pp. 193–265.

ECO, U. (1976) *A Theory of Semiotics*, Bloomington, Indiana, Indiana University Press.

GINSBURG, H.P. (1983) *The Development of Mathematical Thinking*, New York, Academic Press.

GINSBURG, H.P. (1989) *Children's Arithmetic* (2nd ed.), Austin, Texas, Pro-Ed.

HARSTE, J., WOODWARD, V. and BURKE, C. (1984) *Language Stories and Literacy Lessons*, Portsmouth, New Hampshire, Heinemann.

HARSTE, J., SHORT, K. and BURKE, C. (1988) *Creating Classrooms for Authors*, Portsmouth, New Hampshire, Heinemann.

HMSO (1982) *Mathematics Counts*, London, HMSO.

KLINE, M. (1980) *Mathematics: The loss of certainty*, New York, Oxford University Press.

KNORR-CETINA, K. (1981) *The Manufacture of Knowledge*, Oxford, Pergamon Press.

KNORR-CETINA, K. (1983) 'The ethnographic study of scientific work: Towards a constructivist interpretation of science', in KNORR- CETINA, K. and MULKAY, M. (Eds) *Science Observed*, London, Sage Publications, pp. 115–40.

LAKATOS, I. (1976) *Proofs and Refutations*, Cambridge, Cambridge University Press.

LAMPERT, M. (1990) 'When the problem is not the question and the solution is not the answer: Mathematics knowing and teaching', *American Educational Research Journal*, 27, Spring pp. 29–63.

LATOUR, B. (1987) *Science in Action*, Cambridge, Massachusetts, Harvard University Press.

LATOUR, B. and WOOLGAR, S. (1979) *Laboratory Life*, Princeton, New Jersey, Princeton University Press.

LAVE, JEAN. (1988) *Cognition in Practice*, Cambridge, Cambridge University Press.

MEHAN, H. (1979) *Learning Lessons*, Cambridge, Massachusetts, Harvard University Press.

NATIONAL COUNCIL OF TEACHERS OF MATHEMATICS (NCTM) (1989) *Curriculum and Evaluation Standards for School Mathematics*, Reston, Virginia, National Council of Teachers of Mathematics.

NATIONAL COUNCIL OF TEACHERS OF MATHEMATICS (NCTM) (1991) *Professional Standards for Teaching Mathematics*, Reston, Virginia, National Council of Teachers of Mathematics.

NATIONAL RESEARCH COUNCIL (NRC) (1989) *Everybody Counts: A Report to the Nation on the Future of Mathematics Education*, Washington, DC, National Academic Press.

NEILSEN, A.R. (1989) *Critical Thinking and Reading*, Bloomington, Indiana, ERIC Clearinghouse on Reading and Communications Skills.

PALEY, V. (1986) 'On listening to what children say', *Harvard Educational Review*, 56, 2, pp. 122–31.

PEIRCE, C.S. (1877/1982) 'The fixation of belief', in THAYER, H.S. (Ed) *Pragmatism: The Classic Writings*, Indianapolis, Indiana, Hackett, pp. 61–78.

RESNICK, L. (1988) 'Treating mathematics as an ill-structured discipline', in CHARLES, R.I. and SILVER, E.A. (Eds) *The Teaching and Assessing of Mathematical Problem Solving*, Reston, Virginia, National Council of Teachers of Mathematics, pp. 32–60.

RICHARDS, J. (1991) 'Mathematical discussions', in VON GLASERSFELD, E. (Ed) *Radical Constructivism in Mathematics Education*, Dordrecht, Kluwer, pp. 13–51.

ROGOFF, B. and LAVE, J. (Eds) (1984) *Everyday Cognition: Its Development in Social Contexts*, Cambridge, Massachusetts, Harvard University Press.

SCHOENFELD, A.H. (1988) 'Problem solving in context(s)', in CHARLES, R.I. and SILVER E.A. (Eds) *The Teaching and Assessing of Mathematical Problem Solving*, Reston, Virginia, National Council of Teachers of Mathematics, pp. 82–92.

SCHOENFELD, A.H. (1992) 'Learning to think mathematically: Problem solving, metacognition and sense making in mathematics', GROUWS, D.A. (Ed) *Handbook of Research on Mathematics Teaching and Learning*, New York, Macmillan, pp. 334–70.

SKAGESTAD, P. (1981) *The Road of Inquiry*, New York, Columbia University Press.

SMITH, H. (1988) 'Abduction and the signs of expertise', Paper presented at the Annual Meeting of the American Educational Research Association, New Orleans, Louisiana.

STEFFE, L.P., VON GLASERSFELD, E., RICHARDS, J. and COBB, P. (1983) *Children's Counting Types: Philosophy, Theory and Applications*, New York, Praeger Scientific.

WERTIME, R. (1979) 'Students, problems and courage', in LOCHEAD, J. and CLEMENTS, J. (Eds) *Cognitive Process Instruction*, Philadelphia, Pennsylvania, The Franklin Institute Press.

WHEELER, D. (1987) in BERGERON, J.C., HERSCOVICS, N. and KIERAN, C. (Eds) *Proceedings of the Eleventh International Conference of the Psychology of Mathematics Education Group (PME-XI)* (Volume 1), Montréal, Québec, Psychology of Mathematics Education.

WOOLGAR, S. (1988) *Science: The Very Idea*, Chichester, Sussex, Ellis Horwood.

Chapter 17

Reading to Learn Mathematics in the Primary Age Range

Charles W. Desforges and Stephen Bristow

Background

The research reported here arises from a project in which we are exploring the role that 'old' knowledge plays in new learning in the age range 7–11 years. All learners have a range of resources which may be brought to bear on new learning including relevant subject knowledge, general knowledge, meta-cognitive strategies, imitation, scaffolding and social supports. The objective of our research is to develop a theoretical account of how an individual learner brings these resources to bear in the construction of new knowledge in academic subjects. In this present account we discuss the way young children use reading in their pursuit of mathematics learning.

The problem of knowledge use in learning has considerable significance. In practice, the dangers of teaching 'inert knowledge' are well-known (NCC, 1992). The problems of teaching knowledge utilization i.e., of teaching children to use and apply mathematics are equally well-known and proving to be tractable (Desforges *et al.*, 1993). In theoretical terms the relationship between knowledge use and knowledge application has attracted extensive attention (see Singley and Anderson, 1989 for a review) yet there is little to show for this work. As Fodor has observed, '. . . we simply have no idea what it would be like to get from a conceptually impoverished to a conceptually richer system by anything like a process of learning' (in Bereiter, 1985, p. 202).

It has been characteristic of this research tradition to separate the processes of knowledge acquisition (i.e., learning) and knowledge use (i.e., application) for scrutiny. Yet they are inseparable both conceptually and in practice. Knowledge cannot be acquired without the deployment of extant knowledge. And knowledge cannot be deployed without creating at least the possibilities of learning.

Knowledge application has been fostered through strategy training (Pressley *et al.*, 1989), attribution training (de Charms, 1972) open-education movements (Blyth, 1965) and most recently through appeals for a fundamental re-organization of the social contexts for learning (Resnick, 1987). These attempts have not been wildly successful (Bereiter, 1990; Desforges and Bristow, 1992).

The lack of impact is not surprising in retrospect. The striking feature of most of the efforts to promote children's knowledge application is that they are not based on an understanding of children's extant competencies (Desforges and Bristow, 1992). There are almost no studies of how children in the age range

7–11 apply knowledge to learn. Frequently, suggestions for practice are *ad hoc* collections of good ideas culled from a voluminous literature (see Kameenui and Griffin, 1989 for example). Other suggestions come from idiosyncratic views of intellectual functioning (e.g., De Bono, 1976). Yet others are derived from general information-processing theory or are based on the reflections of, mainly, adult experts. All these rationales imply or assume that children's minds are empty of competing or alternative strategies. Interventionist approaches assume that offered strategies are implanted in a vacuum. No interactions with children's knowledge are anticipated.

Our work is an attempt to appraise children's extant approaches to knowledge application, to come to an understanding of the processes they have at their disposal and the manner in which these processes interact with their knowledge as facts, concepts and procedures. We have the longer-term aim of promoting knowledge utilization at the point of knowledge acquisition but our immediate purpose has been to expose and explore the link between knowledge use and learning as a starting point. Our attention has been on children in the age range 7–11 years and on the subjects of mathematics, science and English. In this account we focus on mathematics but we will refer to some of the science work where contrasts prove illuminating.

We have taken a broad view of knowledge to include, subject domain knowledge (concepts, skills, procedures); strategic knowledge (general approaches to learning and problem-solving); contextual knowledge (knowledge of social and working practices) and self-knowledge (i.e., beliefs that children hold about themselves in regard to the curriculum).

As an aspect of knowledge use we were particularly interested in how children interacted with the texts they typically meet in mathematics schemes. The focus on reading in primary mathematics is important in a number of respects. From the point of view of curriculum experience reading opportunities are extensive in mathematics because the professional schemes so extensively used are laden with text-based examples, illustrations and explanations. Reading is, at least potentially, an important way of learning mathematics. Reading has strong curriculum validity in current approaches to teaching mathematics. Secondly, reading is promoted in the primary curriculum as a transferable skill, a cross-curricular activity: it is therefore important to consider the degree to which it does in fact transfer and/or to expose the problems and opportunities in the application of reading skills in the domain of mathematics. And thirdly, reading is a particularly useful research vehicle for those interested in knowledge application. As a process it necessarily recruits 'old' knowledge. As a context it is readily managed: text is easy to design and control in scope and difficulty and there are tried and trusted methods for collecting data on children's thought processes in interacting with print.

Our particular research questions then were, how did the children read the text and in what ways did the text–child interactions provoke or recruit knowledge in pursuit of learning.

The Research

We have worked with a sample of forty-eight children, twenty-four age 8–9 and twenty-four aged 10–11 at the point of data collection reported here. The children were nominated by their teachers who were sympathetic to the aims of the project. We asked for, and believe we got, articulate children who would enjoy our activities. The sample was spread across the attainment range with a slight skew towards higher-attainment levels.

Important features of our research procedure were first that we shared the research objectives with the children. We recruited the children after explaining the objectives. More importantly we persistently reminded them of the objectives throughout the interviews and activities. Secondly we used 'think aloud' techniques in the interviews. Essentially we asked the children to spell out their thinking and decisions for us as they used our learning materials. To help the children use this technique we had first modelled it for them and subsequently reminded them and gave them feedback as they talked through their thoughts. The third, critical feature of our research method, was that in our procedures we deliberately took them to a point of difficulty with a topic in order to create the possibilities for learning. We were not interested in how they ran off familiar procedures. As we worked on the topics (to be described later) we moved the children on until they reached a point at which they could not do the work before them. We then presented them with a learning pack which, we told them, might help them make progress. Our research question was, in what ways did the pupils recruit their knowledge to engage with the proffered learning materials?

Four mathematics topics were explored in this way, namely, area, fractions, ratio and averages. One quarter of the sample (i.e., twelve children) studied each pack. Essentially our approach had a pre-task, learning phase, post-task structure. An interviewer took children individually through the experiences.

The pre-task phase took the form of a set of up to six problems on the topic which progressed in difficulty. Each child was asked to work through these as far as possible. The interviewer's role was to prompt the articulation of thoughts, choices and decisions. The child was forewarned that they would not be able to succeed on all the problems and that when they could go no further they were to turn to the learning pack.

All the materials used were facsimiles of materials to be found in popular mathematics schemes. The children were entirely familiar with the style and layouts with which they were presented. Figures 17.1 and 17.2 show examples of the materials used.

During the children's introduction to the learning pack they were told that they could approach the materials in any way they chose, in any order and in any way they saw fit to help them to learn. They were reminded that we needed them to think aloud whilst they engaged with the pack. They were asked to say why they did certain things, what sense they were making of what they had read or done. During their engagement with the pack they were prompted by the researcher to articulate their thinking. This was achieved by the use of open-ended questions related to the children's actions. For example, 'Tell me why you did that' or, 'You look thoughtful. Can you tell me what you are thinking', or, 'what did you make of what you just read?'

When the children had made use of the learning pack to their own satisfaction they were asked to answer the post-task items. They were asked to comment on the origins of any differences between their pre and post-task answers (for example, why they had put a different answer or how they had come to know that a new answer was necessary). Again, the children were asked whether they thought

Figure 17.1: A Sheet from the Averages Pack

Jan timed herself from home to school each day for a week.

These are the times :

Mon 17 min
Tues 22 min
Wed 19 min
Thurs 21 min
Fri 26 min

Jan wants to work out her average time that week.
To find the average she adds up the 5 times.
Then she divides by 5.
So her average time is $105 \div 5 = 21$ minutes.
When you work out an average like this, it is called the mean.

there was anything for them still to learn. They were then offered the opportunity to do so if they desired.

All the interviews were tape-recorded. Data consisted of tape recordings of the conversations, long-hand notes made by the researcher of the children's actions, and copies of the children's pre and post-task answers together with any written work the children did when using the learning pack including associated workings on rough paper.

Analysis

The particular analysis we focus on here is in regard to the children's interaction with the text of the learning pack. By way of example, the full learning pack for the topic of 'area' is shown in Figure 17.3.

It will be recalled that the children met the learning pack only after they had reached a point of difficulty in problems in the pre-task phase of the study. How did they use the pack to promote their learning? What knowledge, broadly conceived was recruited and brought to bear on the pack? What processes of sense-making were deployed?

Our response to these questions is essentially descriptive. We allowed participating children to articulate their authentic experience of learning and thereby expose how they organized and used their previous experience and knowledge to

Figure 17.2: A Sheet from the Fractions Pack

Peter wants to complete $\frac{1}{5} + \frac{3}{5} =$.

Because $\frac{1}{5}$ and $\frac{3}{5}$ each have 5 as the denominator, I can add them easily.

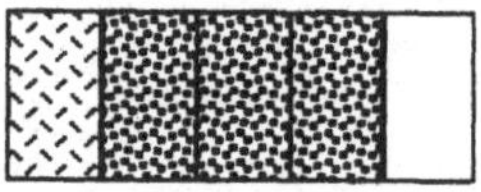

He writes $\frac{1}{5} + \frac{3}{5} = \frac{4}{5}$

Complete.

$\frac{1}{5} + \frac{2}{5} =$ $\quad$ $\frac{1}{3} + \frac{1}{3} =$ $\quad$ $\frac{2}{5} + \frac{2}{5} =$ $\quad$ $\frac{1}{7} + \frac{3}{7} =$

Next, Peter and Gill talk about how to add and take away the fractions $\frac{1}{2}$ and $\frac{1}{3}$.

We could add and take away the fractions if they both had the same denominator.

They both write : $\frac{1}{2}$ is the family name for $\{\frac{1}{2}, \frac{2}{4}, \frac{3}{6}, \frac{4}{8} \ldots\}$

$\frac{1}{3}$ is the family name for $\{\frac{1}{3}, \frac{2}{6}, \frac{3}{9}, \frac{4}{12} \ldots\}$

So $\frac{1}{2} + \frac{1}{3} = \frac{3}{6} + \frac{2}{6} = \frac{5}{6}$ and $\frac{1}{2} - \frac{1}{3} = \frac{3}{6} - \frac{2}{6} = \frac{1}{6}$

learn. We claim validity on the basis of a number of factors. We recruited articulate and highly motivated children to the sample and encouraged their motivation and articulation. Interviews focused on observable actions and were initially non-directive (greater direction only appearing after the children had exposed what they considered to be appropriate learning behaviour). The interviews followed

Figure 17.3a: Full Learning Pack for Area

What does area mean ?

The area is the amount of space inside a shape.

The area of this rectangle

has been shaded :

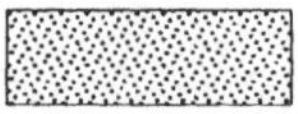

The area of this circle

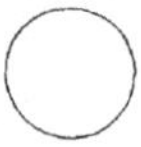

has been shaded :

The area of this triangle

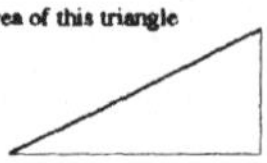

has been shaded :

How to find the area of a triangle.

Here are two ways to find the area of a triangle :

1

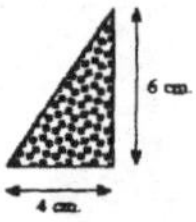

Here is the triangle you want to find the area of.
Its base is 4cm long and its height is 6cm.

To find the area we first have to imagine the triangle to be part of a rectangle the length and width of which are the same as the height and base of the triangle.

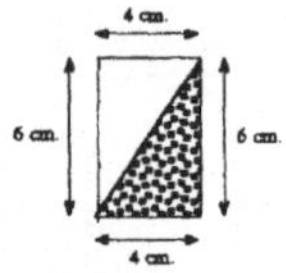

You can see that the white triangle is exactly the same as the shaded triangle (except that it is upside down).

Both triangles have <u>the same area</u> and the area of one triangle is half the area of the whole rectangle.

The area of the rectangle is 6 x 4 = 24 square centimetres.

So the area of the shaded triangle is 24 ÷ 2 = 12 square centimetres.

2 Another way to find the area of a triangle is to use the formula :

$$\text{Area} = \frac{1}{2} \text{ the base x the height}$$

For our triangle, $\frac{1}{2}$ the base is 2 cm.
and the height is 6 cm.

So, using the formula : Area = 2 x 6
Area = 12 square centimetres

How to find the area of a rectangle

Here are 2 ways to find the area of a rectangle :

1

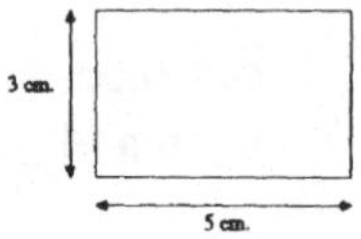

Here is the rectangle you want to find the area of.

If we lay a grid of squares which all have sides 1 cm. long over the rectangle we can count how many whole squares there are :

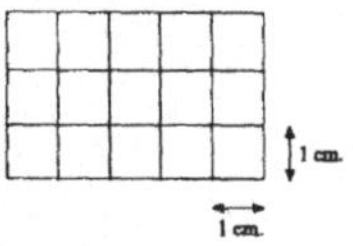

If the rectangle leaves some of the squares incompletely filled, then we need to <u>estimate</u> how many extra squares the incomplete squares amount to.

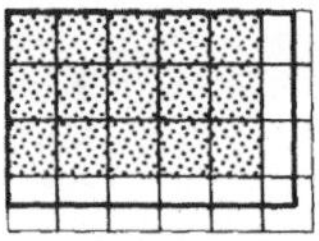

This rectangle is covered by 15 centimetre squares + <u>about</u> 4.5 other centimetre squares.

The area of this rectangle is <u>about</u> 19.5 square centimetres.

2 Another way to find the area of a rectangle is to use the formula :

$$\text{Area} = \text{length x width}$$

So, if like the rectangle at the top of this sheet, the length is 5 cm. and the width is 3 cm. then :

Area = 5 x 3 = 15 square centimetres

Figure 17.3b

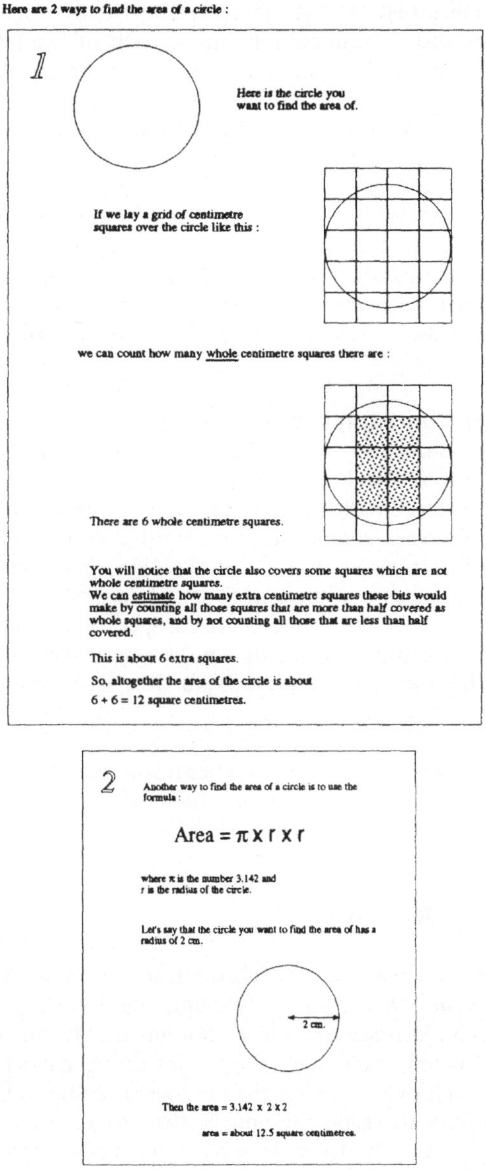

the children's actions and verbalizations throughout. All claims that we make in analysis are directly linked to children's actions.

We have taken several lines in analysis. First, we simply described children's routes through the phases of the task together with their own commentaries and the researcher's questions/prompts. This involved writing out the interviews with the notes of actions integrated into the text. It will be recalled that these routes were self-chosen. Secondly, regardless of routes, we listed, across the sample, children's responses and reactions at each phase of the experience. That is to say, we listed all answers and comments relating to each of the pre-task items in the mathematics task. Then we listed all responses to each section of each sheet in the learning pack and, finally, we listed all answers and comments relating to each of the post-task items in the mathematics task. This we did separately for each of the four topic areas in mathematics.

Using the listings and the descriptions we then searched for themes in the data which would reflect common experiences of, and approaches to, learning with particular reference to the children's use of subject knowledge. Here we used a form of recursive comparative analysis. This is a process in which unfolding descriptions of pattern are constantly tested and refined to take account of all relevant data. Cooper and McIntyre (in press) have described this as a 9-stage process as follows:

1. reading a random sample of scripts;
2. identifying points of similarity and differences amongst these scripts in relation to the research question;
3. generating emergent theories in relation to the research questions;
4. testing the theories against the first sample of scripts;
5. modifying the theories to account for exceptions;
6. testing the theories against a new set of scripts;
7. generating new theories against transcripts already dealt with;
8. carrying all existing theories forward to new scripts; and
9. repeating the above process until all data have been examined and all theories tested against all data.

In this way we identified two main themes in the children's reading of the text. These were: purposes for reading and the level of processing of the text. We discuss these in the following sections.

Purposes for Reading

A common way for children to approach the learning materials on offer to them within the various tasks was to work through them simply because they were there. Many of the children seemed quite content to read through text or answer questions however, when asked why they were doing particular things or what they were hoping to achieve either no answers were forthcoming or their answers indicated an intent only to reach the end of the work. Nathanial (10-years-old) was asked what he was hoping to achieve by working his way through the mathematics learning pack on fractions. He said, 'Nothing really . . . It's just the easiest way of going through them.' This was not an isolated response. It was given by many other children, from both age groups and of both sexes.

Among the sample there are examples of children engaging with the learning packs in a distinctly different way from that identified above. Whereas the learning packs were on some occasions perceived as just more material to work through, on others the learning pack materials were perceived by children as containing something which they needed to know in order for them to be able to complete the learning task successfully, that is, for them to be able to answer the pre-task mathematics items correctly. Such a perception was accompanied by children identifying what we shall refer to as a 'learning need'. To expand on this, at some point during their engagement with the learning materials children would recognize that they needed to know something or find something out, specifically so that they could complete the task to their better satisfaction. This identification of a need to learn something would then, in turn, lead to a particular search for information, or execution of a particular activity. Their engagement with the learning materials would take on a more deliberate air; there would be an intention to learn, and the aim of that intent was, to their way of thinking, to complete the learning activity appropriately. To exemplify this category of knowledge use we shall describe an instance of one child engaged on the fractions learning task.

Colin (10 years) came to this learning task with a rich knowledge base in mathematics. With particular reference to his knowledge of fractions he was able to add and subtract fractions with common denominators and applied this knowledge successfully in answering associated verbal problems. He did not know how to answer questions involving addition or subtraction with mixed denominators, although his protocol for one such problem indicated that he may have had some relevant experience to do with equivalence. For example, he tried to answer the following question: 'Nick ran 4 2/5 km and then walked a further 2 1/4 km. How much further did he run than walk?'

He said, '. . . I'm just trying to get the top one [the numerator] . . . I want the bottom and the top to having something in common.' This procedure, however, was not well established in his repertoire; he was also unable to answer the question 5/8 + 1/3 = .

When he was given the learning pack Colin read through it in page order. In the process he discarded several of the sheets without involving himself in answering the questions because he was confident that he already knew the sheet's content and so had no use for them. The following are indicative of his comments as he did this:

> I know all about this.
> That's easy enough.
> . . . it's just times tables.

When not discarding sheets, Colin's use of the learning pack was characterized by a search for something which might help him answer the mixed denominator problems on the pre-task. For example, on reading one of the sheets he spent some time looking at the arrangement of vulgar fractions printed on some discs (See Figure 17.4).

He articulated his thinking at this point by saying, '. . . I was looking for things in common.' However, he concluded that this sheet was '. . . no use . . .' because '. . . all your top ones [numerators] are smaller than the ones at the bottom [the denominators].' Ultimately Colin came to a sheet which explained equivalence and addition of fractions with mixed denominators (see Figure 17.5). Here he closely followed the written explanation and commented '. . . That helped [he points to '1/2 is the family name for . . .']'. I looked along the lines and found that

Figure 17.4: An Extract from the Fractions Learning Pack

Each disc has been cut in half

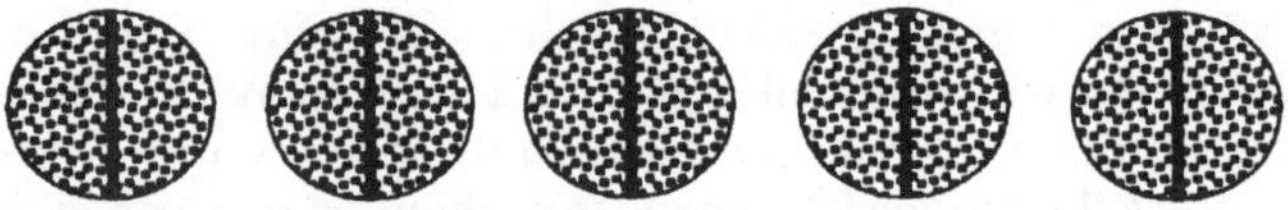

We can count in halves : $\frac{1}{2}$ $\frac{2}{2}$ $\frac{3}{2}$ $\frac{4}{2}$ $\frac{5}{2}$ $\frac{6}{2}$ $\frac{7}{2}$ $\frac{8}{2}$ $\frac{9}{2}$ $\frac{10}{2}$

or we can count in mixed numbers : $\frac{1}{2}$ 1 $1\frac{1}{2}$ 2 $2\frac{1}{2}$ 3 $3\frac{1}{2}$ 4 $4\frac{1}{2}$ 5

Figure 17.5: An Extract from the Fractions Learning Pack

Next, Peter and Gill talk about how to add and take away the fractions $\frac{1}{2}$ and $\frac{1}{3}$.

We could add and take away the fractions if they both had the same denominator.

They both write : $\frac{1}{2}$ is the family name for → $\{\frac{1}{2}, \frac{2}{4}, \frac{3}{6}, \frac{4}{8}.....\}$

$\frac{1}{3}$ is the family name for → $\{\frac{1}{3}, \frac{2}{6}, \frac{3}{9}, \frac{4}{12}.....\}$

So $\frac{1}{2} + \frac{1}{3} = \frac{3}{6} + \frac{2}{6} = \frac{5}{6}$ and $\frac{1}{2} - \frac{1}{3} = \frac{3}{6} - \frac{2}{6} = \frac{1}{6}$

a sixth . . . the denominator is the same there [he points to the ringed 2/6] . . . so I put it as them and I got the answer.' He then returned immediately to the analogous question on the pre-task and transferred the procedure he had just learnt.

Colin treated this task as problematic — as being like a quest. He was involved in doing more than just reacting to elements of the learning pack; he was involved in specific searches and in explicit questioning of what he was experiencing in relation to what he already knew. His activities in the name of learning were aimed towards finding out specific information and his use of the learning pack materials suggested that he perceived them more as a learning resource than as a series of tasks to perform.

Knowledge use which occurred within this category tended to be accompanied by a more explicit presence of executive processes, that is processes by which children monitored and regulated their own thought processes. For example Clare (8), identified a specific learning need when told by the researcher at the end of the interview that one of her answers was still wrong (specifically that 3/10 + 4/10 = 7/20 was wrong). Initially she tried to probe the researcher to help her out, but when this proved fruitless she returned her attention to the learning pack, '. . . I could look in there.' She then instigated a search through the pack for an analogue to the problem (I'm looking if there's some of these . . .'). Like Colin in the previous example, Clare discarded sheets which did not, in her opinion, help and then began to give voice to what can best be described as an 'inkling' she had. She said, 'I have a funny feeling that *that* [she points to the '20' in her answer] is ten . . .' When asked to enlarge on this feeling she added, '. . . because you only add the top numbers . . . We know that this is seven over ten because we don't add these [the denominators].' The significance of this conclusion then struck Clare, 'If that one's wrong then all the others would be wrong.'

Clare's activity is significant because she *used* the learning pack. Rather than merely regarding the pack as more work to get through Clare interrogated the pack with the explicit intention of completing the task to solve particular problems or remedy specific deficits.

Some children dug deeper than this. In asking children to try and learn particular things from the materials provided for the purpose we were alert to the likelihood that children's perceptions about what one does in the name of learning would be different. We have so far indicated two such perceptions: firstly where the task of learning is indistinguishable from the task of getting through the materials and, secondly, where the task of learning is perceived as one necessitating the more appropriate completion of the task through the pursuit of specific procedures. In this section we shall explore a third perception in which children's actions and thoughts in the name of learning seem to have been aimed at deeper levels of understanding. This we refer to as the 'intention to get to the bottom of something'.

In collecting examples in which children demonstrate deliberate attempts to get to the bottom of something we discovered that there are different processes through which these attempts are mediated. We are not yet in a position to be exhaustive about the number of mediating processes through which children try to get to the bottom of something. Two processes seemed salient, namely the use of analogy and the use of hypotheses.

Holly (8-years-old), by way of example, accepted the task of trying to learn about averages from one of our mathematics learning packs. She identified a 'learning need' the moment she read the first question on the pre-task. She asked, '. . . What does "mean number" mean?' She tried initially to interpret the term in relation to her everyday understanding of 'mean', but found this approach both unhelpful and, she thought, inappropriate. ('Well a mean person isn't going to give the right number of things . . . Except that I bet you that isn't what it means.')

She seemed more at ease with the word 'average' which she encountered as she was reading the first learning pack sheet, and illustrated her understanding of the concept with an analogy: 'Average to me means that two things are the same . . . Like you get a weight and you have to pour the salt out 'till it's average.' Here the use of analogy is exemplified in Holly's use of the term 'it's like'. Later she equated the idea of 'average' explicitly with the idea of 'equal to' thus clearly indicating *her* understanding of the concept. ('Average also means equal . . . Like when you pour salt into a jar you say "this is average" or you could say "this is equal" . . .')

In many ways Holly's activities in the name of learning resembled those exemplified in previous sections. She approached each of the learning pack sheets in turn and seemed to accept that doing the work component of the sheets would help her to answer the pre-task questions. ('I think I ought to go onto these sheets . . It's better that you do the whole answer pack as once.' and '. . . The [learning] pack is more for learning things. So I do everything here, I learn lots of things and then I go onto the questions.') However, her approach belied something much more significant in what she did. Rather than concentrating solely on trying to find out how to calculate an average, she seemed to be deliberately addressing (evaluating and updating) her current understanding of the concept of average.

To start with she quickly overcame her everyday confusion with the word 'mean' through her reading from the first card in the learning pack of the sentence 'This is called the "average" or the "mean" number of visitors.' She said, 'Ah! . . . Now I now what "mean number" means . . . It means like average . . . "Mean" is not "getting mean".'

Having removed this anomaly Holly continued to read the sheet and discovered that averages were to do with division. ('Oh! It says divide . . .') Holly, however, was not content just to learn how to calculate an average and proceeded to enlarge on what she thought averages were all about. To do this she chose a real-world context to exemplify her understanding. ('It's like if you have people and they all gave you five pounds . . . And then you suddenly got hold of another twenty people and they all give you five pounds . . . And then you suddenly got hold of another twenty people, and it wouldn't be very fair to give them all one pound when some of them hadn't contributed.') Again the phrase 'it's like' appears in her language illustrating the way in which she was trying to make sense of — to get to the bottom of — the idea of average. Here it would seem that Holly's understanding of 'sharing' and 'fairness' were included in her analogy for the concept of average.

Holly then read the next sheet in the pack:

AVERAGES 2

Patrick is given a different amount of pocket money each week.
For the first week he was given £2. For the second he was given £4, and for the third week he was given £3.
When he was asked how much pocket money he gets a week he found the 'average' mean amount of money he receives.

To find the mean:
(1) Find the total of Patrick's money:

£2 + £4 + £3 = £9

(2) Divide this total by the number of weeks:

£9 ÷ 3 = £3

So Patrick's mean amount of money (or the average) is £3.

Here she felt it necessary to provide for herself some reason as to why Patrick (the boy in the text example) should need to give his answer to the question in the form of an average. ('. . . And he would have to say three pounds because he couldn't go "Oh I get two pound, then four pounds, then three pounds" because everybody would be confused.')

A similar form of reasoning went into Holly's reading of the next sheet, only this time she was prompted to provide another — and this time a very powerful — analogy expressing her understanding of average. She explained, '*It's like* [my emphasis] if you had a pile of . . . different piles of bricks . . . [she points to some wooden blocks stored on a shelf in the classroom] . . . and you went like that with you arm. [She sweeps her arm across the imaginary pile of blocks] So that they all . . . ended up with the same number in each pile.'

It seems to us that Holly was engaged in a quite different learning activity from those we have already exemplified. Holly seemed genuinely puzzled by the terms 'average' and 'mean' and what the terms meant in the context of mathematics. She read the text on the sheets and interpreted their content with the intention of understanding what the terms meant and what purpose the concepts served. This she achieved by interpreting the content of the materials provided for her in relation to her mathematical understanding of addition and division (sharing) and her everyday understanding of human communication (the reasons and purposes for expressing information in particular ways). These interpretations she expressed in terms of explanations and analogies. These, in turn, were then dropped or amended to take account of new situations she encountered and new interpretations she made of those situations.

Different purposes give different directions to learning. Purpose therefore serves as a significant 'knowledge recruiter'. By acting through the mechanisms of selective attention, the goal or purpose brought to an activity works to highlight or to cast into shade, aspects of the task environment. To what degree did children bring different purposes to the text in the mathematics.

We can summarize the above purposes or intentions as follows:

1. an intention to complete a task where the emphasis is on the completion of work;
2. an intention to complete a task where the emphasis is on the identification of a specific learning need; and
3. an intention to get to the bottom of something where the emphasis is on achieving a deeper understanding of the material at hand.

Using these categories as a framework for the analysis, children's responses to each of the sheets on the mathematics tasks were coded. For this analysis the listings of reconstructed data were used and each entry coded. Coding decisions were made on the basis of a child's stated purpose or from an inference made on the basis of a child's actions. The following statements exemplify children's purposes within each category for the mathematics tasks:

Category 1: intention to complete a task where the emphasis is on the completion of work.

Table 17.1: Incidences of Different Purposes Expressed as Pupils Engaged with the Maths Packs

Intention	*Topic* fractions (N = 13)	averages (N = 10)	area (N = 14)	ratio (N = 6)
complete the task as work	44	22	29	4
learn a specific procedure	6	3	13	1
get to the bottom of something	1	5	7	10

Alex Towler (8) responded to a question in the fractions learning pack by saying 'Instead of putting the add in, I took out the plus sign and pushed the number in its place . . . I saw in the example that that's what they were doing, and so I did the same.'

Category 2: intention to complete a task where the emphasis is on the identification of a specific learning need.

Holly Jury (8) approached the first of the sheets in the averages learning pack by saying 'I need to know what to do with them [referring to a set of numbers in a chart]. Like if it said add these together or said multiply these numbers or subtract these numbers . . just some information to tell me what to do.'

Category 3: intention to get to the bottom of something where the emphasis is on achieving a deeper understanding of the material at hand.

Holly Wiles (8) read a statement from the averages learning pack and said 'Aaah . . . Now I know what "mean number" means . . . It means like average . . . Mean is not getting mean. It's like getting like average . . . I don't know who called it mean . . . except that it probably fits in later on.'

Table 17.1 shows the frequencies of different intentions across the different tasks. It clearly reveals that far and away the predominant purpose of the children in approaching the learning materials was simply to get to the end of them as a job of work.

It could be that those children who engaged with the ratio task were more learning-oriented than those engaged on the other tasks. However, case studies we have done elsewhere suggest that it is misleading to label individuals in this way. That is, children often engaged in quite a different way with different subject matter. A more likely explanation, we believe, is that the content and presentational styles of the different leaning packs influenced the purposes which children adopted when faced with trying to learn from them. We believe that the fractions learning pack, for example, was more suggestive to the children of worksheets and work books encountered in their school mathematics experiences and, therefore, their normal working practices for completing such work were brought into play. The ratio task, in comparison, with a greater emphasis on explanation and worked example gave those children who engaged with it little opportunity for exercising their normal working practices and, instead, prompted them to try and make sense of the sheets' content, that is, to get to the bottom of the concept of ratio.

Whatever the case, the predominant intention to 'work the package' i.e., to go through from the beginning simply to get to the end, was, as our illustrations vividly illustrate, a serious barrier to learning since it was associated with no intention therefore no effort to learn.

Figure 17.6: Examples of the Science Text Presented to the Pupils

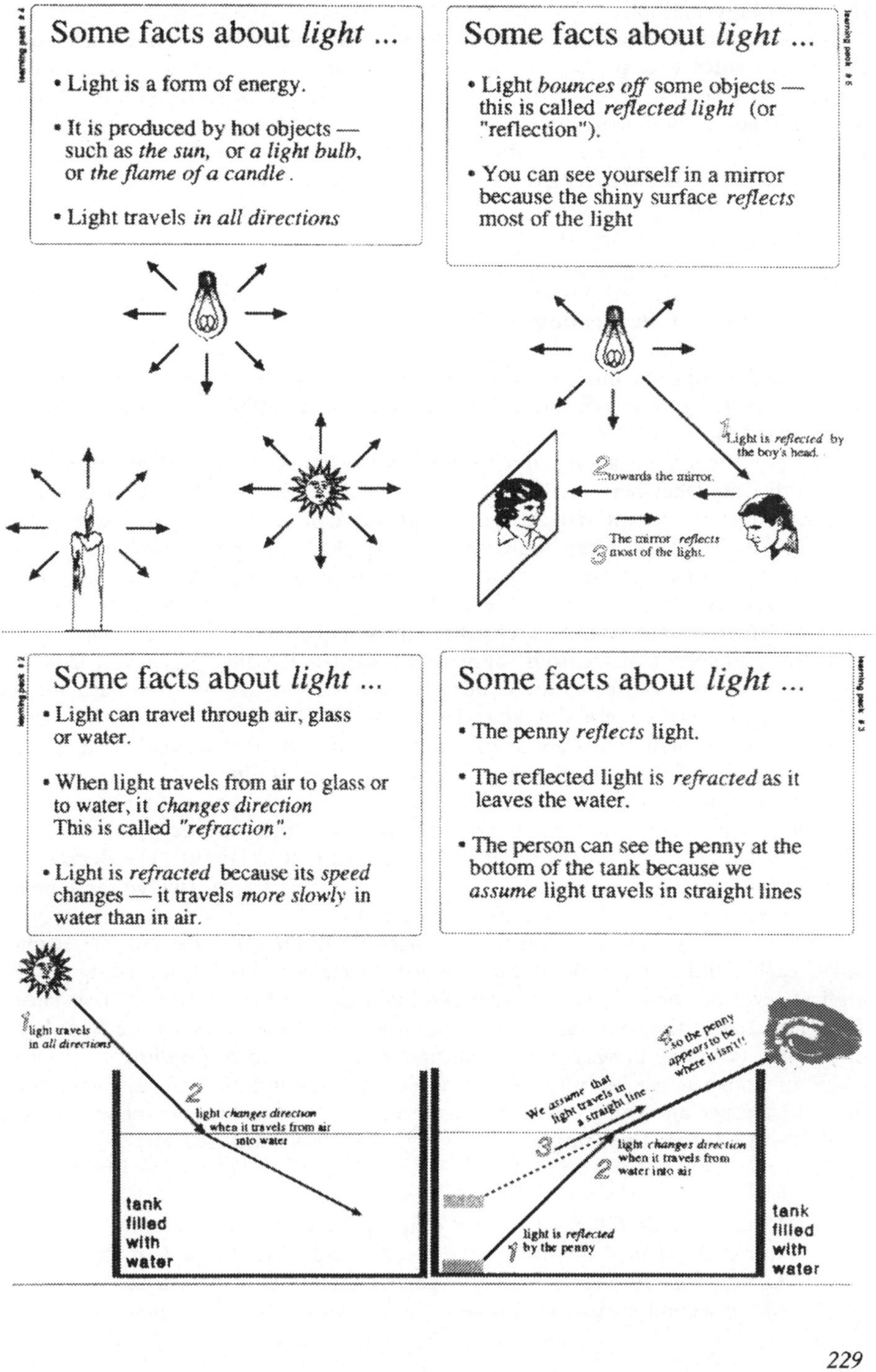

We should note in passing an interesting comparison between the children's intentions for mathematics and those for science. The science text materials are shown in Figure 17.6.

It proved quite straightforward to discern the children's intentions in reading the mathematics text perhaps because in the main they were so ostentatiously completion-oriented. In contrast it proved impossible to identify children's purposes in reading the science text. Despite determined prompting the very children who were clear about their intentions in mathematics were entirely unforthcoming in respect of the science text. Whatever it says about science it highlights that for studying mathematics, text is there in the main simply to be got to the end of, a strong reflection perhaps, of their classroom working practices.

Levels of Processing

How did the children interpret the text? In pursuing this question we used the scale of 'constructive activity' developed by Chan *et al.* (1992). The scale is shown in detail in Table 17.2.

Briefly the scale is an attempt to characterize think-aloud data in response to text. It distinguishes between five levels of sophistication on three dimensions of response: declarative, interrogative and evaluative. In developing the scale, Chan *et al.* (1992) have taken the view shared here that comprehension is an active process of construction involving the reader in interacting with the text rather than merely receiving it. Their scale was developed from the analysis of students' reports of their cognitive processes using a protocol analysis (Bereiter *et al.*, 1988) based on a general constructivist approach (Scardamalia and Bereiter, 1984).

As the texts on offer to the children in our sample in both the mathematics and the science tasks were similar to those use by Chan *et al.* (in that they consisted of discrete statements about the content in question) it seemed an appropriate analytic tool for our purposes: the scale has face validity.

For illuminative purposes in this analysis we have chosen to compare the children's responses to the science text with those to a mathematics text. We chose the area text as the one most similar to the science text in layout. The decision to make this comparison automatically reduces the sample size to twelve these being the children who worked through the area tasks.

Using the listings of reconstructed data for both the area and the science tasks, each child's response to each of the sheets was coded according to the highest level of constructive activity exhibited during the reading of that sheet. Thus for each of the four sheets in the science task and for each of the five sheets in the area task a child was given a rating from 1 to 5 corresponding to the five levels of constructive activity. There were twelve children who engaged with both the science and the area tasks. Of these, five were aged 8–9 and seven were aged 10–11. Across the sample there were seven boys and five girls.

Here are some examples of children's responses for each of the five levels: -

- **Level 1**: Will Tillett (8) read the first paragraph on the second area sheet and said 'That looks quite easy . . It's kind of easy, kind of hard.'
- **Level 2**: Lucy Marsh read the 'partial square' explanation at the bottom of the second area sheet and said 'I didn't know that . . . When you put the

Table 17.2 Levels of Constructive Activity in Learning from Text

Declarative	Interrogative	Evaluative
	Level 1: Prefactual confabulation	
States the topic or repeats fragmentary parts of the text; or gives associations of irrelevant personal knowledge when cued by isolated words.	Asks questions about isolated words.	Reacts to isolated words or brief fragments of text by giving nonanalytic affective responses.
	Level 2: Knowledge/detail retelling	
Makes near-verbatim repetition of the text statement; or elaborates knowledge which is tangentially related to the text when cued by a text proposition; or makes overextended inferences based on personal knowledge when cued by a text proposition.	Asks questions by appending an interrogative word to the text statement.	Acknowledges the text to be true or rejects the text, and makes nearverbatim repetition without giving any reason.
	Level 3: Assimilation	
Paraphrases the text statement to clarify what the text says; or recalls personal knowledge relevant to the text; or paraphrases the text statement and adds simple elaborative inferences to the text.	Asks questions relevant to the text using one's general knowledge.	Disagrees with the text statement and recalls personal knowledge to contradict the text; or agrees with the text statement and recalls personal knowledge related to the text.
	Level 4: Problem solving	
Refers to personal knowledge or previous text statements to explain or interpret new text information.	Formulates text problems by identifying the source of confusion, and hypothesizes to resolve the problems.	Evaluates the text or examines personal knowledge by identifying and resolving the conflict between personal knowledge and text.
	Level 5: Extrapolation	
Constructs inferences and hypothesizes to extrapolate knowledge not already known or shown in the text.	Formulates situationbased problems that go beyond the presenting text to extrapolate world knowledge.	Evaluates conflicting information to resolve inconsistencies leading to the synthesis of a new idea.

Source: Chan *et al.* 1992

Table 17.3: Incidence of Levels of Constructive Activity in Response to Text in Science and in Area

Level	Science	Area
prefactual confabulation	3	22
knowledge telling	26	29
assimilation	9	6
problem-solving	4	1
extrapolation	1	0

grid estimate just to see what it is . . . You could do the ones that are under a half . . put those down, and ones that are half don't count . . . It's just a guess.'

- **Level 3**: Daniel Webb (10) read the middle section of the second area sheet and said, 'The area of this rectangle is 19.5 square centimetres . . . There's no point . . . can't do point in area because the reason is there's either so many square or there's so many square . . . You're asking for the whole squares.'
- **Level 4**: Laura Stoodley (8) read the explanation for finding the area of a triangle on the third area sheet and said, 'Oh, I get how to do that one now . . . Well you have to pretend that it's an oblong, and then you have to find the area of the oblong and then divide it by two.'
- **Level 5**: Lucy Marsh (11) read the statements explaining refraction and said 'It's new to me about the light changes direction because here [card 1] the light goes in all directions, but then it says the light changes direction. So it's a bit odd, because it goes in all directions and then it changes direction.'

The number of times that each level was recorded for both the science and the mathematics tasks was then totalled.

The total for each of the five levels of constructive activity for the science and the area task are shown in Table 17.3.

Two points are worth noting in regard to Table 17.3. The first is the very low levels of response to the mathematics text. Almost 50 per cent of engagements or responses involved 'non-analytic responses' to isolated words or 'repetitions of fragments' of text. The modal response to the area text is characterized by '*ad hoc* agreements with, or dissentions from, the text without reason' or by 'near verbatim repetition'.

We should emphasize that neither the structure of the texts nor the specific contents were unfamiliar to the children. As we stressed earlier, the texts were close facsimiles to popular and readily available curriculum materials. It seems that the children in our sample were hardly reading for meaning where they could be said to be reading at all.

The second point is that the children were more impressive with the science text there being a greater proportion of incidences of level 3 and above. Combining incidences at level 3 and above make the data in Table 17.3 amenable to analysis using X^2. For this distribution X^2 is 15.06 which for 2 d.f is significant at $p < 0.001$. The children perform significantly better with science text than with mathematics text suggesting that the difference in academic subject rather than limitations on the part of the children is strongly associated with the depressed levels of performance on the mathematics text. Children can read with more effect than they appear to deploy on the mathematics task.

Discussion

Before elaborating on the implications of our analysis some cautionary remarks are perhaps in order. The settings in which we studied with our sample of children were not normal classroom circumstances. Our interviews took place in a one-to-one relationship in quiet surroundings. Our materials, whilst close facsimiles of those to be found in commercial mathematics schemes were, none the less not identical to scheme materials. Our data were gathered in closely supportive interviews which had the potential, at least, to be intrusive.

We take the view that none of these factors worked to the detriment of the performance of the children. Indeed the likelihood is that if they had an influence it was to enhance performance. Our design of material cleaned up some of the worst excesses of commercial representations. The quiet atmosphere and the interview's personal attention were both supportive.

Two more problems are worth stating. First our samples were small. In regard to the interactions with text, sample size was down to twelve children. Generalization out of the sample would be unsafe and unwise. Secondly our data took the form of think-aloud protocols. These descriptions of thoughts cannot, of course, be equated with thought itself. Protocols are open to a wide range of reservations and criticisms. Do they constitute 'on-line' descriptions or *post hoc* rationalizations? Everything we know about the limitations of working memory would lead us to conclude that 'talking aloud thoughts' is a contradiction in terms. We take the view that 'talk aloud' records of thoughts are, at best, correlates of thinking in action. This notwithstanding they constitute important data. The ability to communicate thoughts is a significant learning process in itself and in our analysis we need to take our data no more and no less seriously than that.

What do they tell us? They tell us that the overwhelming intention held by the children in our sample (N = 48) when engaging with text in mathematics was simply to get to the end of it. This was overtly claimed in 68 per cent of the cases where an intention could be identified. Put another way, amongst a set of articulate children who were committed to engaging in learning and for whom the possibility of learning had been created by taking them to a point of difficulty which they themselves acknowledged, text was merely there to be got through. We believe these children believed they were pursuing learning in mathematics. In other words, they intended to get through the available text in the name of learning. Their actions in this respect are entirely consistent with Bereiter's (1990) description of a 'school-work module'. Bereiter argues that learners organize their intellectual resources according to their interpretation of the context they are in: their intellectual resources become 'modularized'. His research on reading and writing suggest that many, if not most children adapt to school as a job forming the above named 'school-work module'. Such a module, focusing on activities as work, on completion and production rates for example provides a coherent response to classroom events. It marginalizes learning which becomes no more than coincidental with school work.

Children, in Bereiter's view, develop a school-work module very early in their school careers and this module is extremely difficult to restructure.

This is an interesting perspective in that it is entirely consistent with classroom-observation studies of children's response to activities in general and mathematics work in particular (Bennett and Desforges, 1984; Desforges and Cockburn, 1987). These studies show that it is the teachers who read the schemes, explain the examples and analyse the illustrations: the children are left to do the work. When given the opportunity, as in this present study, to take a learning approach to the text it is perhaps not surprising that the children treat the print as something to be completed in some way. Their completion activity in two-thirds of instances was, in terms of deliberative learning, banal. It was not, however, entirely so.

In 16 per cent of cases there was a deliberate attempt to learn a specific routine and in a further 16 per cent there were deliberate and sustained attempts to garner understanding. It is important to emphasize that we are not here categorizing children but their intentions. Particular children evinced a range of intentions. What we are seeing, we feel, are the effects of general working practices in using text materials in mathematics lessons.

Even given the children's expressed intentions in engaging with the text, the levels of constructive activity were, in our view, surprisingly low. The modal, level 2, response constitutes, at best, a direct readout of the text. It, at least, is consistent with 'getting to the end'. But the prominent occurrence of level 1 engagements could hardly be said to constitute reading at all. This is a particularly interesting point when contrasted with the same children's response to the science text where 'pre-factual confabulation' was rare. Clearly these children can 'read'. Yet they appear not to see the skill as relevant to learning in mathematics. Alternatively, there is something about the context which renders their general reading skill inoperable.

Many if not most of the children in our sample evinced the capacity for logical reasoning and sustained analysis so to find them rendered confabulists says more about the contexts than the pupils.

As our illustrations have shown, and as the decriptors in Table 17.2 systematize, operation at level 1 is associated with the recruitment of fragments of knowledge which do not cohere into a line of argument or of analysis. Operation at level 2 recruits knowledge cued by the text but mainly of a tangential and/or personal nature. If learning is taken to be a constructivist activity there is little at these levels of processing beyond contingent and passing associations on which to build.

Clearly there is provocative material here for teachers of reading on the one hand and for teachers of mathematics on the other. For the former, reading is clearly not, for our sample of children, a readily transferable skill. For the latter, children's core capacities in reading are being wasted in some way. Reading as a constructivist approach to learning is not being deployed here: it is therefore a wasted asset.

Our main purpose however is not to jump to conclusions about the failures and challenges of practitioners. Our main purpose is to understand the observations we have presented: to offer a theoretical account on which practice might be evolved.

We have already made strong allusions to the main parameters of our explanation. The children in our sample, it appear to us, adopt the working practices of the classroom in their approach to mathematics. They equate learning mathematics with doing school work in mathematics. Doing this work does not, it seems, involve reading for meaning or understanding. The text in their schemes, it appears is merely and at best, the continuity material. It is not, for them, an object of knowledge. Left to their own devices they cast their eye over it: pressed by the interviewer and the text begins to afford contingent, unanalysed and unused associations.

Our research leaves open the question of the relationship between levels of constructive active and prior knowledge in the learning process. As we have indicated, constructive activity recruits available knowledge. Low levels recruit contingent associations, higher-level recruit multiple connections and create the possibilities of making new connections. Prior knowledge possibly has its effects on learning through constructive activity. But is the level of constructive activity determined itself by a mass factor of available prior knowledge or can higher levels of constructive activity be promoted on a narrow basis of prior knowledge? Chan *et al.* (1992) attempted to explore the relationship between prior knowledge and constructive activity using instructional techniques. That is to say, they taught a group of children, through modelling, to engage in higher constructive levels in text processing but this did not produce a significant increase in learning from text over and above a control group not so taught. Chan *et al.*, concluded in essence that prior knowledge has its effect on learning only through constructive activity.

The data in Table 17.1 allow us to make an interesting, albeit tentative, comment on this matter. Table 17.1 shows that levels of constructive activity were greatly increased on the ratio task. A note of caution is essential because different children worked on different tasks and this, of course, introduces sources of unaccountable variance. But taken at face value the difference is interesting because the children in our sample were least knowledgeable and least competent with the concept of ratio. The high levels of constructive activity were thus associated with higher levels of relative ignorance. What we cannot say, because we did not collect the appropriate data, is that the higher levels of constructive activity here were associated with higher levels of learning. As we said earlier, the study leaves the question of the relationship between prior knowledge, constructive activity and learning yet to be exposed.

Perhaps safe conclusions for practitioners would be that reading is a potentially valuable process for learning mathematics, that ordinary reading instruction does not appear to transfer well to mathematics text or mathematics working practices and that further reading instruction is unlikely to be helpful unless it is geared specifically to mathematics subject content. Until we know better or different, children need to be taught to read in mathematics.

References

Bennett, S.N. and Desforges, C. (1984) *The Quality of Pupil Learning Experience*, London, Erlbaum.

Bereiter, C. (1985) 'Toward a solution of the learning paradox', *Review of Educational Research*, 55, 2, pp. 201–26.

Bereiter, C. (1990) 'Aspects of an educational learning theory', *Review of Educational Research*, 60, 4, pp. 603–24.

Bereiter, C., Burtis, P.J. and Scardamalia, M. (1988) 'Cognitive operations in constructing main points in written composition', *Journal of Memory and Language*, 27, 3, pp. 261–78.

Blyth, W. (1965) *English Primary Education, Vol 2*, London, Routledge and Kegan Paul.

De Bono, E. (1976) *Teaching Thinking*, London, Temple Smith.

Chan, C.K.K., Burtis, P.J., Scardamalia, M. and Bereiter, C. (1992) 'Constructive activity in learning from text', *American Educational Research Journal*, 29, 1, pp. 97–118.

de Charms, R. (1972) 'Personal causation training in the schools', *Journal of Applied Behavioural Psychology*, 2, pp. 95–113.

Cooper, P. and McIntyre, D. (1993) 'Commonality in teachers' and pupils' perceptions of effective classroom learning', *British Journal of Educational Psychology*, 16, 3, pp. 381–99.

Desforges, C. and Bristow, S. (1992) 'The application of subject knowledge in the primary school: case studies in the use of knowledge of English', Paper presented to the *First European Conference on Educational Research*, Twente.

Desforges, C. and Cockburn, A. (1987) *Understanding the Maths Teacher*, Lewes, The Falmer Press.

Desforges, C., Bristow, S. and Goring, D. (1993) 'Young children's procedures for using subject knowledge to learn', Paper presented to the Annual Conference of the *European Association for Research in Learning and Instruction*, Aix-en-Provence.

Kameenui, E.J. and Griffin, C.C. (1989) 'The national crisis in verbal problem solving in mathematics', *The Elementary School Journal*, 89, 5, pp. 575–93.

NCC (1992) *Using and Applying Mathematics: Book It*, York, National Curriculum Council.

Pressley, M., Goodchild, F., Fleet, J., Zajchoeski, R. and Evans, D. (1989) 'The challenge of classroom strategy instruction', *The Elementary School Journal*, 89, 3, pp. 301–35.

Resnick, L.B. (1987) *Education and Learning to Think*, Washington, DC, National Academy Press.

Scardamalia, M. and Bereiter, C. (1984) 'Development of strategies in text processing', in Mandl, H., Stein, N.L. and Trabasso, T. (Eds) *Learning and comprehension of text*, Hillsdale, New Jersey, Erlbaum, pp. 379–406.

Singley, M.K. and Anderson, J.R. (1989) *The Transfer of Cognitive Skill*, Cambridge, Massachusetts, Harvard University Press.

Part 4

History, Mathematics and Education

The creation of mathematics in Sumer was specifically a product of . . . school. (Høyrup, 1980, p. 45)[1]

In considering the relation between mathematics, history and education, the main issue that springs to mind is to justify the importance of the history of mathematics to school and college mathematics.[2] Thus it is a nice irony to be able to turn the assumed relationship on its head, and to assert with Høyrup that school mathematics provides the origin of the discipline of mathematics, around 3000 BC. Clearly mathematical concepts and inscriptions predate this time. But mathematics was not systematically organized as a discipline until there was a need to teach it to professional scribes, following the invention of writing. Not only was disciplinary mathematics founded there, but Sumerian schooling has a remarkably modern ring to it. In a written disputation between two young scribes (this was a common learning device) one writes: 'When you do multiplication, it is full of mistakes. In computing areas, you confuse length with width.' (Hamblin, 1974, p. 105) Thus Sumeria and the history of mathematics also give us the first written evidence of mathematical misconceptions, some 4000 years ago.

In the modern day, there are many reasons why the history of mathematics is important for mathematics education, and why a historical approach should be incorporated into mathematics teaching. First of all, in the psychology of mathematics education, the parallel between the cultural development of mathematics and the individual's psychological development has been noted since the publication of Haeckel's biogenetic law 'ontogenesis recapitulates phylogenesis'. Of course as Freudenthal said 'We know for sure that this law is not true in a trivial way . . . The young learner recapitulates the learning process of [hu]mankind, although in a modified way.' (Steiner, 1989, pp. 27–8). Understood liberally, the parallel is a central feature of Vygotsky's psychological theory. Similarly, the revolutionary theory of science of Kuhn (1970), also recently extended to mathematics (Gillies, 1992), finds a powerful analogy in Piaget's theory of assimilation and accommodation in cognitive growth. Both are triggered by a contradiction or cognitive conflict. Bachelard's historically inspired concept of an 'epistemological obstacle' suggests that in mathematics *unlearning* is necessary to overcome epistemological obstacles (e.g., multiplication always 'makes bigger') in psychology as in history. Such parallels play an important part in the contributions to this section.

A second area of contact concerns problems and problem-solving. These are central both to the history of mathematics and to mathematics education. In both areas they stimulate knowledge growth. Historical problems, such as the Königsberg Bridge Problem, which stimulated Euler to create Topology (Wolff, 1963), also serve well to introduce students to network theory in today's classroom. A deeper parallel is in the realm of mathematical processes and strategies.

Pappus' specificaiton of analytic and synthetic problem-solving methods is used by psychologists today to distinguish levels of cognitive processing (e.g., Bloom's, 1956). Similarly, methodologists of mathematics such as Descartes and Polya have provided systems of heuristics that continue to drive research and teaching in problem-solving in psychology and mathematics education (Groner *et al.*, 1983; Schoenfeld, 1992). Problem-solving research in mathematics still has much to learn from the history of mathematics, and in the era of computers that history continues to grow and evolve new methods.

Thirdly, the history of mathematics plays the important role of showing how mathematics is socially constructed to meet the demands and needs of society. This supports the contention that mathematics and mathematical knowledge are not objective and neutral products, but that they serve certain historical and sectorial interests. Lakatos has argued that the history, philosophy and practice of mathematics and education are bound up together, and that the 'philosophy of mathematics turning its back on the most intriguing phenomena in the history of mathematics has become *empty*' Lakatos (1976, p. 2). Likewise, mathematics education turning its back on the history or social uses of mathematics also risks becoming empty. But in practice mathematics *is* widely presented as a decontextualized and ahistorical subject. For example, a piece of evidence quoted in the Cockcroft Report (1982, p. 141) states 'Mathematics lessons in secondary schools are very often not about anything. You collect like terms, or learn the laws of indices, with no perception why anyone needs to do such things.' Many school mathematics texts still concentrate on developing students' skills and concepts with little or no reference to either the social context or the history of mathematics. Such procedural and instrumental approaches to mathematics can make it empty and dull. But it is widely remarked that student and teacher attitudes and perceptions of mathematics are important factors in learning. Indeed, much of the research on gender and mathematics talks of the problems caused by the stereotypical perceptions of mathematics as a male domain (e.g., Walkerdine, 1988). Thus an important area of impact of the history of mathematics is on the formation of student (and teacher) attitudes to, and perceptions of, mathematics. At best, this means seeing mathematics as a living, breathing subject with its histories, struggles, triumphs, controversies and panoply of cultural endeavours permeating all of human existence.

These are just some of the issues for the philosophy of mathematics education that a focus on the history of mathematics raises. Not all are treated in the chapters in this section, but some important parallels and reflections are in evidence. This topic looks to be another growth point for research in the future.

Notes

1. The full sentence is: 'So, according to my hypothesis, the creation of mathematics in Sumer was specifically a product of that school institution which was able to create knowledge, to create the tools whereby to formulate and transmit knowledge, and to systematize knowledge.'
2. An issue of *For the Learning of Mathematics* (Volume 11, No. 2, 1991) is devoted to just this.

References

BLOOM, B.S. (Ed) (1956) *Taxonomy of Educational Objectives 1, Cognitive Domain*, New York, David McKay.

COCKCROFT, W.H. (Chair) (1982) *Mathematics Counts*, London, Her Majesty's Stationery Office.

GRONER, R., GRONER, M. and BISCHOF, W.F. (1983) *Methods of Heuristics*, Hillsdale, New Jersey, Erlbaum.

GILLIES, D.A. (Ed) (1992) *Revolutions in Mathematics*, Oxford, Clarendon Press.

HAMBLIN, D.J. (1974) *The First Cities*, The Netherlands, Time-Life International.

HØYRUP, J. (1980) 'Influences of Institutionalized Mathematics Teaching on the Development and Organisation of Mathematical Thought in the Pre-Modern Period,' Bielefeld, extract reprinted in FAUVEL, J. and GRAY, J. (Eds) (1987) *The History of Mathematics: A Reader*, London, Macmillan, pp. 43–5.

KUHN, T.S. (1970) *The Structure of Scientific Revolutions*, Chicago, Chicago University Press (2nd ed.).

LAKATOS, I. (1976) *Proofs and Refutations*, Cambridge, Cambridge University Press.

STEINER, H.G. (1989) 'Relations between historico-epistemological studies and research in mathematics education,' in BAZZINI, L. and STEINER, H.G. (Eds) *Proceedings of the First German-Italian Bilateral Symposium on the Didactics of Mathematics*, Pavia, Italy, University of Pavia, pp. 25–35.

SCHOENFELD, A. (1992) 'Learning to think mathematically', in GROUWS, D.A. (Ed) *Handbook of Research on Mathematics Teaching and Learning*, New York, Macmillan, pp. 334–70.

WALKERDINE, V. (1988) *The Mastery of Reason*, London, Routledge.

WOLFF, P. (1963) *Breakthroughs in Mathematics*, New York, New American Library.

Chapter 18

The Idea of 'Revolution' As an Instrument for the Study of the Development of Mathematics and Its Application to Education

Francesco Speranza

Some Methodological Remarks

'Revolutions in mathematics' is a theme for frequent debates (Gillies, 1992; Ernest, 1992; Speranza, 1992): there are 'revolutionarists' against 'continuists' (the latter is a classical term). In Gillies (1992) the first (and oldest) paper (Crowe, 1975), is continuist; many other contributions, not only in this volume, are 'revolutionarist', even if with some cautions.

In my opinion, a more flexible approach is suitable, for theoretical, a well as practical, reasons. Science does no more claim to find *rerum naturam*, but only to give us some explanatory models; in a similar way, the philosophy of science, in the study of the phenomenon 'science', can give us explanatory models, and, strictly speaking, nothing else. The same event, the same idea can then be studied (also by the same philosopher) from different viewpoints; indeed, a mixed method is perhaps the most useful for a deeper understanding.

For instance, Enriques is considered a continuist:

> Human reason is always the same, in all times [. . .], without this fundamental premise it would be useless to study the history of ideas. (Enriques, 1936, p. 16)

but he writes also, with regard to the science of the seventeenth century,

> To create anew the Democritean spirit and to think again to it, as a living thing, was not possible at Galileo's times, except on condition to overcome Democritus and to make out of his philosophy a richer science, as much as the needs of modern problems overcome the ancient ones. (*ibid.*, p. 340)

'Revolutionary' scientists, on the other end, frequently refer to some famous precedents (Copernicus' recall to Aristharcus, Galileo's to Pythagoras and Plato); the search for the reasons of continuity is useful also when we study the philosophy and history of science by means of revolutionary paradigms.

Both Kuhn (1962) and Crowe, so as other contributors (Gillies, 1992), apply a political metaphor, with the unavoidable consequence that they are led to seek for the 'entity (be it a king, a constitution, or theory) [. . .] overthrown and irrevocably discarded' (Gillies, 1992, p. 3). In the common language, however, we call 'revolution' also a radical change in beliefs and in behaviour of people, as in western countries in the last decades. In science, sometimes there are deep changes in the ways of thinking: Dunmore speaks of a 'meta-level' of mathematics:

> [. . .] the metamathematical values of the community that define the telos and methods of the subject, and encapsulate general beliefs about its nature. (Dunmore, 1992, p. 211)

Incidentally, Dunmore claims that 'mathematics is conservative on the object-level and revolutionary on the meta-level'. In any case, these changes are particularly interesting (let us consider what consequences they can have on teaching). Consequently, also if someone would prefer to use some other term, in my opinion it is suitable to speak also in such cases of 'revolutions'.

Some Parallels Between the Philosophy of Mathematics and the Philosophy of Science

First of all, if we speak of 'revolutions', we must consider also 'normal mathematics' (Mehrtens, 1976). In periods of normal science, scientists work on the grounds of certain 'paradigms', above all on 'solutions of puzzles' (Kuhn, 1962, Chapter 4). Well, also a mathematician works on certain problems that the community considers remarkable, with certain recognized methods, and he is particularly satisfied if he succeeds in solving a 'puzzle' that other mathematicians are not able to solve. On the contrary, if something doesn't work, this lays a blame on the mathematician, not on the theory.

In experimental science revolutions were often the consequence of an anomaly, which the preceding theory could not explain. In mathematics, only few cases are of this kind: among the most noteworthy, the crisis of incommensurable quantities (commonly 'the crisis of irrationals'), which overthrew the proto-pythagorean paradigm 'all is number', and the crisis of foundations, which unmasked every naïve application of a 'self-evident' logic. Perhaps also the non-Euclidean revolution can be connected with a crisis of the research programme for an improvement of Euclid's axiom system.

Noteworthy changes in mathematics are often provoked by a programme of systematization of foundational problems (let us remember that, according to Kuhn, a new theory can simply be at a higher level, which allows it to connect a group of theories at a lower level): obviously, meaningful changes in methods or in concepts must follow. The Cantorian (Dunmore, 1992) and the Fregean (Gillies, 1992) revolutions are of this sort; later on, we shall study Klein's Erlangen programme and Bourbaki's structuralism, both introduced in consideration of an unification of geometry and of mathematics respectively (both had a great impact on mathematical research and on mathematical 'meta-level').

In mathematics, it often happened that the old instruments were no more suitable for solving new problems, and so new ones were introduced. In such cases, many do not speak of a revolution: but (also because of the property of mathematics to change a 'method' or a 'form' into 'mathematical contents') a methodological change often provoked a 'real' and meaningful change. In the next section we shall see two instances.

From 'Methodological' Changes to 'Revolutions'

The case of algebraic symbolism is important (it suffices to compare two texts, of the late sixteenth and of the seventeenth century). Nevertheless it is atypical: the change was prepared nearly 2000 years before, with the use of letters as variables: Aristotle uses letters to denote *terms*, Euclid (but perhaps other mathematicians before him) *points*. Certainly, in both cases there is not a calculus, not even in an embryonal form, with a proper syntaxis, independent from natural language.

But there is another interesting point, concerning the 'slowness' of symbolistic development. A symbolism isn't really accepted, also in a context of natural language, if the fundamental rules of substitution of variables are not applied (Marchini, 1990). In the case of terministic variables, the first that used (in a demonstration about a syllogistic form), a substitution of the kind

$$A \to A, B \to B, C \to A,$$

was Alexander of Aphrodisia (Bochenski, 1956), in the second century AD, approximatively five centuries after Aristotle.

The proposition I.5 in Euclid's *Elements* is the well-known *pons asinorum* ('in isosceles triangles the angles at the base equal one to another') which is proven by means of a complicated construction; on the contrary, if A is the common vertex of the equal sides, it would suffice to apply the substitution

$$A \to A, B \to C, C \to B,$$

and then the proposition I.4; but this simple demonstration occurs only in Pappus (Heath, 1908, p. 254), approximately six centuries after Euclid!

A true 'symbolic system' was developed at the end of the sixteenth century: the consequent development of mathematics is unthinkable without it; the schools that refused to adopt it were soon excluded from the mainstream of science. It was also a (not very bad) epistemological obstacle when Boole conceived his 'logical calculus', which firstly was clearly influenced by the arithmetical calculus. In any case, the way for another chapter of mathematics was open: the study of formal languages.

Now, let us consider the introduction of analytic geometry. Among the authors which studied this case, Cohen (1985) considers it a revolution, whereas Mancosu (1992) advances some objections. Many authors are interested mainly in the 'technical' development introduced by the *Géométrie*. For Dunmore (1992): 'radical as this advance was, it was not accompanied by a revolution at the metalevel.'

I do not agree with this statement, and I try to show here how the method of coordinates influenced 'the concepts of space', upsetting some previous ideas.

The conceptions of space can often be polarized in pairs of opposite points of view: finite–infinite, atomic–infinitely divisible, absolute –relative, and so on. They are normally 'implicit philosophies', to which scientists, teachers, common persons agree, often unconsciously. Typical is the contraposition between those who have in mind figures (or bodies), whilst the space is only their 'container' (Aristotle), and those (as Plato) who firstly have in mind the space (that we call 'independent'), and subsequently plunge in it geometrical or physical objects (Einstein, 1954; Čapek, 1976). According to Erwin Panofsky, Greek-Roman painting is oriented towards the representation of objects; only from the fourteenth century, and especially with the theory of perspective, painters' conception shifted towards the representation of the whole (we can say, independent) space.

Someone observed that Euclid in the *Elements* doesn't speak of 'space': effectively, every figure is considered *per se.* Also the *Géométrie* begins with a determined configuration: Descartes doesn't say 'in the plane we take a system of coordinates'. This approach, which is connected to the idea of 'independent space', can be found only later. Analytic geometry, which rose in a conception of 'space as container', has quickly oriented the mathematicians towards the opposite, now clearly predominant, conception of 'independent space'.

Another meaningful opposition is between the space as a set of points, or as an irreductible continuum. The first idea can be traced back to early Pythagoreanism, the second one to Aristotle, who probably wished to overcome some of Zeno's paradoxes, and to avoid the infinity in geometry (after the crisis of incommensurable quantities, if a line is composed of points, they are necessarily infinite in number). In the *Elements*, this idea is not contrasted: we can say that a line is *in potentia* composed of points. But geometrical entities can be studied by means of coordinates only if we accept the idea that they are composed of points *in actu*.

Again, the infinity of numbers drove mathematicians and physicists to accept the idea of an unlimited space: we can also remark that, with the introduction of perspective, painters had already accepted the idea of a space unlimited *in actu*, because in a painting the points at infinity can be represented.

As the revolution that followed the crisis of incommensurable quantities was the matrix of the Greek rationalism (Pythagoreans, Eleates, Plato), the Cartesian revolution (and the following development of analysis) were at the origin of the 'metaphysical' rationalism of the seventeenth and eighteenth centuries (Descartes, the Cartesians, Leibniz).

Examples of 'Higher-level' Revolutions

As examples of higher-level theories connecting lower-level ones, we shall examine Klein's Erlangen programme and Bourbaki's structuralism. Both have deeply influenced mathematical thought and scientific practice.

Klein presents his programme as a recovery of the essential unitariety of geometry, threatened by its great development during the nineteenth century: after Klein, geometry is the study of spatial properties preserved by the transformations of a given group G. But, as many initiators of revolutions, he tries 'to save the king', in this case the Euclidean elementary geometry: for him, geometrical properties are the properties preserved by the group of similarities (even if 'formally' we could depart from this principle): let us remark that the classical geometries could be considered as it were 'plunged' in elementary geometry. However, there is not even a practical reason why we limit *a priori* G: for instance, the meaning of graphical signs is not invariant under isometries. The mythical 'equality' is replaced by a 'G-equivalence'; geometrical concepts, that an ancient tradition considered the most typical 'platonic ideas', are relative to G.

Moreover, with Klein begins a great, highly 'progressive', research programme: for instance, it allowed to construct new meaningful geometries. In differential geometry, it has been continued by Élie Cartan, with an euristic (the *repère mobile* method) based on the idea of group. Today also, groups are fundamental for geometry (let us think of fibred bundles).

Bourbaki's programme is firstly aimed at founding mathematics on a common basis: the axiomatic method, and above all the ideas of set and of structure (it is noteworthy that many of the founders of the original group came from Élie Cartan's school). It can be considered also as a return (in a modern way) to Descartes's and Leibniz's ideal of a reduction of our knowledge to some 'simple ideas' (in this case, the 'mother structures'). Today, it is impossible to treat at a research level (but often also at an elementary level) a great part of mathematics without 'seeing' the underlying structures: we are accustomed 'to think by means of structures' (as well as non-Euclidean revolution and abstract algebra have taught us to separate the 'syntactic' and the 'semantic' level).

Moreover, Bourbaki's structuralism has radically changed the paradigms of a great part of mathematical research: what problems are meaningful, what methods are accepted by the mathematicians' community, and so on. It is enough to compare a paper of some decades ago with a recent one.

Consequences of Revolutions for Mathematics Education

In traditional mathematics teaching are implicit many ancient traditions: quantitative problems from Egyptian and Babylonian mathematics, pure geometry from Euclid, and so on. As we have seen, many 'implicit philosophies' (whose origin can be traced back to scholastic and extra-scholastic learning) accumulate and superpose in one's mind; an important task is to clarify them, and if necessary to explain their historical place.

The structuralist revolution has a strong influence over the mathematics teaching in western countries; in France, it inspired in the 1970s a famous reform also in secondary schools, but in the 1980s a violent reaction followed.

In my opinion, the role of a revolution (or more generally of a development) in our knowledge over teaching is not a simple problem. To consider a change in teaching as a pure 'technical' fact (e.g., aiming at a modernization of contents) is an error: we are in any case influenced by (implicit or explicit) philosophies. Moreover, when a teacher has accepted a theory, he is accustomed to present it as it was the sole way to treat the subject: pupils have not even an idea of possible alternatives ('it would be a loss of time!').

On the contrary, it is above all necessary to understand the reasons and the meanings of revolutions: they often correspond to 'epistemological obstacles' for the progress of mathematics (and science), and so they can enlighten us about the 'didactical obstacles' that a pupil can find.

For instance, today it is easy to criticize Euclid's geometry ('*A bas Euclide! A bas le triangle!*' was a slogan of the Bourbakist reform of mathematical teaching in France): it can be rebuked for its lack of rigour, or, from an opposite point of view, for its detachment from experience; on the contrary, someone could maintain acritically its superiority (maybe as a reaction to the aforesaid attacks). The correct way is to insert it in a historical-epistemological perspective: so we can appreciate its role in the development of mathematics (possibly connecting it with Plato's theory of knowledge and with Aristotle's scientific methodology).

In Italian high schools will be soon introduced the study of non-Euclidean geometry. If it will be treated with the implicit philosophy and in the traditional style of Euclidean geometry (axioms, statements, proofs), in pupils' minds can enter a great confusion: if Euclidean geometry has been presented as 'proved truth', how can be proposed a theory which says the contrary? Also in this case it is necessary to give an epistemological-historical frame: the acceptance of non-Euclidean geometry destroyed the old rationalist ideal for a sure foundation of knowledge on 'evidently true' principles: in geometry (or more generally in mathematics) there is a 'concrete' level, whose aim is to establish spatial 'true' statements, on the grounds of experience, and then it is fallible; and an 'abstract' level, whose terms have no meaning, and whose axioms are arbitrary.

Usually, 'pure' mathematicians restrict the consequences of non-Euclidean geometry to the trend towards abstraction (likely under the influence of Poincaré's conventionalism, and of the outcomes of the crisis of foundations). But those who must confront mathematics with others fields of experience (didacticians, physicists, engineers, and so on) cannot be satisfied with such a restriction. They need an approach to mathematics which is interested also to more concrete levels, a possibility of an interaction between concreteness and abstraction. Unfortunately, our educational system is constructed on the principle 'from general to particular', or in a modern version, 'firstly the more abstract, and afterwards the more concrete'. An analysis of the development of mathematics could help us to construct a different, more realistic, approach.

References

Bochenski, J.M. (1956) *Formale Logik*, Freiburg-München, V. Karl Alber GmbH.

Bourbaki, N. (1948) *L'architecture des Mathématiques*, in Le Lionnais, F. (Ed) *Les grands courants de la pensée mathématique*, Cahiers du Sud.

Capek, M. (1976) *The Concepts of Space and Times, Introduction*, Dordrecht, Reidel.

Cohen, I.B. (1985) *Revolutions in Science*, Cambridge, Massachusetts, Harvard University Press.

Crowe, M. (1975) *Ten 'Laws' Concerning Patterns of Changes in the History of Mathematics*, in Gillies, D. (1992), pp. 15–20.

Dunmore, C. (1992) *Meta-level revolutions in mathematics*, in Gillies, D. (1992), pp. 209–25.

Einstein, A. (1954) in Jammer, M. *Concepts of Space*, Cambridge, Massachusetts, Harvard University Press.

Enriques, F. and Desantillana, G. (1936) *Compendio di storia del pensiero scientifico*, Bologna, Zanichelli.

ERNEST, P. (1992) *Are There Revolutions in Mathematics?*, PoME Newsletter, n. 4–5, pp. 14–19.

GILLIES, D. (1992) *Revolutions in Mathematics*, Oxford, Clarendon Press.

GILLIES, D. (1992) *The Fregean Revolution in Logic*, in GILLIES, D. (1992), pp. 265–305.

HEATH, T. L. (1908) *The thirteen books of Euclid's Elements*, Cambridge, Cambridge University Press.

KLEIN, F. (1872) *Vergleichende Betrachtungen über neuere geometrische Forschungen*, in GES. MATH. AB. I, Berlin (1921), Springer, pp. 460–97.

KUHN ,T. (1962) *The Structure of Scientific Revolutions*, Chicago, University of Chicago Press.

MANCOSU, P. (1992) *Descartes's* Géométrie *and revolutions in mathematics*, in GILLIES, D. (1992), pp. 83–116.

MARCHINI, C. (1990) *Le sostituzioni e la didattica della Matematica*, Boll. U.M.I., 7, 4 A, pp. 145–53.

MEHRTENS, H. (1976) *T. Kuhn's Theories and Mathematics: A Discussion Paper on The 'New Historiography' Of Mathematics*, in GILLIES, D. (1992), pp. 21–48.

SPERANZA, F. (1992) *The influence of some mathematical revolution over didactical and philosophical paradigms*, Proceedings of the 2nd Italo-German Symposium on Didactics of Mathematics, Osnabruck (Steiner, H.G. and Bazzini, L. Eds)

SPERANZA, F. (1993) 'Contributi alla costruzione d'una filosofia non assolutista della matematica', *Epistemologia*, 16, pp. 265–80.

Chapter 19

Mathematical Practices, Anomalies and Classroom Communication Problems

Anna Sfard

Introduction

A few years ago, in a conversation with a mathematician, I complained about my students' resistance to the seemingly simple idea of a complex number. The mathematician would not accept the claim that the subject might be inherently difficult. He hinted that it could be a problem of an incompetent teacher rather than of incapable students or a difficult subject matter. I insisted that there is a stumbling block, the nature of which was not yet clear to me. I brought lots of evidence. Eventually, the mathematician seemed to be giving up. 'Yes, the problem might be somewhat more complex than I thought', he said. But then, after a thoughtful pause, he exclaimed: 'Ok, it's difficult, but I assure you that I would make the students understand the idea in no time: I'd just write the axioms of a complex field and show that the concept is well-defined.'

I knew then, as I know now, that the mathematician I talked to was not an exception. Teachers are sometimes equally insensitive to profound difficulties students have with seemingly simple concepts. Several years and many hours of reading, researching, and thinking later, I began to realize the reasons of this insensitivity. It may be stated in many different ways, depending on the school of thought to which one belongs: one can say that the students and the teacher don't share one *paradigm*, or that, without realizing, they are engaged in different mathematical *practices*, or that they don't participate in the same mathematical *discourse*. In whatever words one puts the explanation, the message is always the same: the student and the teacher cannot communicate because, although they may be using identical expressions, they live in incompatible mathematical worlds, they are playing according to different rules, and their thoughts go in parallel rather than clash or coincide.

There is no better way to analyse this problem than by scrutinizing the historical development of mathematical knowledge. History is the best instrument for detecting invisible conceptual pitfalls. History makes it clear that the way toward mathematical ideas may be marked with more discontinuities and dangerous jumps than the teachers are likely to realize. If the mathematician I talked to were aware of the turbulent biography of negative and complex numbers, he would probably be more attentive to my classroom horror stories. Similarly, if the mathematics teacher of the French writer Stendhal bothered to look into some original mathematical texts, the latter would have been less likely to complain for the sake of posterity that his difficulties with negative numbers 'simply didn't enter [the teacher's] head' (see Hefendehl-Hebeker, 1991). Indeed, it would be enough to have a brief glimpse at the emotion-laden statements of eighteenth and nineteenth-century mathematicians to realize that there must have been some major disparity between the emergent and established knowledge. Here is a representative sample of relevant utterances, spanning the crucial fifty years in the history of negative and complex numbers: 'A problem leading to a negative solution means that some parts of the hypothesis was false but assumed to be true' (d'Alembert in eighteenth-century *Encyclopédie*), 'It were to be wished . . . that negative roots had never been admitted into algebra or were again discarded from it' (Maseres, in 1759); '. . . it is clear that the square roots of negative numbers cannot be included among possible numbers. Consequently we must say that these are impossible numbers' (Euler, in 1770); The talks about 'impossible numbers, which, being multiplied together, produce unity . . . [are] all jargon, at which common sense recoils; but, from its having been once adopted, like many other figments, it finds the most strenuous supporters among those who love to take things upon trust and hate the colour of a serious thought' (Frend, in 1796).

The point I wish to make here is that such discussions, however fierce, tend to be completely forgotten once mathematicians manage to reconcile themselves with the problematic notions. As follows from the conversation I quoted above, for a modern mathematician nothing could appear simpler than complex numbers. Stewart and Tall (1983) comment, tongue in cheek: 'Looking at the early history of complex numbers, the overall impression is of countless generations of mathematicians beating out their brains against a brick wall in search of what? A triviality' (p. 5). In the classroom, problems are likely to appear at any of the junctions at which mathematicians themselves faltered and asked questions. Indeed, there is no reason to assume that what was difficult to accept for generations of experts will be readily admissible for today's students. To be sure, inventing a new concept is much more demanding a task than trying to learn what may be presented as a ready-made product. Otherwise, how could we hope that the student will be able to acquire the knowledge the development of which lasted for millennia? Even so, there is more than one reason because of which at least some of the difficulties that once troubled mathematicians are likely to reappear in the individual learning.

The last statement will be substantiated in this chapter in the course of an analysis of a change in mathematics. As I shall try to show, mathematics is subject to a permanent tension and conceptual imbalance. The source of the tension lies not so much in a clash between contradictory mathematical theories as in some subtle disparities between the statements of mathematics and the meta-mathematical framework within which these statements are made. The latter usually proves more inert, more resistant to change, than the former. At certain points in the development of mathematical knowledge apparent contradictions would arise which will not be resolved unless the whole system of basic ontological and epistemological assumptions undergoes a substantial change. The major shifts which must occur at such junctions in the thinking of today's students are much the same like those that had to be made by yesterday's mathematicians. Once the shift occurs, the communication between the adherents of the old and the new paradigm becomes seriously impaired, if not impossible. Sometimes, however, inhabitants of different mathematical words are doomed to talk to each other. If these are students and their teachers, such encounters may have unhappy endings.

The task I set to myself will be performed in three steps. Since my analysis of change in mathematics will be inspired by the recent advances in the philosophy of science, it is my duty to begin with an explanation of what might be not entirely obvious: that in the view of current understanding of human knowledge, mathematics and science have much in common both in the way they develop and in the nature of their endeavour. The next step will be to actually use modern philosophy of science — the one which begins with Kuhn (1962) and continues with such writers as Feyeraband, Putnam, and Rorty, as a basis for analysis of the growth and change in mathematics. It will be interesting to see how the conceptual frameworks used by those who investigate mathematical thinking and learning fit into this new vision of science. Finally, I will return to the questions raised in the introduction and will try to show how the suggested analysis of mathematical change accounts for the difficulties in mathematical communication.

The Changing Face of Mathematics: The Loss of Objectivity

In the present section I will draft a background against which my further analysis of a change in mathematics will be cast. Let me begin with a brief account of recent changes in the philosophy of science which evidently had quite a profound effect on the philosophy of mathematics.

Until recently, mathematics was regarded as a province of pure reason, and as such — as something quite different from any other domain of human knowledge and in particular from its nearest neighbours — the natural sciences. The latter had to combine logic with empirical tools in their pursuit of 'the truth about the world'. Even so, natural sciences appeared as a paragon of systematicity, objectivity and methodological purity in comparison to social studies or humanities. The inferiority complex of the latter steadily deepened in nineteenth and twentieth centuries as logical positivism and empiricism advanced the idea of rigorousness in science and tried to formulate strict cannons of the scientific method. As the methodological fence between the 'true' scientific endeavour and what only pretended as such grew higher, it seemed quite natural to try to mould all the domains of human knowledge in the image of the most successful exemplar. The critical comparisons and the attempts at bridging the gap brought, however, a surprising result: rather than transfer the scientific method from physics and chemistry into the social studies and humanities, these efforts ended in second thoughts on the rationality and objectivity of natural sciences. The doubt about the very possibility of ever arriving at a 'God's eye' view of reality — the doubts about the validity of the notion of 'God's eye view' itself — contaminated those regions of human knowledge which, till now, seemed immune to such far-fetched epistemological skepticism. The message of relativity, sometimes interpreted as an eulogy of irrationality — was brought by such writers like Toulmin, Kuhn, Fayerabend, Foucault, and Rorty.

It all began with the disillusionment as to the possibility to ever find an 'objective', 'correct' understanding of historical texts. In the second half of twentieth century nobody would deny anymore that 'that social explanation involves an interpretative or hermeneutical component' (Bernstein, 1986, p. 14). As far as social sciences and humanities were concerned, hermeneutics — once an art of interpretation of sacred and literary texts which broadened gradually into a philosophy of understanding — put an end to the further search after a prescription for 'scientific method' in social studies and discontinued the pursuit of the criteria of objective, ultimate truth. According to the basic tenet of hermeneutical approach, 'understanding means going in circles: rather than a unilinear progress towards better and less vulnerable knowledge, it consists of an endless recapitulation and reassessment of the collective memories.' (Bauman, 1978). Or, to put it in a different language, the thesis of 'hermeneutical circle' presents understanding as a neverending interplay between predictions and their amendments, as an infinite chain of confrontations between the things we try to understand and our own prejudgments. One inevitable conclusion from this dialectical vision of understanding was that 'to understand is always to understand differently' (Bernstein, 1986, pp. 98–9). Since *meaning* does not exist as such and it only realizes itself through the process of understanding, and since the understanding is in a constant flux, visiting the same meaning more than once is just as impossible as stepping twice into the same river. Accordingly, nobody should expect that a theory — any theory — would be any more than one of the many possible interpretations of whatever is to be understood and explained, and as such — that it would ever become a stable, fully defined system of the kind logical empiricism dreamt about.

These claims, if taken seriously, were bound to change all the conceptions about human knowledge. It is hardly surprising, therefore, that what began as a doubt about the possibility of 'scientific method' and 'correct interpretation' in social studies and humanities soon spilled over to natural sciences. Kuhn, one of the precursors of a new philosophy of science, speaks about his own conversion to hermeneutics:

> What I as a physicist had to discover for myself, most historians learn by example in the course of professional training. Consciously or not, they are all practitioners of hermeneutical method. In my case, however, the discovery of hermeneutics did more than make history seem consequential. Its most immediate and decisive effect was instead on my view of science. (Kuhn, in Bernstein, p. 12).

Kuhnian new view of science went beyond Popper's claims about the relativism of scientific theories — it led to a far-reaching revision of practically all the basic beliefs about science. First and foremost, it challenged, in a longer run, the assumption that science aims at a discovery of mind-independent reality. Consequently, the notion of objectivity had to be replaced with the idea of intersubjectivity. Rorty (1991), who calls his philosophy a 'left-wing Kuhnianism,' makes the following exhortation:

> I urge that whatever good the ideas of 'objectivity' and 'transcendence' have done for our culture can be attained equally well by the idea of a community which strives after both intersubjective agreement and novelty — a democratic, progressive, pluralistic community of the sort of which Dewey dreamt. If one reinterprets objectivity as intersubjectivity, or as solidarity . . . then one will drop the question of how to get in touch

with 'mind-independent and language-independent reality'. (Rorty, 1991, p. 13)

With the emergence of a (neo-)pragmatism of which Rorty is the most outspoken protagonist, the criterion of truth and validity was replaced by the ideas of solidarity and of usefulness. People should no longer ask whether anything is objectively true; rather, they are expected to judge knowledge according to whether it can bring them together and whether it can do anything for them.

It is time now to get back to mathematics and to ask what the musing about the nature of natural science has to do with the discipline always considered as an exclusive (or almost exclusive) domain of the analytic thought. Mathematics was able to keep for millennia its superior status of a stronghold of objective truth. Even this well-guarded bastion of certainty could not, however, remain unaffected by the upheavals that changed social and natural sciences beyond recognition. When the distinction between analytic and synthetic became blurred, when the idea of 'given reality' was denounced as untenable, when scientists were saying their 'farewell to reason' (Fayerabend, 1987) and to objectivity, the epistemological foundations of mathematics were bound to become subject to scrutiny and revision. To be precise, mathematics had its own 'public image' problems even before Popper, Kuhn, Fayerabend, and Foucault revolutionized the vision of science. The face of mathematics did not remain the same after the famous crises in foundation at the turn of the century. The antinomies and paradoxes which infested logic and set theory discredited the view of mathematics as a domain of ultimate truth and ended in 'the loss of certainty' (Kline, 1980), namely in the renouncement of the absolutist view for the sake of fallibilism (see a comprehensive account of the relevant developments in Ernest, 1991). From the denial of certainty, however, mathematics had still a long way to go until it arrived at doubts about objectivity similar to those that can be found in natural sciences. After all, questioning the possibility to reach an objective truth is not the same as challenging the very idea of objective truth.

The already blurred difference between sciences and mathematics became even more fuzzy in the hands of Lakatos (1976). Not only did Lakatos ultimately deny mathematics the right to the title of infallible, he also indicated that mathematical and scientific methods bear a family semblance. In fact, Lakatos did to mathematics what Popper did to science: has shown its relativity, its constant need for revision and rebuilding. With the help of ingenuous examples he has made it clear that the purely deductive nature of mathematics is but an illusion, and above all — that mathematics is quasi-empirical in that like Popperian science, it endlessly proceeds from hypothesis to its denial and to the reformulation of the hypothesis. Following Polya (1945) who once stated that 'mathematics in making is an experimental inductive science' (p. 117), Lakatos stressed the importance of informal mathematics — this mathematics which bears the closest semblance to empirical science.

Lakatos remained, however, relatively silent on the question of objectivity, so readily tackled by the philosophers of science. For some reason, philosophers of mathematics seem more reluctant than their scientific colleagues in renouncing the dream of objectivity. As Ernest (1991) put it, even though 'we accepted the fallibility of mathematical knowledge . . . the fact remains that the objectivity of both mathematical knowledge and the objects of mathematics is a widely accepted feature of mathematics, and must be accounted for by any philosophy of mathematics' (p. 49). It seems that for all the concessions which have already been made for the sake of less mind-independent, more human-oriented approach, mathematics unrivalled ability to create consensus evokes more wondering than the visibly controversial science. Thus, while some thinkers opt for the banishment of the term 'objective' from any account of human knowledge (see e.g., Johnson, 1987), the philosophers of mathematics tend to reinterpret the notion rather than abandon it altogether. Referring to Popper's definition, according to which objective knowledge is 'the world of the logical *contents* of books, libraries, computer memories, and suchlike' (Popper, 1979, p. 74), Ernest redefines *objective* as *intersubjective* or *publicly shared*:

> I shall use the term 'objective knowledge', in the way that differs from Popper, to refer to all knowledge that is intersubjective and social. I wish to count all that Popper does as objective knowledge, including mathematical theories, axioms, conjectures, proofs, both formal and informal. One difference is that I also want to include additional 'products of the human mind' as objective knowledge, notably the shared (but possibly implicit) conventions and rules of language usage. Thus I am referring to publicly shared, intersubjective knowledge as objective, even if it is implicit knowledge, which has not been fully articulated. This extension is very likely unacceptable to Popper. (Ernest, 1991, p. 46)

This approach to the question of objectivity of mathematical knowledge is a cornerstone of *social constructivism* — Ernest's special synthesis of current trends in the philosophy of mathematics. Radical constructivists, who stress with even greater force that there is no other site for knowledge than human mind and at the same time are very much concerned with the social dimension of knowledge construction, admit the existence of 'consensual domains' but prefer to bypass the notion of objectivity by talking about 'institutionalized knowledge' and about meanings that are 'taken as shared' (Cobb, 1989). To sum up, whether the notion of objective knowledge is banned or reinterpreted, one may certainly say that in mathematics, exactly like in science, the loss of certainty is now being followed by the loss of objectivity, or at least the loss of objectivity as we knew it.

The comparisons between mathematics and natural science did not go much beyond those made by Lakatos — they barely transcended the adaptation of Popper's ideas to the domain of mathematics. Only very recently some attempts have been made at using Kuhn's theory of scientific revolutions (Kuhn, 1962) as a point of departure for reflections on the deep structure of change in mathematics (notably Kitcher, 1984; see also Griffiths, 1987; Cobb, 1989; Otte, 1990; Ernest, 1992). In what follows I will try to take this project a little farther. Certain modifications made by post-Kuhnian writers, as well as those introduced by Kitcher (1984) in an attempt to fit Kuhn's ideas into the context of mathematics will serve as a basis for further elaborations. At this point, a word of caution seems necessary: for all that has been said about the similarities between mathematics and natural sciences, one must remain aware of substantial differences between the two. Many distinctive features still keep mathematics and science quite apart. Even so, I find Kuhn's ideas potentially useful in the analysis of discontinuities in the growth of mathematical knowledge.

Anna Sfard

The Change in Mathematical Practice

Fitting Kuhnian Theory to Mathematics

Any attempt to summarize Kuhn's theory is doomed to turn out highly personal. Kuhn's seminal book *The Structure of Scientific Revolution* (1962) has spurred a great deal of discussion, the echoes of which can still be heard. In the course of the dispute, many interpretations have been given to Kuhn's theory and to its basic notions. The terms 'paradigm', 'natural science', 'anomaly', 'crisis', and 'scientific revolution' were scrutinized, construed, and reconstrued so many times that it is hardly possible to suggest a definition without evoking somebody's protests (which shows that reading text is a hermeneutical task even when the text in question is itself a product of a hermeneutical effort!). In the following brief summary of the basic tenets of Kuhnian framework I will try to remain closer to the original Kuhnian text than to the works of interpreters.

The claim about the relativity of knowledge leads Kuhn to concentrate on discontinuities in the development of science. According to his theory, periods of relative calmness and more or less cumulative growth of scientific knowledge are disrupted, time and again, by certain special events which, rather than add new bulk of data, bring a far-fetched reinterpretation of the previously accumulated findings. The former were called 'periods of normal science', the latter got the title of 'scientific revolutions'. According to Kuhn's own definition, normal science 'means research firmly based upon one or more past scientific achievements, achievements that some particular scientific community acknowledges for a time as supplying the foundation for its further practice' (p. 10). The body of this accepted knowledge is expounded and perpetuated through textbooks. As well established as it may be, the scientific practice is bound to undergo a far-reaching change, sooner or later. Scientific revolution is the name given by Kuhn to these conceptual upheavals which de-legitimize approaches hitherto widely accepted and turn the contents of even the most revered of textbooks into obsolete.

Kuhn puts much effort into an explanation of the mechanism of scientific revolutions. His insightful interpretation of such events like the victory of Copernician astronomy over that of Ptolemy or transition from Aristotelian to Newtonean mechanics brings a totally new outlook at the reasons and the results of change, and makes it clear that periodical conceptual transformations are both inevitable and beneficial (some time later, Fayerabend (1975, p. 24) will take this claim to the extreme by saying that freedom from restrictions, such as those imposed by an established scientific 'rules of game', is a precondition of progress and that 'uniformity impairs [the] critical power' of science). To give a thorough account of scientific revolutions Kuhn asks three basic questions: What is it that changes in the course of revolution? Why and how does it change? What are the final outcomes of such change?

To tackle the first of these queries Kuhn invokes the concept of *paradigm*. This is probably the most widely known but also the most abused component of Kuhn's theory. Kuhn himself defined paradigm as a specific instance of a particularly successful, comprehensive, and convincing scientific achievement, such as Copernician or Newtonean theories; or, to put it in Kuhn's own words, it is an 'accepted example of actual scientific practice — example which include law, theory, application, and instrumentation together' and thus 'provides models from which spring particular coherent traditions of scientific research' (p. 10). Kuhn's followers, however, silently agreed on a much broader interpretation of the term, according to which paradigm encompasses practically all the components of scientific endeavour, namely all the beliefs, both implicit and explicit, about the object and method of scientific inquiry shared by a given scientific community. However the word is construed, it remains closely related to the idea of normal science: paradigm is what shapes, guides, and restricts the scientific endeavour in the periods of calmness and steady growth. Paradigms is what divides all the human knowledge and actions into scientific and non-scientific.

To answer the question concerning the reasons and the course of revolution, Kuhn introduces the concept of *anomaly*. Revolution begins with a *discovery*, namely a 'novelty of fact' or with an *invention* — a 'novelty of theory'. Discovery, says Kuhn, 'commences with the awareness of anomaly, i.e., with the recognition that nature has somehow violated the paradigm-induced expectations that govern normal science'. In other words, anomaly arises when the accepted theories cannot account for a certain observation, or, to put it even stronger, it is an appearance of data which do not fit into the space delineated by the existing paradigm. The awareness of anomaly is equally instrumental in inducing scientific inventions. To be sure, since no theory is sufficiently equipped to give clear-cut answers to all the possible questions, and since the acceptance of a certain paradigm must therefore be to some extent an act of faith (as opposed to a deliberate rational choice), the appearance of anomaly does not necessitate revolution — it only creates favorable conditions for such event.

Finally, in Kuhn's account revolutions are only those major changes which have far-reaching, penetrating effects on all the aspects of scientific endeavour, changes that alter the conceptions, perceptions, and methods so deeply that they lead to an emergence of a paradigm incommensurable with the one which is to be replaced. The historical examples surveyed in his book show that scientific revolutions

> necessitated the community rejection of the time-honored scientific theory in favor of another incompatible with it. Each produced a consequent shift in the problems available for scientific scrutiny and in the standards by which the profession determined what should count as admissible problem or as a legitimate problem-solution. And each transformed the scientific image in ways that we shall ultimately need to describe as a transformation of the world within which scientific work was done. (Kuhn, 1962, p. 6)

Kuhn's conception of the growth of scientific knowledge was not less revolutionary than the events he baptized as revolutions in science. It would be in point now to ask whether this radical change in the vision of science can have any implications for the philosophy of mathematics. The most serious attempt to look at the latter through the Kuhnian lens, made by Kitcher (1984), will be reviewed in the next section. Here, I will briefly mention some recent, less comprehensive, studies.

As was already noted, profound changes have occurred in the philosopher's vision of mathematics in the last several decades. As a result of these changes mathematics is more and more often compared to science. Even so, until recently, one of the basic beliefs about mathematics was that unlike physics, biology or chemistry, it is a cumulative science. Saying this is tantamount to an assertion that growth of mathematical knowledge is free of discontinuities, and such conviction rules out the possibility of mathematical revolution. After Lakatos, this view is no longer uncritically accepted. Lakatos (1978) conveyed his disagreement through ironic account of the popular view: '[history of mathematics] is still regarded by many as an accumulation of eternal truths; false theorems are banished to the dark limbo of pre-history or recorded as regrettable mistakes, of interest only to curiosity collectors.' More recently, Crowe (1988) presented a similar position by listing the claim about the cumulative nature of mathematics among his 'ten misconceptions about mathematics and its history'.

That the development of mathematics was not as smooth and undisturbed as generally believed was, in fact, noted already by Bourbaki:

> Since the earliest times, all critical revisions of the principles of mathematics as a whole, or of any branch of it, have almost invariably followed periods of uncertainty, where contradictions did appear and had to be resolved. (Bourbaki, 1950)

The recognition of the existence of the singular points on the curve of mathematical growth leads in an obvious way to the conjecture that the development of mathematical knowledge might in fact be marked with far-reaching conceptual changes, much like those observed in the history of science. In the light of all that is known about historical controversies and about the debates led by the finest of mathematical minds, it would be only natural to check whether mathematics could be subject to revolutionary changes like those known from the history of science. The few writers who already made some attempts at answering this question came up with similar verdicts (the semblance may be easily overlooked because of the differing ways in which the authors express their opinions). All of them seem to sustain that the notion of revolution can only be applied to mathematics if given an interpretation that differs significantly from the one offered by Kuhn in the context of sciences. A representative example would be that of Griffith (1987), who doubts 'whether Kuhn's notion of a paradigm applies to mathematics in the way it does to other sciences'. Pointing to the fact that the acceptance of non-Euclidean geometry did not lead to rejection of the Euclidean theory he concludes that 'Mathematical paradigms seem to be coherent tracks of thinking that yield fruitful results, and they can live side by side with other paradigms, even within the mind of the same mathematician.' With a conception of paradigm so dramatically different from the one proposed by Kuhn (Kuhnian paradigms do exclude each other by definition), Griffith must compromise on a very 'mild' version of mathematical revolution: 'the growth of an important paradigm in mathematics induces a "revolution" in the sense that the thought habits, language, and ways of questioning of later mathematicians will differ from those before paradigm.'

Ernest (1992), in his discussion of Griffith's positions, suggests to focus on 'shifts in background scientific and epistemological context' rather than on 'the replacement of one mathematical theory by other'. This makes him more sympathetic toward the idea of a stronger version of revolution in mathematics:

> some radical changes or global restructuring of the background epistemological and scientific context of mathematics can be described as Kuhn-type revolution. Such changes result in a profound re-orientation of mathematics, which can lead to as much 'incommensurability' as is found in science. (Ernest, 1992)

It becomes obvious that a question *what changes* must be answered before we define the nature of change and start analysing the significance of its results. In other words, the preliminary task is to decide what aspects of mathematics, if any, are liable to revolutionary changes. As *revolutionary* I will regard only such shifts that produce an incompatibility, shifts as a result of which the pre- and post-revolutionary communities are no longer able to communicate with each other. There is little doubt that history of mathematics witnessed many such events. Pythagoreans, for whom the idea of incommensurability was unacceptable since it meant the end of the world structured by — or maybe even built of — integers, would not arrive at mutual understanding with those who admitted the existence of irrational numbers; Cardan's followers, even though they knew what would be lost if they refrained from the use of square roots of negative numbers, would nevertheless have no common language with nineteenth-century mathematicians who decided that the outcome of this 'impossible' operation was a legitimate number; Sacceri, who was just one step distant from the discovery of the independence of Euclid's Fifth Axiom but recoiled at the last moment unable to face the consequences, would not be able to communicate with Gauss, Bolyai, and Lobachevsky.

It seems that when bringing the example of Euclidean and non-Euclidean geometries in order to pinpoint the difference between mathematics and science, Griffith somehow missed the point. When introducing the concept of paradigm, Kuhn tried to answer the question what aspects of science are liable to revolutionary changes — to the changes which produce incommensurability. Such approach, whatever the actual definition, entails the impossibility of 'peaceful coexistence' of different paradigms. It also implies that the move from one paradigm to another is one-way and irreversible. With such understanding of the term, Griffith's statement that 'mathematical paradigms . . . can live side by side with other paradigms' sounds like a contradiction in terms. No one theory, however comprehensive and successful, can be regarded as a paradigm (when Kuhn gave the name paradigm to Newtonean and Einsteinean frameworks, he did not mean the theories themselves but rather all the aspects of mathematical practice associated with these theories). Incidentally, if this principle was not observed, then what Griffith said about mathematics would also be true about science: at a closer look it becomes obvious that science does allow a plurality of outlooks similar to that attained in mathematics by the acceptance of different axiomatic systems. In a sense, science seems even more tolerant than mathematics toward incompatible theories; its history knows more than one case of dual theoretical approaches to the same phenomena. Einsteinean framework, in spite of its being apparently incompatible with Newtonean theory, did not overthrow the later but rather incorporated it as being a useful approximation; the corpuscular and wave theories of subatomic phenomena live side by side in physics, and this coexistence is sanctioned by Bohr's principle of complementarity. Indeed, many more examples could be brought to show that the question of what is the object of revolutionary changes — of those changes that lead to incommensurability of the old and the new scientific or mathematical worlds — deserves a deeper thought. This is exactly what will be done in the next section.

The Object of Change: Mathematical Practice

The question of the object of change in mathematics was tackled by Kitcher (1984) in a detailed way. According to Kitcher, Kuhn's notion of paradigm grew from the latter's intention to divide history into large, clearly defined segments on the one hand, and from his wish to emphasize that science does not advance by mere modifications of its statements, on the other hand. While Kitcher tends to endorse Kuhn's second claim, he casts doubt on the first. In other words, while he agrees that there is much more to the development of science than local amendments of theories, he rejects the idea that history of either science or mathematics can be viewed as a chain of quiet periods of accumulation interrupted with acts of radical reorganizations. It seems obvious, says Kitcher, that tensions and anomalies are ubiquitous and they are part and parcel of any scientific and mathematical endeavour. From here it follows as an immediate conclusion that our knowledge is in a constant flux rather than in a smooth process of growth marked with clear-cut discontinuities.

When it comes to mathematics, Kitcher's dissatisfaction with Kuhnian concept of scientific paradigm leads him to suggest a replacement: what changes, often quite dramatically, in the course of mathematical progress is the entire *mathematical practice*. This last concept is defined with an almost mathematical precision as a quintuple composed of a language (L), a set of accepted statements (S), a set of reasonings (R), a set of questions selected as important (Q), and a set of meta-mathematical views (M). As mathematics develops, each of the five components may undergo more or less radical changes, changes that stem from internal discrepancies within any subset of the quintuple. The peaceful coexistence of incompatible theories, such as Euclidean and non-Euclidean geometries, is made possible by an appropriate adjustment of the language and the set of admissible questions. This is where, according to Kitcher, the illusion of the purely cumulative nature of mathematics takes its roots: thanks to the reorganization of the whole framework, 'one-time rivals' turn into equally legitimate alternatives.

In the remainder of this chapter I will use Kitcher's concept of mathematical practice as a point of departure for further reflections on mathematical change. I shall begin with a new attack at the question of the relative resilience of mathematics and its apparent immunity to restructuring and to sharp discontinuities. Kitcher's reinterpretation of the cumulativity of mathematics is descriptive rather than explanatory. What lacks in Kitcher's theory is an epistemological background detailed enough to account for the unique way in which mathematics guards itself against too radical, too abrupt a change. The required background information relates to the nature and sources of tensions which eventually bring the change.

The Setting of the Tension: Between Mathematics and Meta-mathematics

Mathematical change would not take place without a good reason. The reason, according to Kitcher, is a lack of equilibrium between different components of mathematical practice. The lack of equilibrium creates tension which, sooner or later, has to bring a shift and to change the whole system, often beyond recognition.

Let us have a closer look at the setting of the tension, on which Kitcher does not give a detailed account. The emergence of the new geometries is a useful example. The main point that must be made here is that the tension which preceded the reconciliation of the Euclidean and non-Euclidean geometries did not reside between the concurring theories but between S and M component of mathematical practice — between the dramatically enlarged yet still consistent system of mathematical statements on the one hand, and the meta-mathematical framework that rejected the possibility of incompatible axiomatic systems on the other hand. To put it differently, Sacceri would not be able to communicate with Lobachevsky not because of their disagreement on the logical argument that led to the crucial discovery, but because of the kind of conclusions which each one of them was prepared to accept. Indeed, in order to bring the Euclidean and non-Euclidean geometries into a 'peaceful coexistence' mathematicians had to revolutionize their most fundamental epistemological premises. No longer could they sustain that mathematics in general, and geometry in particular, deal with a reality in which existence and properties of objects are subjected to laws similar to, and as restrictive as, those governing the physical world. The most basic of these laws, the law of contradiction and of excluded middle, seemed to be violated. In the abstract universe of mathematics, claims about existence and non-existence of objects would no longer be mutually exclusive and the truth of such statements would stop being a matter of either 'yes' or 'no'. From now on, a mathematician could claim that since the existence of certain entities proved inherently non-decidable, it would be equally admissible to regard these entities as existent and as non-existent. If applied to physical objects, these utterances would be declared 'meaningless' or 'impossible' and those who made them would likely be suspected of being out of their minds. Paradoxically, the acceptance of such statements in mathematics was made in the name of sanity: the tension between the evidence of logic and the confining epistemological premises could only be alleviated through modifying the latter.

To sum up, the most vulnerable, most problematic point of mathematical practice seems to be the junction between the meta-mathematical component and all the other ingredients. At this point it is necessary to make explicit the meaning of the term 'meta-mathematics', as used in the present context. After foundational movement made an attempt to give a mathematically rigorous treatment to the philosophy of mathematics, the word 'meta-mathematics' is often understood as tantamount to mathematical logic. This is not the meaning in which it will be used in this chapter. Here, meta-mathematics should be interpreted as a broad philosophical framework which includes all the epistemological and ontological assumptions, whether explicit or tacit, of those who engage in the given mathematical practice. To put it in a more detailed way, meta-mathematics encompasses beliefs about the object and methods of mathematical inquiry, about the nature, sources and conditions of existence of mathematical objects, and about the admissible forms of knowledge representation.

The meta-mathematical framework clearly reveals much more inertia than mathematical theories themselves and this is probably the main reason of the tension ever present in mathematical practice. While the M component remains relatively stagnant, the set S of mathematical statements grows quickly thanks to the efficient mechanism of mathematical reasoning (R). The latter never remains idle and hence it often acts as a forerunner. The history of mathematics is replete with examples showing that logic and mathematical deduction have often covered new area before the existing epistemological restrictions were able to justify the conquest. Much has been told on the centuries-long debates about acceptability of negative and complex numbers, the nature of functions, the foundations of analysis and the meaning of new findings on Euclid's Fifth Axiom. All these disputes seemed to result from the lack of congruency between the ideas in question and the existing system of fundamental beliefs about mathematics. Very often, generally held convictions about what counts as mathematically legitimate initially barred the way of new concepts and theories. 'Logic sometimes makes monsters', complains Poincaré (1899; quoted in Kline, 1972, p. 973), meaning that under a sole guidance of rigorous mathematical reasoning one can arrive at ideas in the face of which 'common sense recoils'. The fact that basic epistemological and ontological beliefs is much more inert than the other components of mathematical practice accounts for a tension which eventually would lead to a change of the entire system.

The Source of Tension: An Anomaly

What has been said in the last section about the setting of the tension does not yet explain the particular resistance to change typical of mathematical practice in general and of meta-mathematical beliefs in particular. In order to find what is missing I shall turn now to a more basic question of the origins of abstract mathematical ideas.

The main point I wish to make here is that all our thinking has perceptual sources and mathematics, for all its abstractness, is no exception. The claim about bodily origins of human thinking was thoroughly expounded by Lakoff and Johnson in their three books (Lakoff and Johnson, 1980; Lakoff, 1987; Johnson, 1987). According to Lakoff and Johnson's non-objectivist theory of meaning and understanding, the bodily experience constitutes our imagination and even the most abstract of our ideas are shaped by our senses. The mechanism which transfers 'the bodily' into 'conceptual' is that of a metaphor: our basic perceptual sensations produce mental schemes which, when detached from their origins and metaphorically transformed, give rise to abstract concepts.

This last claim may seem surprising when made in the context of mathematics. Mathematical concepts seem so abstract, so distant from anything perceptual, that at the first glance it is difficult to see any bodily connections. In a series of interviews with mathematicians and through a careful reading of mathematical texts I tried to examine this issue as thoroughly as possible (Sfard, 1994). My search brought much evidence that, indeed, be a mathematical concept as abstract as it may be, it still remains in a close relationship with our sensory-motor experience. Mathematical language is perhaps the most immediate and most revealing source of such evidence. Mathematicians refer to abstract concepts, such as numbers, sets or functions, as if they were talking about physical objects that can be manipulated and transformed; they describe these objects in such physical terms like 'open' and 'close', 'increasing' or 'decreasing', 'compact' and 'smooth'. All this leaves little doubt as to the metaphorical nature of the mathematical universe.

At a closer glance, our meta-mathematical beliefs will also show clear signs of being a direct product of the way we conceive the physical reality.

An important point to make here is that the metaphorical nature of our abstract thinking is a doubly-edged sword: the metaphors, and our perceptual experience through them, both create and restrict our imagination. It is the restrictive power of metaphorical thinking which is of relevance to the present discussion. Since our meta-mathematical beliefs are usually implicit, and often outright transparent, they may be more affected by such restrictions than any other component of mathematical practice. What seems absurd in the light of our perceptual wisdom can hardly be accepted by our imagination. Thus, using Kuhn's terminology, one may say that the tensions in mathematical practice result from *anomalies* which appear at these junctions in the development of mathematical knowledge where the statements of mathematics violate the expectations produced by our perceptual experience. History knows many examples of such anomalous situations, situations in which the judgment of logic seemed to contradict the evidence of senses. The resistance to non-Euclidean geometry, for instance, is easily understood in the light of perceptual implausibility of the existence of more than one parallel to the given straight line through the given point. Another, more subtle point here is that of our *freedom to decide* about the number of parallels. In the physical world, the question about existence of objects can only be given one answer: either yes or no. In the post-Lobachevskyan mathematics the existence of mathematical objects was to become a matter of choice. A testimony of a mathematician whom I recently interviewed brought another example of the same type of anomaly. The mathematician said that he 'could not put up with the independence of the continuum hypothesis', since this implied that the question of existence of a certain transfinite number could not be answered. 'It was my deep conviction that this set either exists or not', he explained. Another example of anomalous situation has already been mentioned in this chapter: Pythagoreans could not meet the challenge of incommensurability because they perceived the world as built of countable pieces and thus describable in terms of whole numbers and their ratios. The evidence of logic — implication of the theorem on right-angled triangle — clearly contradicted this conviction.

In all the above examples, the source of an anomaly was an incongruity between Platonist framework and the logical consequences of certain mathematical assumptions. The historical debates dissolved in gradual loosening of realist constraints and in the emergence of alternative philosophies of mathematics, such as intuitionism and formalism. Even so, deep in their hearts most mathematicians still seem to be Platonist (or, to say it in the way Davis and Hersh (1983) put it, 'mathematicians are Platonists on weekdays, and formalists on Sundays' (p. 321)). At a closer glance, the claim about experiential roots of mathematical abstraction implies that Platonism is more than one of the possible philosophies of mathematics; in a sense, Platonism is a default state of mind of a person engaged in mathematical activity. Indeed, all the mathematicians I talked to testified to their deep sense of the 'reality' of abstract objects, even though some of them emphasized that this was only a matter of feeling and not of a sound philosophical belief. The following statement by René Thom (1971) forcefully summarizes this points:

> Everything considered, mathematicians should have the courage of their most profound convictions and thus affirm that *mathematical forms indeed*

> ***have an existence that is independent of the mind considering them.*** (Thom, 1971)

The persuasive power of Platonism is not surprising in the light of the claim on experiential roots of mathematical thinking. The Platonist realm of ideas speaks to the mathematician's heart and reason thanks to its being in so many respects like a physical reality: mind-independent and ruled by laws which are clearly dictated by perceptually-based intuition. In the light of this, we may certainly expect our students to behave like naïve Platonists and be exposed to the same kind of difficulties which were described in the above examples. The next section will be devoted to a detailed analysis of one particular kind of anomaly which arises when evidence of logic violates Platonist expectations.

The particular inertia of meta-mathematical framework and its frequent bias toward Platonism may now be easily accounted for in the light of what has been said about the origins and the nature of our beliefs about mathematics. First, these beliefs are directly tied to the most robust of our mental structures, to the structures that originate in perceptual experiences; and second, they are *implicit* and deeply hidden, and thus are not readily amenable to rational revision.

The Resolution of Tension: A Change in Meta-mathematical Framework

The claim about perceptual roots of mathematical abstraction implies that changing meta-mathematical assumptions requires overcoming primary intuitions. These intuitions are so fundamental, so tightly connected to the perceptually based cognitive structures, that they are almost never a subject to reflection, and their role in the creation of an anomaly is often completely overlooked. Some events in the history of complex numbers aptly illustrate this point. Evidently, the mathematicians who objected to the notion of a square root of negative number were not aware of the primary sources of their doubts. They justified their resistance by pointing to certain contradictions which, in their opinion, were internal to mathematics while, in fact, the argument that has been used involved extra-mathematical assumption about the nature on numbers. Euler, for instance, stated in his 1770 *Algebra* that 'The square roots of negative numbers are neither zero, nor less than zero, nor greater than zero.' In his opinion, this was a sufficient reason to call these new numbers 'impossible' and to exclude them from mathematics. After all, incomparability with other numbers was in a direct contradiction with the basic conception of a number as a God-given object that represents a quantity. Euler, unable to make revision of this meta-mathematical assumption, concluded that the only way to save mathematics from the apparent contradiction was to ban square roots of negative numbers from the realm of 'possible numbers'.

In order to accept complex numbers as legitimate mathematical objects or to put up with the logical implications of independence of Euclid's Fifth Axiom mathematicians had to revolutionize the deepest and most elusive of their beliefs. Since it was practically impossible to bring such hidden convictions into the open, the only way to arrive at the necessary change was to say 'Go on; faith will come to you' (Jourdain, 1956). In other words, recognition of the square root of –1 as a fully-fledged number was not a matter of finding a fault in the deductive reasoning, nor did it require a better logical argument. Such recognition, being dependent on a change of meta-mathematical assumptions, might only be an act of faith. As such, it could not be expected to happen overnight, not even in a few years. Switching allegiances and changing the awareness requires much time.

One can hardly be surprised, therefore, that Sacceri, although at the face of it endowed with almost all the necessary information, was unable to do what Gauss, Bolyai, Lobachevsky, and Riemann did more than a century later. The change caused by the work of the nineteenth-century mathematicians was revolutionary in Kuhnian sense: the mathematical world of Lobachevsky was incommensurable with that of Sacceri. The idea of a unique God-given geometrical universe was overthrown and from now on mathematicians would be able to choose at whim one of several mutually exclusive systems. It was definitely not easy to get used to such a dramatic increase in the degree of freedom — the acceptance of which was, in a sense, a declaration of independence of mathematics from the physical world. To become true citizens of this new free world, the mathematicians had to relinquish what seemed till now one of the basic laws of rationality. This law, with its roots in human perceptual experience, was so obvious and 'natural' that it was practically impossible to get hold of it and to submit it to a rational scrutiny. Therefore, saying that mathematicians just replaced one epistemological premise by another, although technically correct, would not make justice to the inherent difficulty of the enterprise. Emancipation from conceptual constrains that shape the awareness can only be attained through a painfully gradual process.

One question should be addressed before this section is closed: To what extent does the mathematical change fit the Kuhnian description of *revolution*? The existence of mathematical change cannot be doubted anymore and, as was explained above, such change would often be as radical as a Kuhnian revolutionary change is supposed to be. However, the course of mathematical change defies Kuhn's description. The transformation of mathematical practice is not an abrupt, clearly delineated series of events and it does not consist in a simultaneous shift in all the components. There are no sudden upheavals — the permanent lack of balance has a gradual, continuous effect. At any given moment, many different kinds of tension are present in the system. Since the different components are interrelated and mutually dependent, the regulation of these tensions cannot happen in a linear order. Rather, it will be carried out over time through the combination of little shifts which would give an overall effect of a steady flow. In this way, the continuous transformations will remain almost imperceptible and mathematical practice will preserve its unity through the change. Usually, one would need a historical perspective to realize the magnitude, if not the very existence, of the shift that has taken place over time.

Anomaly of a Missing Object: A Case Study

In a search for an illustration of the claims made in the former sections, I aimed at anomalies which once brought major innovations in the philosophical infrastructure of mathematics and which are likely to reappear in today's classroom. These considerations led me to look at the development of numerical systems.

The turbulent story of number concept is well documented in literature (see, e.g., Novy, 1973; Kline, 1972, 1980; Cajori, 1985; Fauvel and Gray, 1987; Kleiner, 1988; Boyer and Mertzbach, 1991) and it seems indeed to have a great potential as a source of insights about anomalies. What appears to be the most striking common characteristic of the many ways in which new kinds of numbers entered the historical scene is the great deal of distrust with which they were invariably greeted. Let me make a quick survey of objections raised by different writers at different times.

The difficulties with the idea of an *irrational number* were already mentioned before. Greeks, who discovered incommensurability as early as the sixth century BC, never put up with the notion and in order to solve the problem of length which could not be expressed as a pair (ratio) of integers kept the concept of number and that of (continuous) quantity separate. Even as late as the sixteenth century 'the problem of whether irrationals were really numbers troubled some of the very same people who worked with them' (Kline, 1980, p. 113). The idea of *negative number* was brought to Europe by Arabs and became truly indispensable when subtractions of bigger from smaller quantities started to pop out from Cardan's algorithms for the solution of cubic and quadric equations (*Ars Magna*, 1545). For almost three centuries this useful notion was a source of doubts and disagreements. The list of individuals who expressed their concern about the legitimacy of the notion, is long and impressive. It includes such names as Vieta, Descartes, and Pascal. The difficult history of the *complex numbers* has already be mentioned many times in this chapter. Kleiner (1988) calls it 'tortuous'. Kline (1980, p. 119) says that " 'the debate about the meaning and use' of the 'imaginary' numbers was truly 'strident' ". All this started in the middle of the sixteenth century, when Cardan, in his *Ars Magna*, showed that extracting square roots from negative quantities may be useful for solving cubic equations. Among those who doubted the meaning and legitimacy of complex numbers were such leading mathematicians of the sixteenth, seventeenth and eighteenth centuries as Cardan himself, Bombelli, Descartes, Newton, Bernoulli, Leibniz, and Euler.

The story of numbers is particularly appropriate for my purpose as it abounds in examples of an important kind of anomaly — the one which appears when a well-known process, applied to a certain kind of mathematical object, does not produce any known object. On the face of it, there is nothing simpler than to bring the missing entity into existence by stipulation — a procedure fairly common in today's mathematics. In fact, the idea of teaching complex numbers in an axiomatic way, suggested by the mathematician whom I quoted in the introduction to this chapter, was meant to be a classroom version of this procedure. Judging from history, however, such simple solution often proves too simple. According to Kuhn,

> The commitments that govern normal science specify not only what sort of entities the universe does contain, but also, by implication, those that it does not. (Kuhn, 1962, p. 7)

This statement will remain in force if we substitute an expression 'metamathematical beliefs' instead of Kuhn's 'commitments that govern normal science': even if a new mathematical construct seems logically sound, its acceptance would often be obstructed by tacit epistemological and ontological assumptions. This chapter will be devoted to the analysis of the special type of anomaly which arises when new mathematical objects seem to be necessary to give meaning to a certain useful procedure, but the current system of meta-mathematical beliefs bars their way into mathematics. As we have seen, such anomaly, which will be called here an 'anomaly of a missing object', can be observed time and again throughout the successive generalizations of the notion of number. Let me count the ways in which the concepts of negative and complex number violate Platonic framework firmly grounded in the perceptual experience.

Expectation 1: Processes are performed on objects and produce objects
It happened more than once in the history that mathematicians felt obliged to manipulate non-existing outcomes of one operation to perform another useful operation. Such was the plight of Cardan's followers who tried to use the effective algorithms for solving cubic and quadric equations. In order to do that, they often had to put up with the necessity to add, multiply, and even take square roots from the products of such subtractions as 2–5 or 5–15 that, in fact, did not seem to have any product (when the algorithms to solve an equations produced 'sensible' (real) solutions but led to computations involving square roots of negative numbers, Cardan insisted that one should go on with the calculations 'putting aside the mental tortures involved'). This situation certainly violated one of the most basic expectations rooted in our perceptual experience — belief that process is performed on objects and produce objects. We cannot perform mathematical process on nothing, just as we cannot perform physical manipulation on non-existent bodies. On the other hand, if a process proves useful even though it seems to have no sensible product, we have good reasons to persist in using it. We become entangled in a quandary that was called elsewhere 'the vicious circle of reification' (Sfard, 1989, 1991, 1992): we must anticipate a certain object as a condition for the later acceptance of the existence of this object.

Expectation 2: processes and objects are ontologically different entities
The aforementioned term 'reification' refers to the mechanism through which mathematical objects come into being. At a closer look, what really counts in mathematics are the processes that are performed first on material objects (e.g., counting) and then on other, lower-level processes. Mathematical objects are but metaphors which allow us, when necessary, to refer to the mathematical processes as if they were permanent, self-sustained, object-like entities. Talking about the negative number –3 or about the complex number i is but an alternative way of referring to the operations 1–4, 2–5, etc., and to extraction of the square root from –1. As can be seen from these two examples, 'process-object duality' is inherent in the notion of number as well as in its representations. Our perceptually constrained imagination can hardly cope with this 'split personality' of mathematical concepts. Whereas in the physical reality process and product seem to be two ontologically distinct entities, in mathematics they both have the same incarnation. When one says to a pupil 'There is a number $\sqrt{-1}$, or, better still. 'There is a number i such that $i = \sqrt{-1}$', one clearly uses metaphors coming from the physical world. An expression 'i is a square root of –1' relies on a tacit assumption, rooted in our bodily experience, about the separateness of processes (in this case extracting of the square root) and their products (the number i). Thus, our expectations are violated when we are compelled to talk about an output of a certain process but are not able to point to any free-standing entity, the existence of which can be asserted independently of this process.

The fact that our perceptually rooted intuition revolts against process–object duality of abstract concepts found its expression in the historical attempts to separate the operational and the structural ingredients of mathematical constructs by designing different representations for the processes and for their products. In the case of complex numbers, this representational solution indeed seemed to have been of some help: Euler's symbol *i* for a product of the operation $\sqrt{-1}$, Hamilton's ordered pairs (a, b) instead of the operational a + *bi*, and Argand-Gaussian plane in which a complex number is represented by a point, evidently had a considerable impact on mathematicians' attitude toward complex numbers and helped in the eventual acceptance of the concept.

Expectation 3: a given type of objects must have a given set of properties
New abstract objects may be openly at odds with the beliefs about the nature of the concept of which it purports to be a new instance. Being a product of numerical computations, the square root of a negative quantity must be a number. Historically, the concept of number grew from the processes of counting and measuring, and as such was supposed to express quantity. Complex numbers clearly violate this expectation: they could not be ordered in a way which would preserve the order of real numbers — they could not measure magnitude. For three centuries, the idea of number without magnitude never occurred to those who struggled with the notion of complex number. Unaware of their being prisoners of a certain bodily-based metaphor — metaphor of a material quantity — mathematicians projected the properties of real numbers onto square roots of negative numbers and were led into such paradoxes as 'a square root of −1 is both greater than infinity and smaller than minus infinity'. What was a result of an incongruity between the statements of mathematics and the meta-mathematical framework seemed to them as happening within the boundaries of mathematics itself and was interpreted as a threat to its inner consistency.

To make this point clearer, let me illustrate it with another example. As a mathematical metaphor for material quantities, numbers were expected to comply with all the laws the quantities seemed subjected to. A belief that a part is always smaller than the whole is rooted in our perceptual knowledge of material quantities. No wonder then, that the concept of transfinite number, introduced by Cantor, brought his creator into despair when it became clear that the cardinality of infinite sets may be equal to the cardinality of some of their subsets.

Expectation 4: mathematical objects are mind-independent
The very act of introducing new objects may violate the deeply rooted intuitive convictions about the human (as opposed to divine) rights to bring new mathematical entities into existence. Leibniz's imaginative description of the square root of a negative quantity as 'amphibian between being and not-being' bespeaks the ontological nature of his doubts. Euler operated on complex numbers freely and introduced the symbol *i* which helped mathematicians in getting used to them; however, he did all this not without qualms — a fact which found its expression in the statement that square roots of negative numbers 'are called imaginary or fancied numbers, because they exist only in imagination'. These words reveal that for Euler, the new numbers had different ontological status than all the others: his insistence that the former were just figments of the human mind indicates that the latter were conceived by him as something 'real', and were believed to have an objective existence of a sort.

This doubt about the human right to bring new abstract objects into existence is but another manifestation of the fact that our mathematical imagination is rooted in the bodily experience and that perceptual metaphors force mathematicians into a 'Platonic frame of mind'. The unfulfilled expectations effectively blocked the way of complex numbers into the world of legitimate mathematical entities for a long time. The name 'imaginary numbers' persisted even after a geometric model of the complex field was constructed and the theory of complex numbers was recast into a consistent axiomatic structure.

The example of complex numbers makes it clear that frequently observed anomaly of a missing object could not be resolved without a major modification of a meta-mathematical framework. Paraphrasing Kuhn we may say that

> Assimilating a new sort of *mathematical object* demands more than additive adjustment of theory, and until the adjustment is completed — until the mathematician has learned to see *mathematics* in a different way — the new *object* is not a *mathematical object* at all. (Kuhn, 1962, p. 53, my emphasis)

Kuhn's referred to 'scientific facts', 'scientists', and 'nature', and these terms were replaced here, respectively, with 'mathematical objects', 'mathematicians', and 'mathematics'.

What had to change in the vision of mathematics if complex numbers were to become legitimate mathematical objects? To begin with, mathematicians had to give up the basic metaphor which originally brought the concept of number into existence. This was the only way in which they could reconcile themselves with the idea of a number which is not a measure of a magnitude. Also, they had to admit themselves the right to introduce new objects on the sole basis of logical consistency. Hamilton's work on complex numbers was an important step toward both these ends. In presenting complex numbers as mere ordered pairs of real numbers governed by a set of arbitrary rules, Hamilton detached the concept of number from the metaphor of material quantity. By doing so he also brought a more general meta-mathematical message: a message of freedom from commitments external to mathematics itself. Hamilton immediately exercised the newly acquired right for free creation in his work on quaternions. Gradually, introducing new objects by fiat turned into a common practice. At the face of it, consistency of a definition, along with such non-measurable considerations like the beauty and richness of the idea, became the only criteria for its acceptance or rejection. In fact however, mathematical creation will always be mitigated, at least to some extent, by our perceptual experience which both creates and limits human imagination.

Masters and Apprentices in Mathematics: Can They Communicate?

I began this chapter with the issue of communication between the teacher and the student. In the example I brought I tried to show that the 'masters' and the 'apprentice' may live in separate mathematical worlds, worlds so dramatically different that there is no possibility of real exchange.

In an attempt to account for the existence of different mathematical worlds — the fact that is often overlooked by those who tend to view mathematics as monolithic and unique — I went so far as to the general philosophy of human knowledge. I analysed the mechanisms which brought about the realization that creation of knowledge — any kind of knowledge — is a hermeneutical task and as such cannot be expected to produce either certainty or the kind of knowledge which could be called 'objective' in the traditional sense. All this is true even in mathematics, which, like any other discipline, seems to be in a constant flux. A permanent tension between the body of mathematical statements and the system of tacitly accepted epistemological and ontological assumptions keeps the system in a constant motion. This tension stems from the fact that the meta-mathematical framework, being rooted in our primary perceptual intuition, is much more inert than all the other components of mathematical practice and does not change at the pace which would ensure a full equilibrium. Since ontological and epistemological discrepancies are the most difficult to detect, the teacher and the student may be unaware of the fact that they participate in different mathematical practices and that their mathematical imagination is shaped by different metaphors. Their inability to communicate with each other will then be interpreted as the student's failure to learn or the teacher's failure to teach, or both.

The pupil's fundamental problems with such ideas as negative or complex numbers tend to be overlooked by the teacher mainly because the latter's own implicit beliefs make him or her oblivious to the possibility of somebody having a different ontological stance. Indeed, those who stand at the top of a mountain might have forgotten the changing landscape they could observe themselves while climbing the hill for the first time. Another circumstance that helps in concealing ontological difficulties is the fact that a student may become quite skilful in manipulating concepts even without reifying them. This seems a plausible explanation for the phenomenon observed with a particular clarity by Kuchemann in the large-scale CSMS investigation carried out among 13–15-years-old pupils. The study has shown that when operations involving negative numbers were concerned, older children tended to do significantly worse than the younger. Kuchenmann conjectures that 'this fall-off in performance . . . may . . . be due to the older children having forgotten the meanings that were given to the integers when they were first introduced.' I shall venture an even more far-fetched explanation. It is possible that these students never really overcame the anomaly which they faced when exposed to the idea of negative number for the first time, and thus they never reified the concept. In other words, the negative number was never a fully-fledged mathematical object for them, and therefore their ability to manipulate the integers was purely mechanical. As such, it was fragile and quick to disappear.

By the same token, it requires more than just observing a student manipulating complex numbers to realize the existence of difficulties rooted in his or her meta-mathematical beliefs and in the metaphor he or she developed for the concept of number. One of the few works which have shown that such problems do exist is the study by Tirosh and Almog (1989). The authors' attempt to probe learners' hidden conceptions led them to the conclusion that 'the students are reluctant to accept complex numbers as numbers, and [they] incorrectly attribute to complex numbers the ordering relation which holds for real numbers.'

The natural question to ask now is what can be done to overcome the deeply rooted problems that obstruct the communication between students and their teachers. In the foregoing discussion I have pointed out to more than one reason why the task may be extremely difficult. Our success seems to depend on our ability to change the most hidden, most elusive, and most robust of student's beliefs.

Rather than suggesting any solution, let me begin with an easier task: with a critique of certain existing, commonly employed, methods of teaching. I have already suggested that introducing complex numbers as abstract pairs of real numbers is bound to be as ineffective as it was for generations of mathematicians. Some empirical support for this can be found in Vinner (1988). Why the axiomatic method cannot be pedagogically effective should now be quite clear. Indeed, no mutual understanding can be expected between pre-Hamiltonian students, unable to relieve themselves from the grip of number-as-quantity metaphor, and post-Hamiltonian teacher who is no longer fettered by anything except the abstract structure of field and who feels free to introduce new numbers by stipulation.

Teachers and textbooks are often paying a lip service to the idea of 'meaningful learning' when they motivate the introduction of complex numbers with a need to find a solution for the equation $x^2 + 1 = 0$. This kind of argument will not help the student who, being a captive of the metaphor of quantity, feels neither the need nor the right to talk about numbers the square of which is negative. The idea of an existence of such number violates all the meta-mathematical expectations I listed in the last section. Those who suggest this solution seem to be unaware of the subtle difference between equation $x^2 + 1 = 0$ used to introduce complex numbers, and the equation $x^2 - 2 = 0$, with which irrational numbers are usually ushered into mathematics. The idea of a solution to $x^2 - 2 = 0$ is in tune with the metaphor of a number as a quantity: after all, if a diagonal of a unit square is to have length, this length must be a root of $x^2 - 2$ (this does not mean that the introduction of irrational numbers is easy; also in this case there might be a metaphor which would obstruct the learning: the metaphor of a number as a discrete quantity or as a pair — ratio — of such quantities).

After the critical survey of the existing methods, I cannot escape the central question any longer: If all this cannot help, what can? Like before, when we were dealing with the nature of difficulty, let us turn to the history. It seems that the way ideas developed over time is not only a perfect lens for detecting invisible conceptual pitfalls, but also a valuable source of ideas about how these pitfalls may be overcome.

Inspired by historical example, Kleiner (1988) suggests to use as a motivation for complex numbers the special instance of a quatric equation which has real solutions but, if solved with the help of Cardan's algorithm, produces roots of negative numbers as intermediary results. Thus, Kleiner proposes replacing the argument of a *need* with the argument of *usefulness*: the candidacy of the square roots of negative numbers for citizenship in the world of numbers is worth considering not because of a necessity to solve a hitherto unsolvable equation (why should such thing be necessary?), but because the new numbers might be useful for purposes that transcend the borders of mathematics. Judging from the history, this suggestion may be helpful. While the direct claim about an arbitrary need to solve the unsolvable does not seem to convince anybody, the argument of usefulness (and consistency!) proved effective over time. As de Morgan put it in 1831,

> We have shown the symbol $\sqrt{-a}$ to be void of meaning, or rather self-contradictory and absurd. Nevertheless, by means of such symbols, a

> part of algebra is established which is of great utility. It depends upon the fact, which must be verified by experience that the common rules of algebra may be applied to these expressions without leading to any false results. (de Morgan, 1831, in Kline, 1980, p. 155)

Whatever the advantages of the idea of usefulness, giving an explicit argument in favour of the existence of the new kind of number cannot make the concept instantly accessible. After all, coping with anomaly involves replacement of one metaphor by another and requires a deep conceptual change. As can be learned from history, transformations of this kind always take time and require persistence and belief in the necessity of change. It was mathematicians' stubborn refusal to abandon the problematic square roots from negative numbers which, after centuries of struggle, brought the reconciliation with the idea of a complex number. As Jourdain (1956) put it,

> Mathematicians thought, then, that imaginaries, thought apparently uninterpretable and even self-contradictory, *must* have a logic. So they were used with a faith that was almost firm and was only justified much later. (Jourdain, 1956)

The 'technique' of persistence was deliberately adopted by many mathematicians. For example, Kitcher (1988) tells us about 'Leibnitz's exhortations to his followers to extend the scope of his methods without worrying too much about what the more mysterious algebraic maneuvers [those involving square roots of negative quantities among them] might mean.' And he concludes,

> the Leibnitzians seem to have thought that the proper way to clarify their concepts and reasonings would emerge from the vigorous pursuit of the new techniques. In retrospect, we can say that their confidence was justified. (Kitcher, 1988, p. 307)

There are reasons to expect that what worked for generations of mathematicians has a chance of success in today's classroom. Thus, rather than require an immediate grasp of a new concept, we should encourage the students to use it in as many contexts as possible *in spite* of the initial doubts about its meaning — in spite of the inadequate meta-mathematical framework and notwithstanding the violated expectations. de Morgan formulated this pedagogical advice explicitly:

> We are not advocates for stopping the progress of the student by entering fully into the arguments for and against such questions as the use of negative numbers, etc., which are inconclusive on both sides; but he might be made aware that the difficulty does exist, the nature of which might be pointed out to him, and he might then, by the consideration of a sufficient number of examples, treated separately, acquire confidence in the results to which the rules lead. (De Morgan, 1831, in Kline, 1980, p. 155)

The most important part of this recommendation is de Morgan's insistence on *using the new concept before all the conceptual difficulties are resolved.* Whether and when the student should deal explicitly with the epistemological and ontological quandaries is another question. One thing, however, seems unquestionable: the teacher himself or herself cannot effectively cope with the student's difficulties if he or she is not able to predict the existence and understand the nature of the student's doubts. Thus, in the light of the example I brought in the introduction to this chapter, de Morgan's request that people are 'made aware that a difficulty does exist' should probably be applied to teachers before it is directed toward the students.

Teachers' awareness of the differences between their own meta-mathematical beliefs and students' tacit epistemological and ontological assumptions are the first step toward better classroom communication. Should the differences be dealt with explicitly? Not necessarily. The incommensurability of teachers' and students' words may impede any direct attempt to bring about a mutual understanding. Fortunately, there is another lesson to be learned from history: the fact that generations of mathematicians were able to persist in using ideas which were not entirely clear to them shows that there is a possibility of an effective learning even when the things that are done cannot be fully justified. The 'master-apprentice' mode of learning, based on participation in doing rather than on explicit dialogue between the teacher and the student, seems to be a promising pedagogical option:

> Learning . . . can take place even when coparticipants fail to share a common code. The apprentices' ability to understand the masters' performance depends not on their possessing the same representation of it, or of the objects it entails, but rather on their engaging in the performance in congruent ways. Similarly, the master's effectiveness at producing learning is not dependent on her ability to inculcate the student with her own conceptual representations. Rather, it depends on the ability to manage effectively a division of participation that provides for growth on the part of the student. Again, it would be this common ability to coparticipate that would provide the matrix for learning, not the commonality of symbolic or referential structures. (Hanks, 1991, p. 21).

Rather than try to make his or her meta-mathematical principles explicit, the teacher should aim at providing a student with an appropriate experience. The student is more likely to be converted to a new mathematical practice by living in a new mathematical world than just by talking about it.

References

BAUMAN, Z. (1978) *Hermeneutics and Social Sciences: Approaches to Understanding*, London, Hutchinson.

BERNSTEIN, R.J. (1986) *Philosophical Profiles*, Cambridge, Polity Press.

BOURBAKI, N. (1950) 'The architecture of mathematics', in LE LIONNAIS, F. (Ed) *Great Currents of Mathematical Thought*, New York, Dover Publications, pp. 23–36.

BOYER, C.B. and MERTZNBACH, U. (1991). *A History of Mathematics*, 2nd ed., New York, John Wiley.

CAJORI, F.A. (1985) *History of Mathematics*, 4th ed., New York, Chelsea Publishing Company.

COBB, P. (1989) 'Experiential, cognitive, and anthropological perspectives in mathematics education', *For the Learning of Mathematics*, 9, 2, pp. 32–42.

CROWE, M.J. (1988) 'Ten misconceptions about mathematics and its history', in ASPRAY, W. and KITCHER, P. (Eds) *History and Philosophy of Modern Mathematics*, Minneapolis, Minnesota, University of Minnesota Press, pp. 260–77.

DAVIS, P.J. and HERSH, R. (1983) *The Mathematical Experience*, London, Penguin Books.

ERNEST, P. (1991) *The Philosophy of Mathematics Education*, Basingstoke, The Falmer Press.

ERNEST, P. (1992) 'Are there revolutions in mathematics?', *Philosophy of Mathematics Education Newsletter*, 4/5, pp. 14–18.

FAUVEL, J. and GRAY, J. (1987) *The History of Mathematics — A Reader*, London, Macmillan Education; Milton Keynes, The Open University.

FAYERABEND, P. (1975) *Against Method*, London, Verso.

FAYERABEND, P. (1987) *Farewell to Reason*, London, Verso.

GRIFFITH, H.B. (1987) 'Looking for complex roots', *Journal for Research in Mathematics Education*, 18, 1, pp. 58–75.

HANKS, F. (1991) 'Foreword', in LAVE, J. and WENGER, E. (1991) *Situated Learning*, Cambridge, Cambridge University Press.

HART, K. (Ed) (1981) *Children's Understanding of Mathematics 11–16*, London, John Murray.

HEFENDEHL-HEBEKER, L. (1991) 'Negative numbers: obstacles in their evolution from intuitive to intellectual constructs', *For the Learning of Mathematics*, 11, 1, pp. 26–32.

JOHNSON, M. (1987) *The Body in the Mind: The Bodily Basis of Meaning, Imagination, and Reason*, Chicago, The University of Chicago Press.

JOURDAIN, P.E.B. (1956) 'The nature of Mathematics', in NEWMANN, J.R. (Ed) *The World of Mathematics*, New York, Simon and Schuster, pp. 4–72.

KITCHER, P. (1984) *The Nature of Mathematical Knowledge*, Oxford, Oxford University Press.

KITCHER, P. (1988) 'Mathematical naturalism', in ASPRAY, W. and KITCHER, P. (Eds) *History and Philosophy of Modern Mathematics*, Minneapolis, Minnesota, University of Minnesota Press, pp. 293–325.

KLEINER, I. (1988) 'Thinking the unthinkable: The story of complex numbers', *Mathematics Teacher*, 81, 7, pp. 583–91.

KLINE, M. (1972) *Mathematical Thought from Ancient to Modern Times*, New York, Oxford University Press.

KLINE, M. (1980) *Mathematics, the Loss of Certainty*, New York, Oxford University Press.

KUCHEMANN, D. (1981) 'Positive and negative numbers', in HART, K.M. (Ed) *Children's Understanding of Mathematics: 11–16*, Newcastle, John Murray, pp. 82–7.

KUHN, T.S. (1962) *The Structure of Scientific Revolutions*, Chicago, The University of Chicago Press.

LAKATOS, I. (1976) *Proofs and Refutations*, Cambridge, Cambridge University Press.

LAKATOS, I. (1978) *Mathematics, Science, and Epistemology*, Cambridge, Cambridge University Press.

LAKOFF (1987) *Women, Fire and Dangerous Things: What Categories Reveal About the Mind*, Chicago, The University of Chicago Press.

LAKOFF and JOHNSON (1980) *The Metaphors We Live By*, Chicago, The University of Chicago Press.

NOVY, L. (1973) *Origins of Modern Algebra*, Leyden, The Netherlands, Noordhoff International Publishing.

OTTE, M. (1990) 'Intuition and logic', *For the Learning of Mathematics*, 10, 2, pp. 37–43.

POLYA, G. (1945) *How to Solve It*, Princeton, New Jersey, Princeton University Press.

POPPER, K.R. (1979) *Objective Knowledge* (*Revised edition*), Oxford, Oxford University Press.

RORTY, R. (1982) *Consequences of Pragmatism*, Minneapolis, Minnesota, University of Minnesota Press; Brighton, Sussex, Harvester Press.

RORTY, R. (1991) *Objectivism, Relativism, and Truth*, Cambridge, Cambridge University Press.

SFARD, A. (1989) 'Transition from operational to structural conception: the notion of function revisited', in VERGNAUD, G., ROGALSKI, J. and ARTIGUE, M. (Eds) *Proceedings of the Thirteenth International Conference of PME*, 3, Paris, Laboratoire PSYDEE, pp. 151–8.

SFARD, A. (1991) 'On the dual nature of mathematical conceptions: Reflections on processes and objects as different sides of the same coin', *Educational Studies in Mathematics*, 22, pp. 1–36.

SFARD, A. (1992) 'Operational origins of mathematical notions and the quandary of reification — the case of function', in DUBINSKY, E. and HAREL, G. (Eds) *The Concept of Function: Aspects of Epistemology and Pedagogy*, MAA Monographs Series.

SFARD, A. (1994) 'Reification as a birth of a metaphor', *For the Learning of Mathematics*, 14, 1, pp. 44–54.

STEWART, I. and TALL, D.O. (1983) *Complex Analysis*, Cambridge, Cambridge University Press.

STRUIK, D.J. (1986) *A Source Book in Mathematics, 1200–1800*, Princeton, New Jersey, Princeton University Press (originally published in 1969).

THOM, R. (1971) 'A modern mathematics: an educational and philosophical error?', *American Scientist*, 59, pp. 695–99.

TIROSH, D. and ALMOG, N. (1989) 'Conceptual adjustments in progressing from real to complex numbers', in VERGNAUD, G., ROGALSKI, J. and ARTIGUE, M. (Eds) *Proceedings of the Thirteenth International Conference of PME*, Paris, Laboratoire PSYDEE, Volume 3, pp. 221–7.

VINNER, S. (1988) 'Subordinate and superordinate accommodations, indissociability and the case of complex numbers', *International Journal of Mathematics Education, Science and Technology*, 19, 4, pp. 593–606.

Notes on Contributors

Stephen Bristow is a research officer at the School of Education, University of Exeter, England.

Stephen I. Brown is a professor at the Graduate School of Education, University of Buffalo, New York, USA.

Tony Brown is a senior lecturer in mathematics education at the Manchester Metropolitan University, Manchester, England.

Raffaella Borasi is associate professor of education at the University of Rochester, New York, USA.

Kathryn Crawford is senior lecturer in mathematics education at the University of Sydney, Australia.

Charles W. Desforges is professor of primary education at the School of Education, University of Exeter, England.

Paul Ernest is reader in mathematics education in the School of Education at the University of Exeter, England.

Stephen Lerman is senior lecturer in mathematics education in the School of Computing, Information Systems and Mathematics at the South Bank University, London, England.

Philip Maher is senior lecturer in mathematics at Middlesex University, London, England.

John Mason is a professor of mathematics education at the Open University, Milton Keynes, England.

Michael Otte is professor in mathematics education at the Institute for the Didactics of Mathematics, University of Bielefeld, Germany.

David Pimm is senior lecturer in mathematics education at the Open University, Milton Keynes, England.

Falk Seeger is a researcher in mathematics education at the Institute for Didactics of Mathematics, University of Bielefeld, Germany.

Anna Sfard is a professor of mathematics education at the Hebrew University of Jerusalem, Israel.

Marjorie Siegel is associate professor at the Teachers College, Columbia University, New York, USA.

Erick Smith is assistant professor in mathematics education at the University of Chicago at Illinois, USA.

Francesco Speranza is professor of mathematics at the University of Parma, Italy.

Leslie P. Steffe is a professor of mathematics education in the College of Education at the University of Georgia, USA.

Heinz Steinbring is a researcher in mathematics education at the Institute for the Didactics of Mathematics, University of Bielefeld, Germany.

Dick Tahta is a retired lecturer in education from the University of Exeter, England.

Robert S.D. Thomas is professor of applied mathematics at St John's College and the University of Manitoba, Canada.

Ron Tzur is a researcher in mathematics education in the College of Education at the University of Georgia, USA.

Ernst von Glasersfeld is a research associate at the Scientific Reasoning Research Institute, University of Massachusetts, USA.

Index

9 780750 705707